I0493677

An open letter
"TO SELECTED ACADEMICS"
4

Written By Peet (P.S.J.) Schutte

ISBN-13: 978-1499515381

ISBN-10: 1499515383

Indicating Singularity

Please note that
This is a direct Follow on
An open letter
TO SELECTED ACADEMICS # 3

This book uses the flowing work as reference :

© KOSMOLOGIESE EN ASTRONOMIESE TEGNIKA
An open letter

TO SELECTED ACADEMICS
ISBN 0-9584410-9-X

An open letter **TO SELECTED ACADEMICS** is THE ACADEMIC PROLOGUE AND AN ACADEMIC INTRODUCING LETTER TO ACADEMICS PRESENTING A NEW COSMIC THEORY AS IS STATED IN **MATTER'S TIME IN SPACE THE THESES,** ISBN 0 – 6 2 0 – 2 7 0 4 1 – 1 which consists of the following seven books in title:

1) **A Cosmic Birth...Dismissing Nothing** I.S.B.N. 0-620-31609

2) AN OPEN LETTER ON: **XEPTED ASTRONOMICAL MISTAKES** ISBN 0-9584410-1-4

3) AN OPEN LETTER ON: **INTER GALACTICA SPACE TRAVEL** ISBN 0-9584410-2-2

4) AN OPEN LETTER ON: **CORRECTING COSMOLOGY** ISBN 0-9584410-5-7

5) MATTER'S TIME IN SPACE: **THE HYPOTHESIS** ISBN 0-9584410 –6-5

6) AN OPEN LETTER ON: **" STARSSTUFFN'** ISBN 0-9584410-3-0

7) **" SEVEN DAYS OF CREATION"** ISBN 0-9584410-4-9

© KOSMOLOGIESE EN ASTRONOMIESE TEGNIKA
An open letter

TO SELECTED ACADEMICS
ISBN 0-9584410-9-X

THIS LETTER IS THE ANNOUNCING OF THE BOOK IN SEVEN VOLUMES CALLED **MATTER'S TIME IN SPACE:** THE THESIS ISBN 0-9584410-8-1 VOLUMES 1-7

This letter was the letter that was sent to near enough eighty Universities through out the world in regard of announcing a new cosmic theory. It now is turned into a separate and individual commercial book.

TO WHOM IT MAY CONCERN,

An open letter **TO SELECTED ACADEMICS** ISBN 0-9584410-9-X Is THE ACADEMIC NOTIFYING OF

MATTER'S TIME IN SPACE: THE THESIS ISBN 0-9584410-8-1 Written by PEET SCHUTTE

$$(4(\Pi^2)$$
$$\Pi$$

Earth $\overset{\Pi}{\bigcirc} \xleftarrow{(\Pi^2/4 \times 10) + \Pi} \overset{\Pi}{\bigcirc} \longrightarrow \overset{\Pi}{\bigcirc}$ Moon

Behind this principle is the sound barrier and the reason why aircrafts break the sound barrier. It is all a relation in different positions of singularity. From the time, I wrote the first few pages I was on a quest to find a more suitable person to take over the work. I, more than any one else, know my limitations and limitations they are. I knew that anyone of the more than one thousand five hundred academics I eventually contacted held more knowledge in the tip of his (or her) little finger, than I have in my entire body. I tried in desperation as I tried in vain, to convey my message to the right person that could see what I could see. It all came to nothing because the academics all sat on the same mighty Sear Tower far too high to even notice me pointing at the cracks down below. From where the Official Policy Protectors sat, they did not notice me. Those to whom I drew some attention be it personal, by mail or on the Internet saw me as a nonsense proclaiming nonsense.

From where I stood, I could see the mighty tower they sat on. I could also see what construction held that mighty tower together. I could see how much that mighty tower was leaning over like the tower of Pizza. I could see how the tower will fall one day if not soon then later, because I was at the foundation of the tower. From where I stand, at the very bottom, I see the foundation of this enormous Petranos Towers collapsing from the misconceptions it holds as a base. Down at the very bottom where I am, I could see whatever other one holding a position in this tower could not see. Those that are at the very top, ARE so high, so very secure, they would not even hear me or take notice, and yet they are the only ones that can do something about the inclining tower.

After attempting for seven years to make myself heard to get the High and Mighty to notice the insignificant me down below, so small in relation to their greatness, I decided to show the world what holds them that high. I decided to show everyone how great "they" are, especially about the greatness of their misconceptions that put them at such a dizzy height.

To every one of the Official Policy Protectors, I say this: Every opportunity you had, you thoroughly rubbed my nose in the knowledge that we are not in the same class. I accept that fact as I accept my academic qualifications being so very poor. I shall never be your next-door neighbour or the bloke living down the road from your house, because I am not in your league and I shall never be. I do not begrudge you your academic position, your mighty achievements or the height you have reached in your sphere. I shall never enjoy your company as an equal, because I can never be your equal. Your brainpower puts you light years ahead of me, and for that reason I do not even wish to have the honour of your company.

All I ask, is listen to a mere mortal, a mindless illiterate compared to you, one sod down here at the bottom where you are at the very top and that may just see things you cannot see from the height you hold. I do not wish to join your company for I shall not fit. I never had or have any ambition to fit either, because I am quite happy being in the sub-minor league. All I ask is to be heard. So many times you, honoured members of the clan of Newtonian High Priests did not even attempt reading my book that I sent to you. You did not even try to pretend I had a point, therefore you merely threw my book away in disgust. To you my illiterate arrogant views about science being wrong for the past few hundred years are the epitome of a mindless person. You saw me as a totally mentally underdeveloped excuse for breathing and you could not bare my company because for me to have my view such as my view is in rejecting the view of an establishment centuries old. I can understand your disgust, but that does not change my point and that does not change the incorrectness of Newtonian science!

Before you throw the book down in total disgust, first answer the following argument and if you can answer it truthfully then throw the book down. If you cannot answer it, go on reading the book and you may just set your thinking mind in motion. Hear this from a mindless: you may have the ability to learn and afterwards reflect on that which you learn, but you cannot think, and I do not know who is the most mindless, me without education or you with education and without reason.

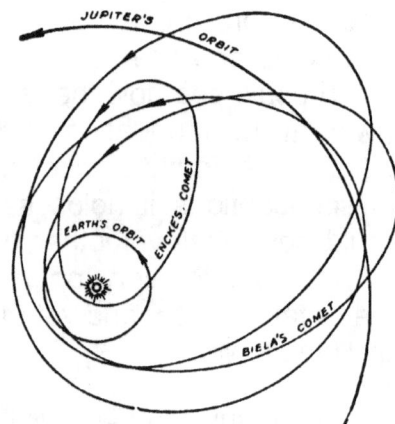

You have had an interest in science for as long as you can remember. You have an inquisitive mind stronger than most others around you. You have an investigative nature because that is why you are reading this book. You are more intelligent than most and that is why you are where you are. You can grasp most things much better than many others because that is what makes you intelligent. With all that attribution you still wish me to believe you when you tell me you never, not once, asked yourself why the comet misses the Sun by almost an astronomical mile? You are addicted to Newton!

Not accepting the truth about your condition is all part of your addiction to Newton. That is part of all addicts where those addicts would rather fight the truth than fight their condition. All addicts, (your condition included), biggest fight presently is a fight against accepting what they have become. It is time that you confronted your fears head on and now is as good as any time there ever will be. Hit the iron now when it is still red hot. Accept your addiction to Newton and admit to yourself out loud where every one can hear you that the comet does not hit the sun in spite of Newton's claim on fame being $F = G(Mm)/r^2$. That will be your first step on your long walk to recovery.

You are addicted to Newton. You do not understand Newton because if you did understand Newton as far as cosmology goes, then you would reject Newton. You were force-fed by your superiors on Newton at the time and during the time when you were a student and being force-fed you either had to grow addicted to Newton or die from Newton. Being where you now are it means you are not dead. That means you took the second option by becoming addicted to Newton. You saw comets come and you saw comets go. You did not question the reason why the comet went because of your addiction to Newton. You accept Hubble's expanding yet you go about looking for a critical density in the hope you do not have to abandon your Newtonian addiction. You know the Big Bang cannot be possible while Newton's mass annihilates the radius between masses because then the Big Bang had to be the Big Crunch and from where we are there is only one enormous out explosion that had no Newtonian implosion. The enormous force that were suppose to contract the small radius that was in place during the Big Bang never imploded on the mass by nullifying the radius through the tremendously large mass and small radius being present at the time. All the facts modern day science accepts does not stroke with Newton. Is it because subconsciously everyone is silently admitting about Newtonian incorrectness?

Now I come and I tell you your addiction is killing you. As all addicts do, you are not willing to kill your addiction although you do realize that your addiction to Newton is senseless, stupid and all together wrong. The second choice facing all addicts including you is to hate the messenger that comes to take the drug from you. You will hate me, as the messenger. You will be willing to kill me being the messenger of evil because you see me as the evil that wished to part you from your addiction, which is Newton. You now find the bizarre feeling to put your anger distrust and vengeance on me and on the book you are reading. I have seen many academics stop reading at the part you now have arrived at. Brave yourself and throw out you fears by reading the rest. It will eventually cure you. Newtonians declare that a force brought about by the content of the mass of a body will therefore pull objects closer where the pressure coming from the weight brought about by the mass of the bodies, will lead to heat.

If a cylinder is pumped with air, the pumping is a force. The force comes about because the force is that of the intentional action of the only force in the Universe, the force of life. The more pumping there is and the longer the pumping will last, the hotter the air will become, and therefore the hotter the walls of the cylinder will get, as it transmits the heat from a point where the heat is most abundant to a point where the heat is least abundant. Heat flows from hot to cold.

After pumping stops, the heat on the inside will reduce in value up to a point where there is equilibrium between the heat on the outside of the cylinder and the heat on the inside of the cylinder. The heat reached equilibrium with the event of time. Please for the sake of sanity, do not reply that it is the molecules in the air tank that is bumping against each other and through that collision, friction causes the heat. That is as much Newtonian rubbish as one can ever find. Should you insist on that being your answer, then ask yourself what will calm the molecules down afterwards where they get so calm, there is no more heat in un-equilibrium. Did the molecules take drugs, or did the force calm them by telling them gentle nighttime stories. I had so much bullshit thrown at me wherever Newtonians defended their Master it sickens me. I may be uneducated, but I am far from mindless!

No molecule can ever, ever touch another molecule because the electrons guarding the outside are equally negatively charged, and will therefore reject any contact or coming closer to one another or with another molecule.

When we look at the Earth, we find the coldest region on the outside of the atmosphere, where the least mass, weight and gravity is. As the circle grows smaller, the molecules become more, the mass becomes more, and this will increase the weight that brings about heat from pressure. This is not my say so this is Newtonian science.

The increase in mass, bringing about an increase in weight, puts most pressure at the centre to be the hottest. There is this force in matter that pushes and pulls, until everything is boiling hot, and that is gravity.

There must be a point, where all the matter has finally found a position to bring about the overheating that occurs in the centre. The heat is energy and cannot be manufactured, but has to come from somewhere. The heat cannot be lost, because heat will transfer to a colder region to bring about equilibrium. The heat cannot continuously come from nowhere because at a point, the molecules will settle in their individual positions, find equilibrium and maintain the heat balance in that spot. It cannot produce heat from nowhere, on a continuous basis, because heat is energy and being energy it must come from somewhere as much as it is going somewhere. Either the atoms lose their mass to heat and the mass becomes heat in order to generate heat on a continuous basis, or the heat must stop, decrease and become as cold as outer space because no further heat comes about because the heat is exhausted.

If the matter as much as mass becoming weight established heat by applying continuous pressure, the Earth should decrease in size as mass turns to heat with the consequential loss of mass and gain in heat. The Earth must then deflate and be a pretty small place by now.

You may say there is a lot of mass-producing a lot of pressure becoming a constant flow of heat, but that will mean some of the mass must have disappeared to heat because four thousand five hundred million years is a pretty long time. In four thousand five hundred million years, some size diminishing should show, or the Earth should have reached a point of equilibrium by now where the heat supply is completely exhausted because again four thousand five hundred million years is pretty much enough time I would say for matter to have cooled down by now.

Yet, the flow of heat maintains in the Earth, very much uninterrupted and in no way showing signs of decrease. If the force of gravity is manufacturing heat, from where does it get its raw material and where is the manufacturing plant? If weight and pressure leads to heat, then the heat should have transferred altogether to outer space, the coldest place we know at minus 276°C. Four thousand five hundred million years have come and gone since the Earth became the form it has now.

On the other hand, if the force continued its pushing and pulling it applied when it formed the Earth, the Earth should by now have incinerated. What force of gravity is required in getting enough pressure to push hydrogen dust into a solid iron formation and release that formation as a molecule in being as solid as the Earth is. Such an effort requires a lot of pushing and compressing. If that pushing and pressure continued as much as it must have had it had to increase with the demise of particle space that is separating the molecules? Keeping that in mind as a natural law, everything on Earth should be covered in flames by now. When you go with the argument that there is not enough matter to produce such pressure, then there was not enough matter from the start to get the place as dense as it currently is. The matter did not decrease, the force therefore had to become weaker with time and the force is so weak now, it does not collapse the Earth into a smaller space than it was say three thousand five hundred million years ago. However there are very little signs of that, in fact it seems the whole thing is getting bigger, with all the lakes losing their depth, and the mountains rising, including the volcanic activities establishing new islands.

If you cannot state why the heat has not reached a point of saturation or has decreased to equilibrium with outer space, it proves you are not thinking. In that case read on ... it may arouse your thinking ability once again. This I say, not in arrogance because I, of all people should know how poor my abilities are. I sat day after night, night after day, breaking my brain to find answers, or only clues to the questions in hand. I knew at each point I arrived, that the answer is right in front of me, but through the darkness of my personal ignorance, I could not see what I knew there was to see. From that feeling of incompetence I tried once more on every occasion to contact any person with more brains than I but every time I was luckless to energise interest and had to continue with my personal incapacity.

Every time I contacted an Official Policy Protectors in desperation for help, they ignored me flat. Every message I sent telling whomever I contacted, that there is no such a thing as gravity, it is all a medieval hoax, Newton is altogether completely wrong with his gravitational laws and the Bible is one hundred percent correct about creation, they would not even reply or at least respond. Well... to them I say this: I still maintain that there is no such a thing as gravity: it is a hoax.

Obviously, the obvious is that the Newtonian Order of High Priests would lead every body to believe that Newton and Einstein's findings are flawless. All these mentioned discrepancies are known to "Xepted science", yet they keep the charade going on about other planets they are about to find, only to mislead the public and milk their tax money, in the name of research. If it were not for funding and an effort to provoke general interest in skimming tax money, then why would they deliberately spread such malice?

Scientists know about this discrepancy in the Newtonian laws, yet there is never any mention about it.

The so-called "evidence of the existence of planets" is based on just as laughable principle. Allow me to explain:

The findings which science base their proof on about the location of other planets are the gravitational pull.

At first the way in which the facts are presented does not sound that unfamiliar in an argument, and one tends to accept it without a second thought. When given the second thought, the blatancy in the matter leaves one breathless.

In the evidence, the stars and "planets" are presented to be in a tug of war. This one can see in any sketch about the solar system. How this supposedly works is that the one star first pulls the other system closer with the force of gravity, and then it is the other systems turn to apply gravity and jerk the first system to its side. What they do explain in explicate detail is that they do not understand the first thing about the matters they pretend to understand.

The Official Policy Protectors know very well that all children play this game, and therefore every one will associate this explanation with familiar events in their past, and no further questions will be asked. Let us examine this principle with obvious general knowledge about space flight and how this applies in outer space. An Astronaut is capable of lifting four tons of equipment by self-propulsion and relying on human muscle. He (or she) can perform this action effortlessly. However, what is impossible to perform in outer space is to correct his position when he is not secured to a stabilized object. Every person knows it can be life threatening when an astronaut loses his grip or connection to the spacecraft. In this, the question is; how can two heavenly bodies have a tug of war under such conditions? There is obviously some stabilizing factor on Earth one do not find in outer space, and that is not gravity because such an answer is avoiding the issue

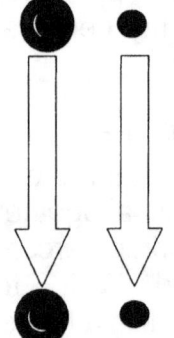

In the past so many Official Policy Protectors dismissed my rejecting the Newtonian claim about mass being a factor in falling objects. Some even went as far as refusing to read my book beyond that statement, that being on page four of a one thousand seven hundred page document, claiming I do not understand Newton, but no one can explain why oxygen is a gas, and yet oxygen is more massive than boron and carbon. If mass were a factor of the essential being the claim of Newton, then oxygen would fall to the earth faster than boron and not float in the air as air. Yet boron is a solid and oxygen is the gas.

The reason why the **mass of the smaller object does not apply**, is because **it is not the object** that is **drawn** to the Earth, but **the space-time in which the mass of the object finds itself in,** that is being **drawn towards the Earth**. The most obvious proof that something applies movement to something directing the movement towards Earth is the simple pendulum. If gravity was a force, one should get the same result by applying a pull spring to the pendulum. With the spring connected to the bottom of the pendulum, both time as well as space would compromise. However, we know that the result proves the opposite, which proves that there is no force applied to the pendulum. On one of my many crusades I met one of the most influential academics on astro-physics in Africa.

I tried to explain the pendulum swing to a high ranking man of much academic importance, doing great work on behalf of NASA in South Africa but with me not knowing his superiority on the matter of the pendulum, I took this man on, on one of his specialties. This man was apparently an expert on research or lecturing or what ever on matters about the pendulum. Of course, as usual, the very first thing he asked (as all Newtonians do) before even asking a person's name, was at which university did I study and what my academic qualifications was. By replying that I have never been at any university for longer than a few hours in my life, and therefore my academic qualifications was less than zero, this highly rated person of high standings was less than impressed to spend any of his valuable NASA paid time with me. To complicate the whole aspect of my un- welcome visit was that I tried to explain to him being the expert on the pendulum that he is, about the pendulum. Boy, was the man annoyed with me.

He was very polite and very civilized about the whole issue, but his annoyance with MY EFFORT ABOUT EXPLAINING THE PENDULUM TO THE EXPERT ON EXPLAINING THE PENDULUN WAS MORE THAN HE COULD BE CIVILISED ABOUT. There is a point where a person gets a little too civilized to be true and at that point where a person gets a little too polite to be civilized about a topic. Any person can sense such a point the point when one realizes that is the point of limits. I also realized that what ever I had to say on any matter concerning all relevant matters was as good as never said as far as our Professor Doctor, was concerned. As a matter of fact that Professor from the University of Potchefstroom is one of the most exceptional men to walk this planet. I gave him a copy of The Thesis and after reading only four pages from a book containing over two thousand pages he was able to draw a conclusion and condemn my work. Of course the reason was the usual: I did not understand Newton obviously because of the lack of education on my part.

You may ask yourself: *"What was me (the un-welcome person) trying to say?"* This is what I am saying:

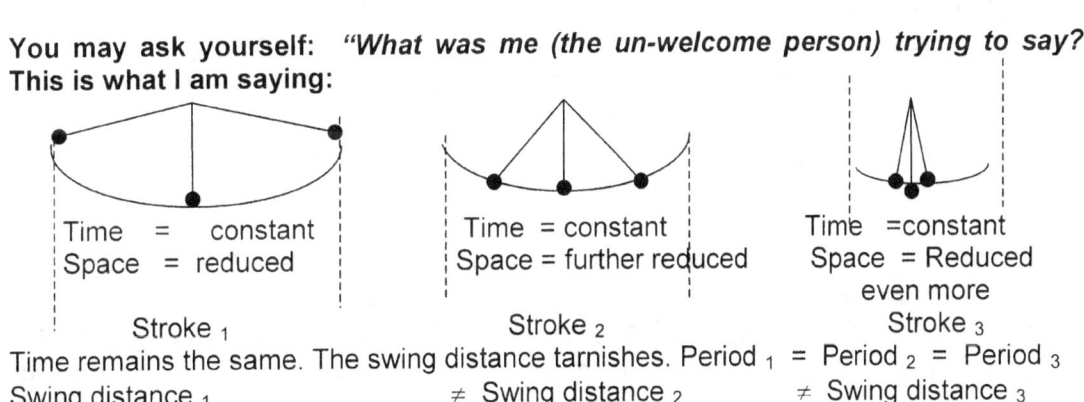

Time = constant	Time = constant	Time = constant
Space = reduced	Space = further reduced	Space = Reduced even more
Stroke $_1$	Stroke $_2$	Stroke $_3$

Time remains the same. The swing distance tarnishes. Period $_1$ = Period $_2$ = Period $_3$
Swing distance $_1$ ≠ Swing distance $_2$ ≠ Swing distance $_3$

In the pendulum principle that brought Galileo his everlasting fame, the pendulum swings at an even interval. As the **pendulum swings**, the **space tarnishes** while the **time (period) remains** the same. This is the principle on which all clocks work. What is it that Newtonians are missing for three hundred and something years about the pendulum? Newtonians are not seeing the very best example there is to indicate singularity outside singularity

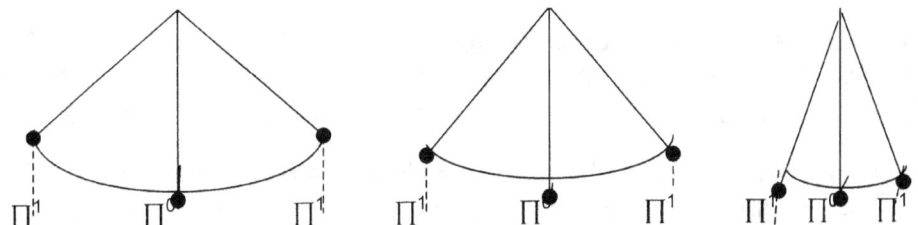

The pendulum indicates the very point of singularity the earth holds and the marks on both sides where singularity deviates in space, giving time to that singularity diverting.

If gravity, which is a force, did apply, then it would be as if a spring was fixed to the bottom of the pendulum and to an unmovable object below the pendulum. With the applying of the springtime will not remain at equilibrium but will tarnish in a vector with the declining of space. Something is holding time steady to the demise of space.

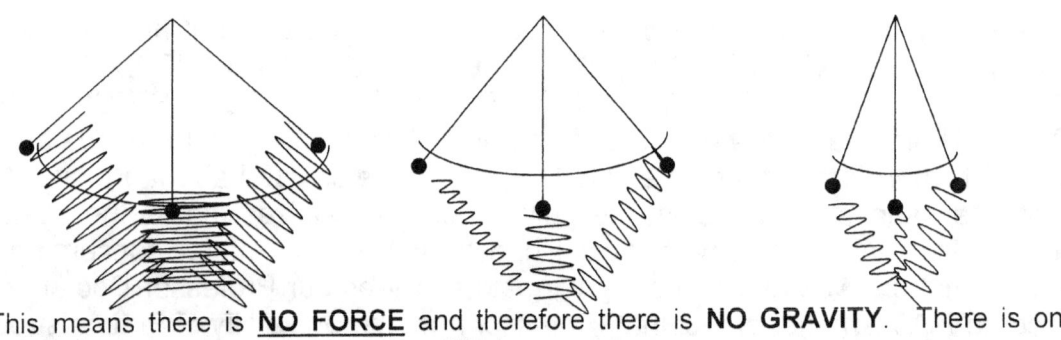

This means there is **NO FORCE** and therefore there is **NO GRAVITY**. There is only **space** (stroke) **time** (period) = **space-time.**

When a spring of 9,81 Nm is mounted to a pendulum, which is an equal force to that of gravity, both the time period and the swing distance would equally be affected, but to a lesser degree as the swing distance declines.

ALL SCIENTISTS GO INTO FRENZY BECAUSE GALILEO WAS PUT IN HOUSE ARREST FOR TEN YEARS! HOWEVER, THESE VERY SAME SCIENTISTS ARE STILL HAVING GALILEO'S WORK KEPT IN HOUSE ARREST AFTER ALMOST 350 odd years. HOW DO THEY EXPLAIN THAT? Galileo introduced the best devise indicating space-time and half a millennium onwards, nobody but me can see it.

What is space-time exactly? Einstein was the first to explain the existence of space-time, but Galileo was the first to indicate space-time and Kepler was the first to pin point the position of space-time. Einstein made one big error in judgment. In his all to well-known formula $E=MC^2$, he relates to space-time as if space had a factor of one and time was the altering factor. According to the Einstein / Newtonian Order of High Priests, we live in a total dark, totally flat, and single dimension Universe.

Why would it be a total dark Universe? According to Einstein, the speed of light is the same as the speed of time. Should that be true, photons have to freeze in time, and must be unable to move through space in time! (I shall elaborate on this in due time.) This proved how far the greatest Newtonian outside Newton really were off the mark

Look around you and see all structures (**space**) are different in size. Therefore, **space** cannot have a factor of none converting to one and back to none, but relate to the size of the object, whether it is an atom or the cosmic Universe.

Time (C^2) is at an even factor as all things in the Universe relate to the same time, (although not the same duration of time). By implying that $E = MC^2$, he puts R^3 at a relative value of one throughout the Universe. Space can hardly disappear, but can compromise under abnormal star growth.

Every round object has a point establishing a very centre, a middle dividing one side from the other. That division determines the space from one side away from the other side. At one point there must be a point that does not fall on either side of the divide. Such a point will still be a circle, because from that side the circle divides into two sectors.

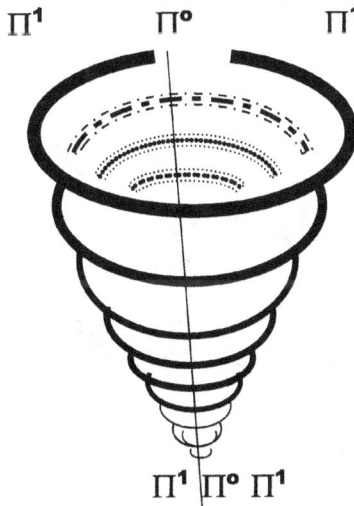

Π^1 Π^0 Π^1

Π^1 Π^0 Π^1

Every solar structure is spinning around an individual axis while the whole lot is spinning around a mutual axis the Sun provides The spin that shows on the different planets is the most crucial aspect of their orbiting the Sun. Calculating a circle involves two aspects where the one is either the radius or the diameter that is double the radius. The other is the factor Π. **$\Pi \times D^2 / 4$ = circle and $\Pi \times r^2$ = circle** The point of singularity cannot be in space at large because space is not there and secondly what ever is there spin to slowly to have a connection with singularity directly. The pendulum indicate the very point where all the Universe conjuncts placing space in relation to the time-Zero singularity as indicated through the position the Earth maintains individual singularity parting from cosmic singularity The pendulum is a direct measure of space-time flowing but to this moment where I write this, I still have to find one Academic that understand my connection. Since there is no Newtonian thus far capable of seeing the comparison I draw between the moving of space in the time it takes such space to move I shall repeat it once more in the hope one might see the light.

Take the pendulum.

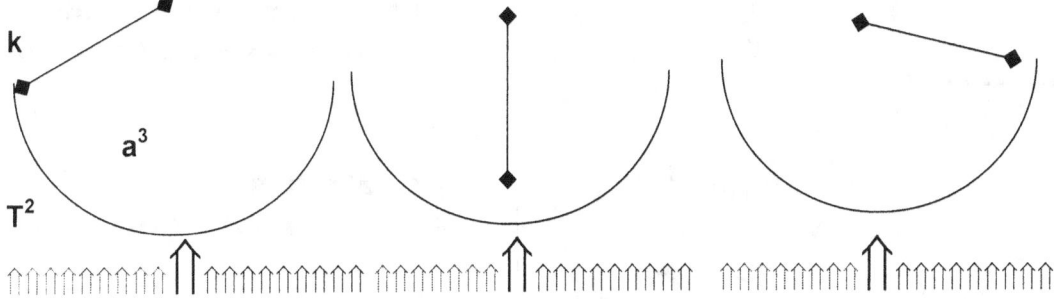

k

a^3

T^2

Every time the pendulum arm crosses to the other side it indicated the most important factor.

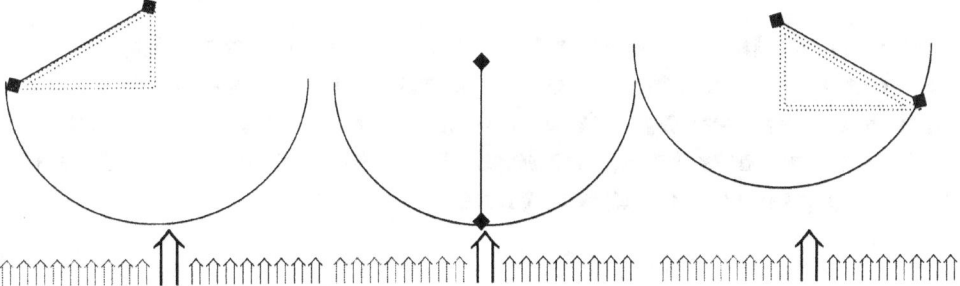

Every swing the pendulum arm does it brakes through the factor holding the universe in place.

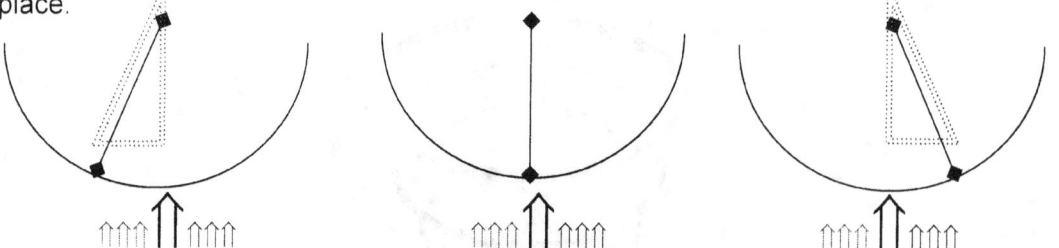

The pendulum not only crosses the singularity the earth dictates at that given time and the pendulum not only points at the factor maintaining space-time on earth.

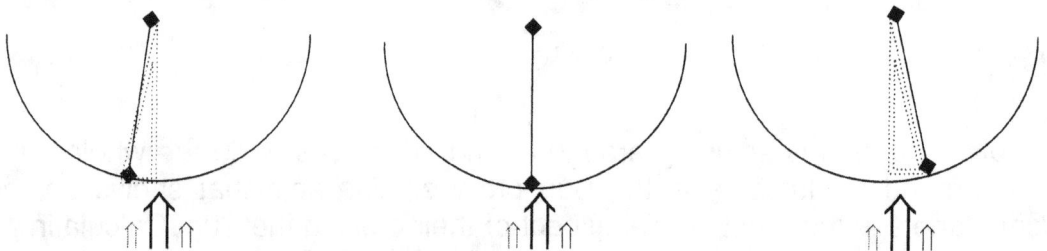

The spin of the top placed the top in a relevancy with time by producing space-time

The relation of time in 3 points in space

In the center runs a line called the axis line.

The line does not show any influence on managing the top when the top is motionless and bounded by the Earth gravity. However the sooner a motion sets in that is adequately strong enough to support the independence of the top the top generates enough gravity to sustain and independent attitude in relation to the Earth.

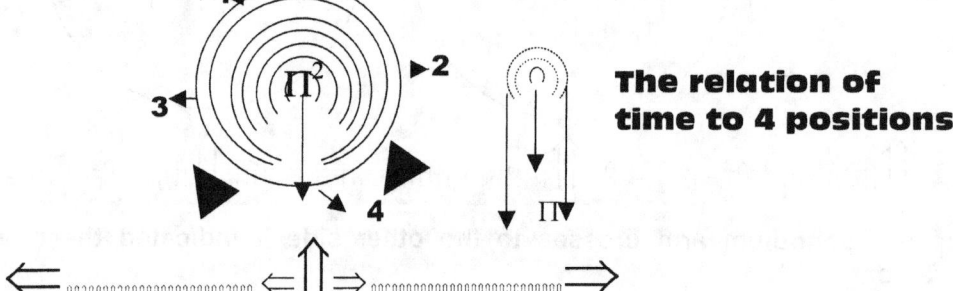

The relation of time to 4 positions

The spinning sphere activates the seven points, which places gravity in relation to a centre. Outside the centre there are five sides by dimension. The sphere has seven points of which four is spinning. The four spinning stands related to the gravity of spin, which are Π^2.

As the top hits the ground after being thrown with spin it starts to move around in small circles while rotating the axis. It spins in a vigorous manner as if the top is suddenly too energetic to stand still and that is precisely what happens. This surging finds a new dynamic and is a most important rule in cosmic principles.

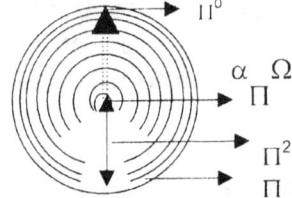

Nothing said so far is high tech or mind bending complicated. All the above arguments from the first page to this point reached are simple and there's only ordinary primary school mathematics involved that every scholar should know. One does not need a brain fitting Einstein to come to these conclusions but just thinking about everyday issues.

With all the excitement and no where to take it the extending of the drive line runs down the singularity inwards towards the newly established governing singularity that keeps the whole job erect.

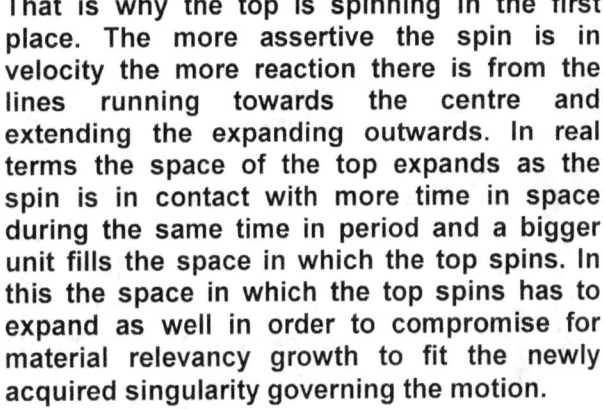

That is why the top is spinning in the first place. The more assertive the spin is in velocity the more reaction there is from the lines running towards the centre and extending the expanding outwards. In real terms the space of the top expands as the spin is in contact with more time in space during the same time in period and a bigger unit fills the space in which the top spins. In this the space in which the top spins has to expand as well in order to compromise for material relevancy growth to fit the newly acquired singularity governing the motion.

The support that the spinning top finds in the established governing singularity keeps the top spinning in an upright stance, only supported by the singularity that takes charge of the spinning space-time

The time activates a line from which space begins. The line forms a space that turns the top, but just as important is the fact that the line extending singularity activates a dimension in time in which the top can spin.

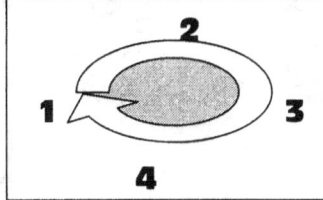

It is the time that relates to the spin of the space filled with material that allows the position the top has when spinning. When the co ordinance between the space spinning and the timeline keeping the balance falters that the top looses independence and fall.

The issue is that the rotation activated the time component where four points set about by motion. This formed the basis of the Coanda principle as the rotating top then became liquid although the wood is a solid. The point on which the rotation turns becomes the solid. This is evident in the fact that the center is activated and not activating the spin. As the spin reduces its dynamics the turning still fight for balance

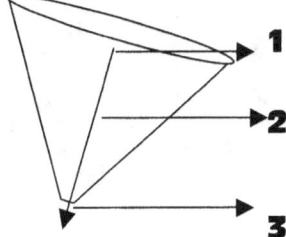

More spin increase both lines that force gravity by the increase of T^2 extending k, k^{-1} as well as a^3. The space wants to exceed its boundary because suddenly the motion allows the space to become extended. The gravity line running to the centre wants to extend for the same reasons and so does the gravity line running towards the liquid that should be there and that should be enforcing this sudden living up to better standards.

With all the excitement and nowhere to take it the extending goes down to singularity keeping the whole job erect. That is why the top is spinning in the first place. The more assertive the spin is the no reaction there is from the lines running towards and extending outwards. It should in real terms expand the top because the motion creates more space per time unit filled and the space has nowhere to go but the excite the newly established centre singularity.

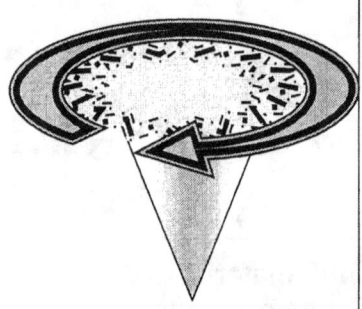

The spin under normal conditions can only come about as a result of more heat. With that aside the spin normally caused by heat will bring on a linear gravity running towards the centre of the top. This is then a product of $k = a^3/T^2$. But to counter this (Newton's law on action and reaction), another balance comes about $k^{-1} = T^2/a^3$ that centres the material in line with the progressive spin and the extending of the motion that should be because of a liquid heat adding to the material.

The support that the spinning top finds keeps it upright and performing as if in a fighting mood. However, again I have to press the point that it is life that initiates the motion and for this motion to start as a natural flow of events requires a lot of nourishing by the independent singularity that starts to drive the object when this process is indicating the birth of a newly developing star in the cradle of a galactica. However it can only be gravity that is able to fight gravity by extending the Earth gravity and by extending the Earth gravity we find some part of the Roche limit also applying.

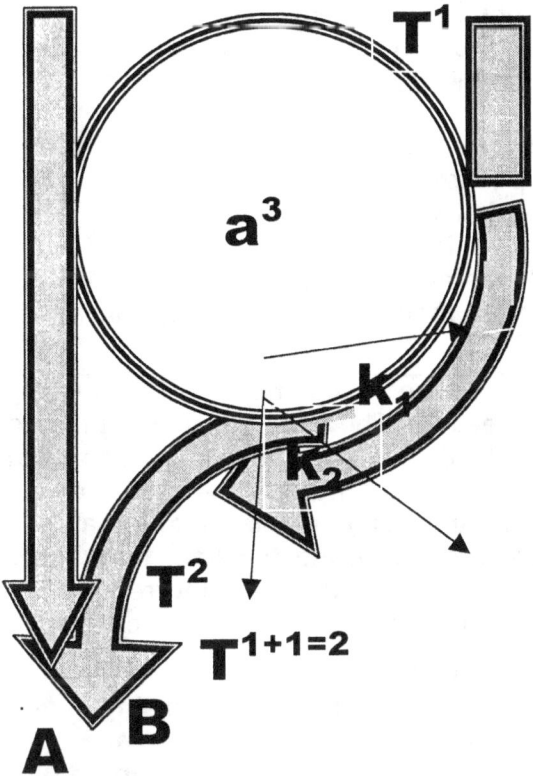

Normally water will run down to the centre of any gravity point, as A shows. By allowing the flowing water to come into contact, an object of a specific form the flow will divert (B) from the normal line and follow the contour of the object presented. For that to take place there is one condition that has to come about.

The Coanda effect is proof of gravity coming about through space forming motion. In the case where water diverts the normal directional flow the space that translates to the motion is deflecting singularity with the flowing water charging the motion. In the centre of the object having the round form, singularity is duplicated and by transferring Π to form Π^2 and the motion of the water creates a line of gravity that pushes the flowing water to follow the direction that the newly gravity applies to the water. This again proves Kepler's statement of $k = a^3/T^2$ that specifically states that space (in this case the object transferring singularity to a new position within the round object) and with the motion of the water redirects the gravity flow of the water to new space in new time. Only Kepler can explain the phenomenon but only when Kepler stands alone, correctly interpreted and divorced from Newton's opinion about Kepler's statements.

When standing still the body holds a relevancy of $7(3\Pi^2)$. That is the contact in gravity motion

In motion producing more of the relevant contact with time in relation to the Earth increase from Π^0 to $5\Pi^0$. The relevancy increases by $7(3\Pi^2)\Pi^0$ to $5\Pi^0$. It is because more material contacts more space or time in time duration as motion produce more space over time in time.

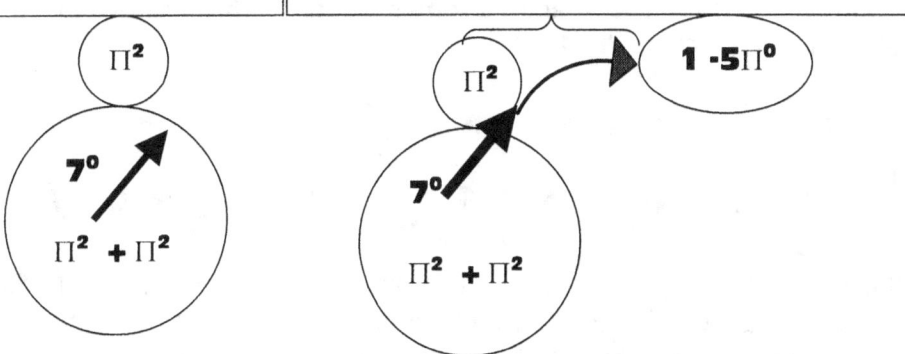

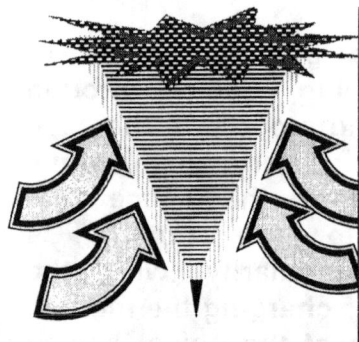

The heat that should supposedly under cosmic law drive the spinning top will come from the governing singularity accumulating the heat in concentration by the contraction or cooling ability the top singularity acquired. But in this case the spin is a result of life's ability to manipulate space-time and lead cosmic events. The heat that would establish such a drive in motion in real cosmic terms would require a lot of nourishing and sustaining from a large number of maintaining atoms that produce a large flow of space-time.

With sufficient energy the top gets into a fighting mood making the top very reluctant to give up this newly established freedom. The behavior now attributed to the top is normally the manner how a star develops in the galactica cocoon and how the fledgling star gains its birthright to leave the nest of the cradle of the galactica. The atoms form a sum total of space-time displacement that can support the generating of the required gravity in securing the heat that would unleash such a drive. Such singularity in governing come to life and release the new star from the blanket of heat that covered the star up to the time of release.

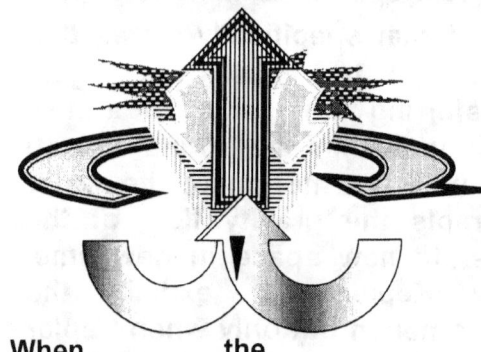

The example we can gather from the top shows how desperate a governing singularity can become when starved of motion and how such an exited singularity can put up a fight for life and independence. The top is in a fight for independence while the Earth is restraining the independence. The fight goes on until the earth suppresses the last bit of motion that the top has and the top uses the last motion it has to defy the Earth's control.

When   the motion exceeds the level of the Earth gravity, the top shows an eagerness to rise to higher levels of independence in the same manner that an electron reaches into higher rings of energy because the top with motion is in an electron relation with the Earth filling the proton role and the atmosphere being in the neutron role.

Let's quickly establish events as they translate singularity from a dot to a controlling entity that is commanding space-time through the establishing of a separate individual drive. The motion comes about which proves to be that which generates the gravity that drives the individuality in the top.

In the sphere centre is a spot that has to be there mathematically by measure of $(\Pi r^2) / (\Pi r^2) = \Pi^0 r^0 = 1$. In order to provoke the line into action, motion is required just as Kepler indicated where the space becomes equal to the motion and the motion is equal to the space $a^3 = T^2 k$

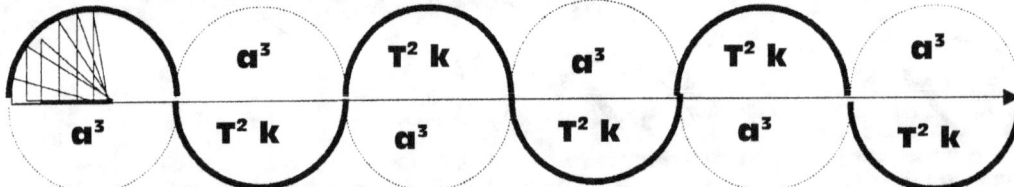

It is as if one then must claim in affect that Kepler held $a^3 = T^2 k = 0$. If the Sun and the Earth have a rotating relevancy of zero either the Sun has gone away or the Earth stopped existing. One cannot claim there is a wheel and then remove the spokes because according to you taste, too do not like the spokes

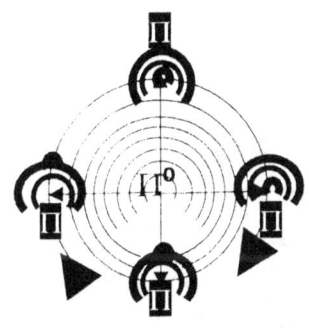

With everyone of the four rotating points duplicating the value of Π in relation to the centre $\Pi°$ at a measure of $\Pi / 2$ and where Π^2 is the establishing of Π by the motion thereof therefore $\Pi^2 / 4$ became a limit in relation to the developing centre. One has to remember that the star of the present takes characteristics of the form from the era before space was a factor.

As the absolute master of motion **Newton** should have placed emphasis on the motion aspect when he as a young man that saw an apple fall from a tree he made a brief calculation but he used the mass instead of the motion while Galileo proved that mass has nothing to do while the falling occurs. Seeing this he jotted down a formula and chucked it away. Newton however insisted on mass in spite of the clear evidence brought by Galileo to the contrary of mass playing a part. However most surprising to me is that most Newtonians are not only incapable of seeing the facts my way but they get sometimes pretty unpleasant in a very coldish pleasant way about my view. If mass had a major part, then the more massive must fall quicker because the mass will provide the drive and the drive will excel the velocity. While it is true that all things fall equally, then mass has no part to play while the dealing is occurring and that is in spite of all the Newtonian abstinence about the matter.

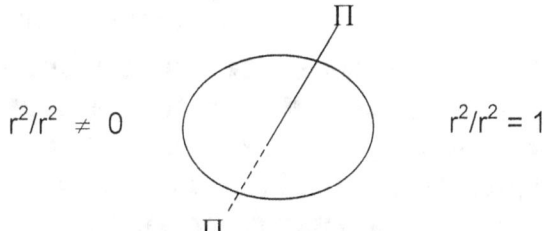

$$r^2/r^2 \neq 0 \qquad\qquad\qquad\qquad r^2/r^2 = 1$$

The relativity remains one, eternally zero. Therefore, dJ/dt cannot be zero dJ/dt can be eternal or infinitive or (dJ/dt = 1 but dJ/dt $\neq$ 0

In this matter I am disputing Newton's honesty. He placed the relevance on mass when he was a young man and retracting his former claim would have tarnished his reputation as a genius. His glory was worth more to his mind than what the truth was. As young man he drew instant fame by claiming mass as the driving force and when as an older man he found he had to retract the first genius, his fame seeking would have left him a scar on his reputation. In that I can forgive the man for the man was human and as Cecil John Rhodes said, all men have a price by which the man can be bought. Newton's academic genius was his all-important vice. The problem is that the incorrectness stuck with science fore almost four hundred years on and no brilliant mind since than was able to make the Galileo mass connection. What happened to the many wise that walked the path after Newton had gone to better grounds. Where is the honesty in those that were supposed to search for the unblemished truth? Gravity is motion and mass is the restraining of the motion of gravity. The top shows the truth. Let us reflect once more

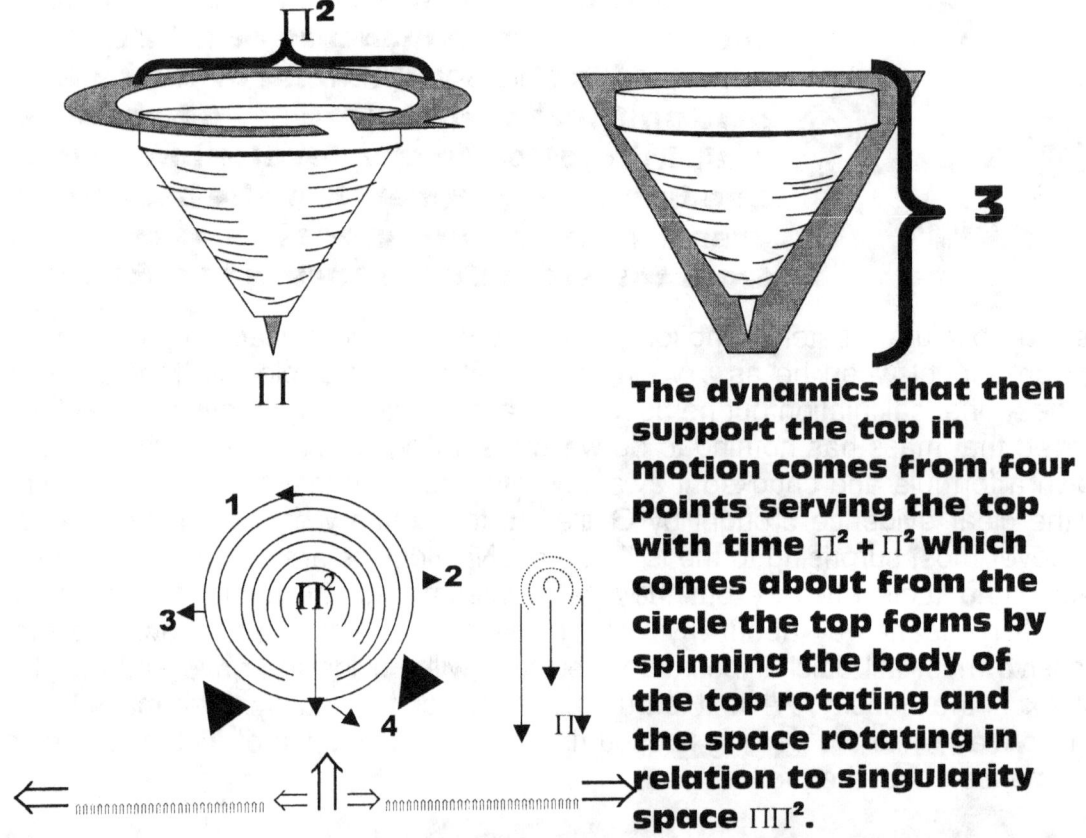

The dynamics that then support the top in motion comes from four points serving the top with time $\Pi^2 + \Pi^2$ which comes about from the circle the top forms by spinning the body of the top rotating and the space rotating in relation to singularity space $\Pi\Pi^2$.

Then there is the time aspect of three positions in space standing related to singularity in the centre which is formed by the seven points singularity hold in the spin as a unit. This is the position that the spinning body of the top forms with in time of from the past onto the present and into the future by motion. This is the layer of heat that immediately surrounds the top and forms the space that time holds in relation to the space the material holds $\Pi\Pi^2$ relating to the position in time $\Pi^2 + \Pi^2$.

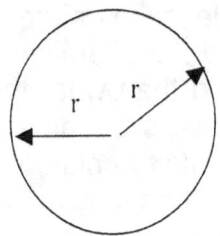

The radius r runs from the circle outwards, from a circle centre point towards Π, the value of the circle. In the centre of the circle, there is a point where the radius starts. It runs outwards from that point in all directions towards the circle Π. Technically, there then has to be a point where r is infinite and not zero, an absolute infinite. However, the circle therefore remains Π. The circle does not disappear; it remains there for all to see. It is only the radius that almost disappears into the infinite, but it does never become zero!

What is it the Newtonians fail to see? If an electron is orbiting around an atom, the inside of the atom must be a circle. If the atom was not a circle, it then had to be a cube. The electron cannot rotate around a cube; therefore, the inside of the atom is a circle. The cosmos is one big atom imitating all atoms as all atoms produce one Universe

In a circle, there is a radius that initiates the circle. The calculation of such a circle is $\Pi \times r^2$.

$$\frac{\Pi r^2}{r^2} = \Pi$$

If one removes the radius from the circle, the circle remains, only holding the value of Π. By removing the value of r, Π becomes singularity with no place to be. Singularity is the place where there is no space to be in place. However, Π remains because once r receives the slightest of space Π will find space. Then the circle will grow to Πr^2 and r would determine the space. Without space, there is no r but there is a circle with the value of Π.

Singularity is in every single rotating object, be it the proton or the combining effort of all particles in the Universe. That is what light and the photon is. It is concentrated heat that the Sun (or any other generator of electricity) concentrates to connect the concentrated heat to singularity where the heat receives either temporary connection to singularity or a small piece of individual singularity. All spinning matter has the point where the spin is still there but the radius is to small to measure by any means. That point is standing still in relation to the rest of the spin. In relation to that logic I do not accept Newtonian science holding the radius of the spinning object unrelated to the spin, whether the spin is applying or not.

Applying Newton's second law F=ma
One arrive at the formula
GMm / r^2 = m ($\omega^2 r$)
By replacing ($\omega^2 r$) with 2Π / T we obtain Kepler's third law
This law predicts that $T^2 = a^3 r$

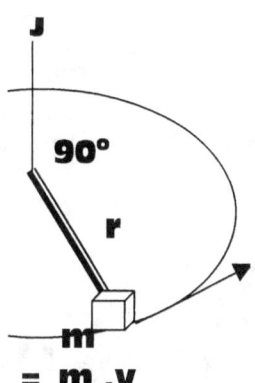

The mass (m) multiplying the speed (v) forms a new value J AND THEREFORE j CONTINUOUS TO IMPLY J = I ω

J = r X p where p = (v =r x ω)

J = r.m.v = m.r^2 .ω = I. ω and becomes interpreted as J = I ω

This establishes that r = dJ / dt

Since this is the absolute crux that Newtonian science pivots around I feel it is important enough to return to the whole issue once more in similar detail.

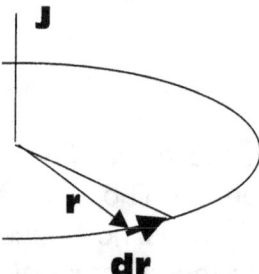

r = dJ / dt In the case of planets in orbit around the Sun r forms a value of zero because dJ / dt = 0.

Since Newton became an institution forming the King bee of the academic cartel world wide The Brainy Bunch had Newton's vision written in the minds of the future generations almost at gunpoint…well definitely at an academic gunpoint.

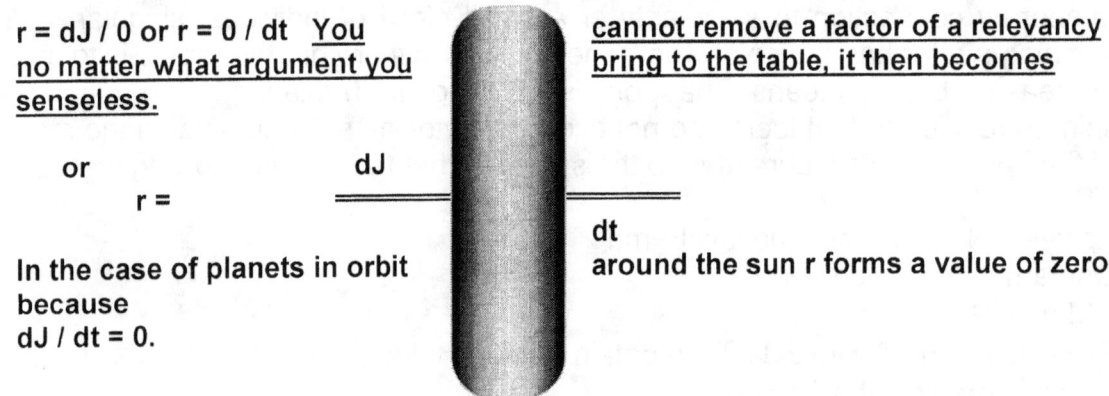

r = dJ / 0 or r = 0 / dt <u>You</u> <u>no matter what argument you</u> <u>senseless.</u>

<u>cannot remove a factor of a relevancy</u> <u>bring to the table, it then becomes</u>

or

r =

dJ

dt

In the case of planets in orbit because dJ / dt = 0.

around the sun r forms a value of zero

I am not the brightest in the world that I admit, but one thing no one can do, not even if you are the one and only Isaac Newton, is that you cannot place any relevancy in a relevancy and then claim it not to be in a relevancy because such a relevancy does not suit your taste.

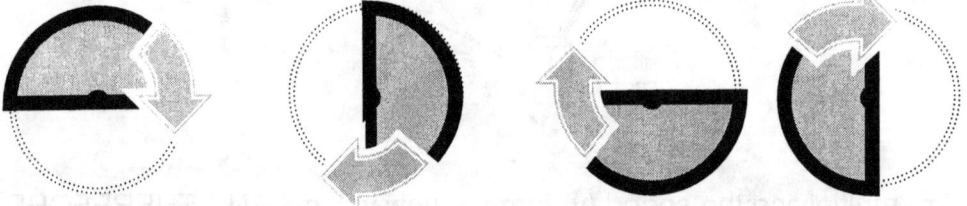

I wonder where would one put the zero part on the spinning wheel and what part must be excluded from the wheel. What Newton suggests, is a wheel has one side on top and no side at the bottom. While the wheel is spinning one may not remove the one side and then claim there is no attachment between the top and the bottom. That would mean in a graph the top is not connected to the bottom because a wheel spinning is a graph moving against time. It is the principle all driving is done and not the least electricity.

Every quarter of a rotating body is opposing the opposite sector directly and completely.

You cannot put something in relation to another object and then decide there is no relevancy in the relevancy

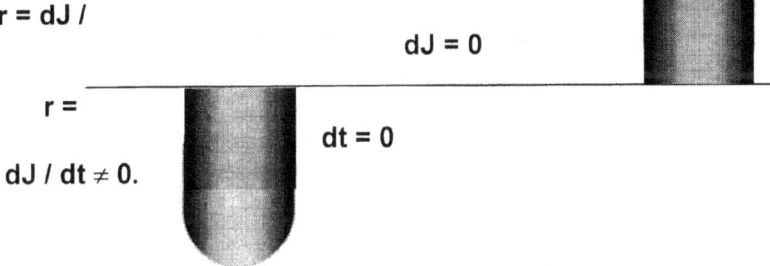

r = dJ /

dJ = 0

r =

dt = 0

dJ / dt ≠ 0.

If dJ = 0 then dt = 0 That is a mathematical principle, much larger than even Newton

Quite the very opposite is true and the rotating wheel in fact is the moving wave.

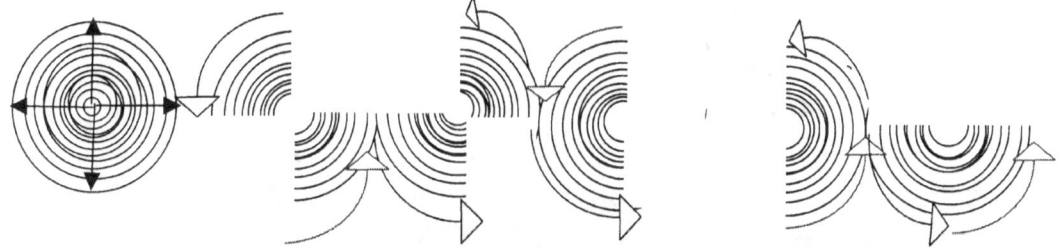

The graph which is a cornerstone of mathematics and which is used extensively in various calculating procedures would not function because at the end of a cycle there will be no cycle.

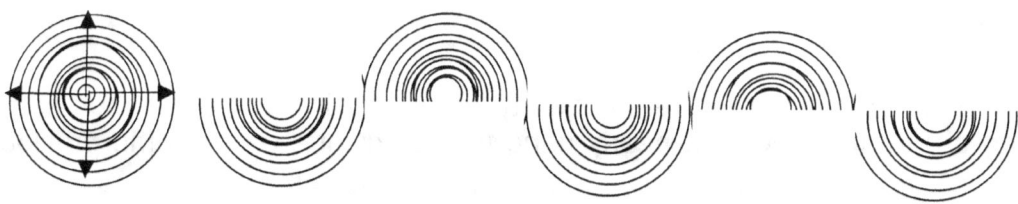

That proves that every rotating object holds the form of a wave and mostly it

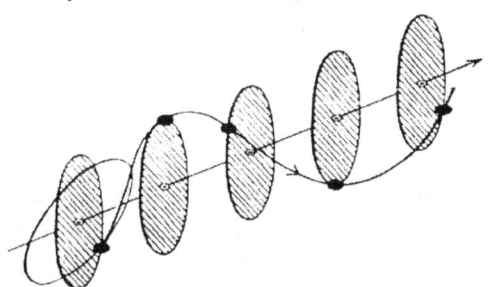

etiolates Newton's claim totally that no relevancy exist between the rotation and the axle which is pointing towards motion. That proves that while spinning, an object holds an absolute relevancy of one in order to maintain equilibrium as to ensure rotating motion. In both cases, one of the factors then does not exist. Such a claim is incoherent, because you proclaim that a circle has no radius, or a radius has no circle. When calculating a circle, you multiply either the square of the radius by Π, or the quarter of the diameter at a square by Π. Newton's claim suggests that a wheel in rotation will return to the same spot it had previously, as it does not affect the spin. It is the same process, which science uses to try and locate unidentifiably small cosmic objects. In that case they call it the gravitational pull.

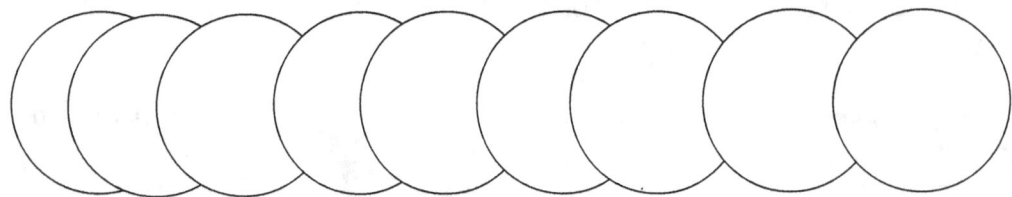

The fact that rotation does not produce work is impossible since rotation brings about motion changing the principles of the location.

One do seem to get the impression that little changes in line with the rotation will bring some forward motion and some returning to the original position.

Even by using half a wheel would still bring considerable confusion but one can clearly see that Newton's presumption does not quite match reality

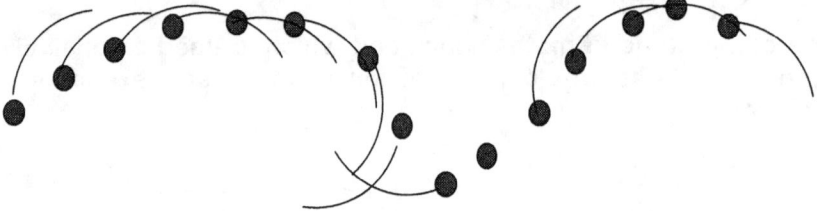

Shortening the arch changes the complexity considerably as one can then see a changing of the arch does not nearly bring the return of the dot to the previous spot.

When placing pointing arrows indicating the direction the line of movement, it becomes clear that there is a complete mismatching and the cosmos changes as rotation progresses. The behaviour, which I describe, is a flow of space through the line of time. Electricity is charged in this manner. The same generated force keeping the top upright is what is used to generate electricity. The flow of a charged conductor through excited space-time brings about the flow of a current.

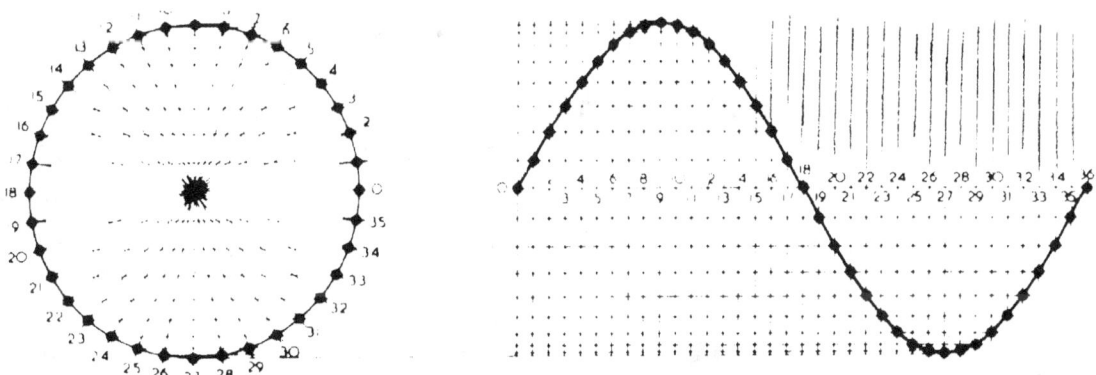

The motion the top has is directly a mirror image of what we find in the sine wave, which is used to calculate electricity. There is the line or Earth or singularity from which space-time diverts.

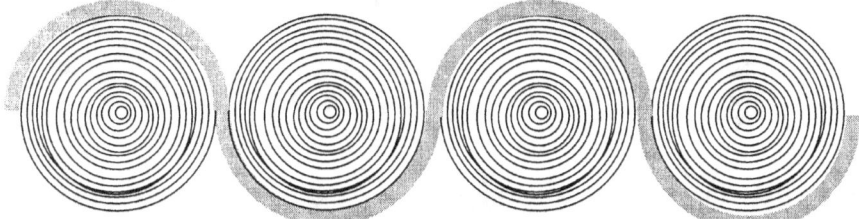

From   such   a   relevancy there then   must   be   four different values relating to singularity and since the atom has a proven relevancy of $(\Pi^2 + \Pi^2)\, \Pi^2 \times \Pi \times 3$. In the motion that replace the motionless, the motion made the motion less an atom by putting the object through the commitment of motion into an independent Universe where every aspect of the Universe becomes an Individual atom that maintains its independence as all atoms do.

It started with a dot, because that is the only form, size and dimension mathematical logic will allow our brain to accept. From the one dot had to come a second dot and a third dot. The dynamics of such a dot is smaller than we can understand because such a dot is in negative relation to what we see Π to be, and the deeper we delve in finding the smallest fragment where space started, in the spot where time is still eternal as much as we can accept eternity to be. The reason why we should first locate the spot is because we can only work from that point forward. By working forward we have to work backwards to locate where we are heading. The cosmos started at a point and where such a point is, we will find the Universe. Every one knows where the Universe is, because we can see where the Universe is, but if we can see where the Universe is, then we should find the centre of the Universe in that spot. Einstein theoretically positioned the point of beginning at a place he indicated where singularity should be. With the cosmos the size it is and space so large compared to our smallness we have no chance in finding the centre of the Universe. The Universe started where singularity is and singularity is the sure indicator of the Universe. With all spinning objects holding singularity we then have located singularity in as much as finding the centre of the Universe. The Universe started with a dot forming. That answer arrive from taking mathematics back to a point of being the smallest possible position, far smaller than we may be able to calculate form.

Even the electron serves the line of time, in the same manner. As the Earth spins through time by repositioning space in time singularity is re-applied, repositioned and re-aligned with the entire Universe in the manner I describe. The relation of the proton moving has to effect the following location of the electron since the electron is relevant to a position in space in time by a continuous motion through time.

Route the electron follows

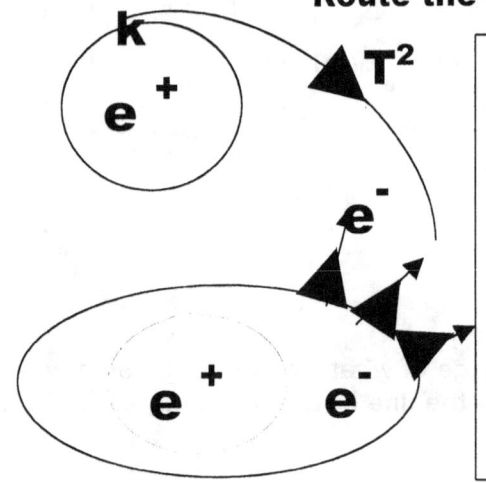

There is **k** that forms the distance between the proton and the electron while the electron is spinning T^2 around the proton k^0. While all this action is going on, we think of the atom as being very still and satisfied with being a small part in a lump of metal we call iron. It could be any element but I use iron just as an example this time. The lump of iron is as motionless on earth as anything can be while being.

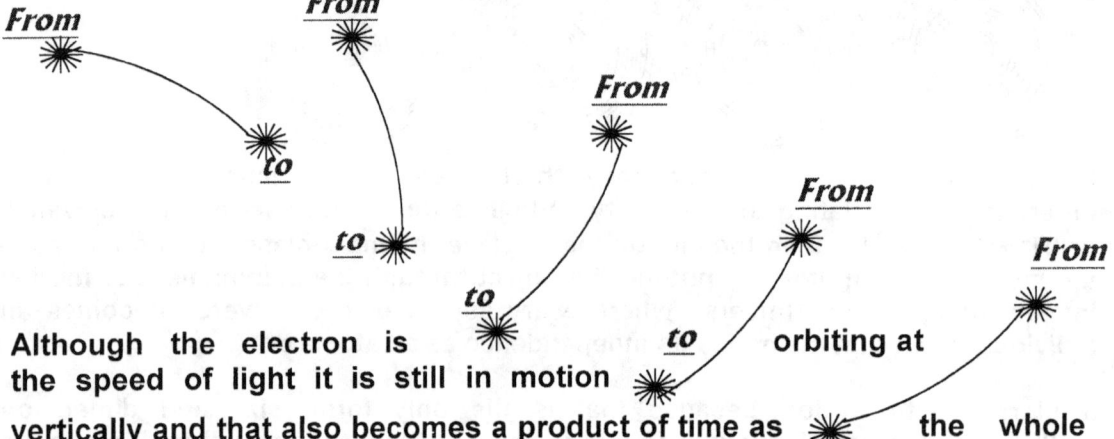

Although the electron is   orbiting at the speed of light it is still in motion vertically and that also becomes a product of time as   the whole structure is repositioning the relevancies, it had a moment

before to what it will have the next moment. Such   motion will again have an influence on the relation in the   position the electron forms with the rest of the Universe while the lump of metal is now travelling as a spacecraft destined to other galactica. It is if we use the logic those intellectuals calling them Academics show and those Super-Educated that advocate how we may travel to far away galactica while we go on skipping the nearby galactica that is only two to twenty million light years away. Since the electron is duplicating by motion the motion links the electron to a time constant. The time constant is linked to the speed of light but time as such, is part of the speed of light. The faster we take the electron to go straight in the motion man produces, the less time there will for the electron be to circle around the atom. If we make **k** bigger in relation to increased motion, the smaller will T^2 produce a usable space.

Even by coming erect through motion the generating of this stance finds its roots in the relocating of the rotating (T^2) in relation to the alignment with the line (**k**) in relation to space-time a^3 in time-space T^2k. The generating of the top and of gravity and electricity is provided in the very same manner by the Coanda principle.

By the motion and the singularity the top evokes a graph form where the graph runs along the line of time.

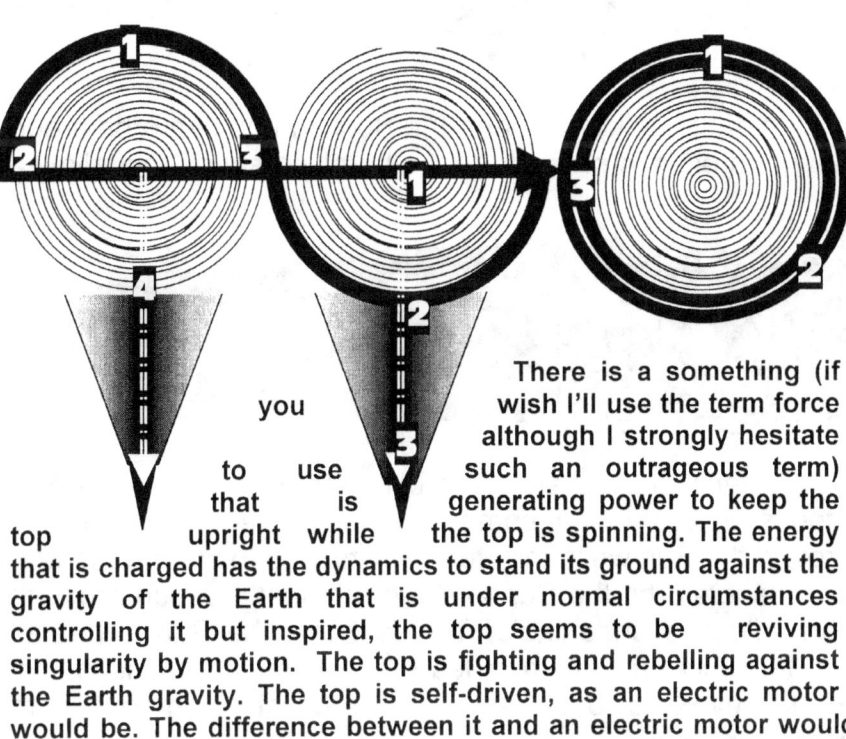

The balance is a control of motion that is established as a flow of space-time supporting the ends **(4)** holding time while this generates the space **(3)** in singularity containing and creating the space **(3)** in which the spinning takes place.

There is a something (if you wish I'll use the term force although I strongly hesitate to use such an outrageous term) generating power to keep the top upright while the top is spinning. The energy that is charged has the dynamics to stand its ground against the gravity of the Earth that is under normal circumstances controlling it but inspired, the top seems to be reviving singularity by motion. The top is fighting and rebelling against the Earth gravity. The top is self-driven, as an electric motor would be. The difference between it and an electric motor would be the origin of the source from where the energy comes which drives the spinning top. The top stands upright as individual as any self-propelled object can be. Although gravity is retaining the motion of the top, it is not contradicting the motion. It is not combating but is merely suppressing the motion. What we would think of as air restriction is no restriction because from the restriction comes support that keeps the top standing on a very thin needle edge. The top should tell us so much about cosmic laws, if we would only listen and learn and not tell cosmic laws what we think nature should tell us.

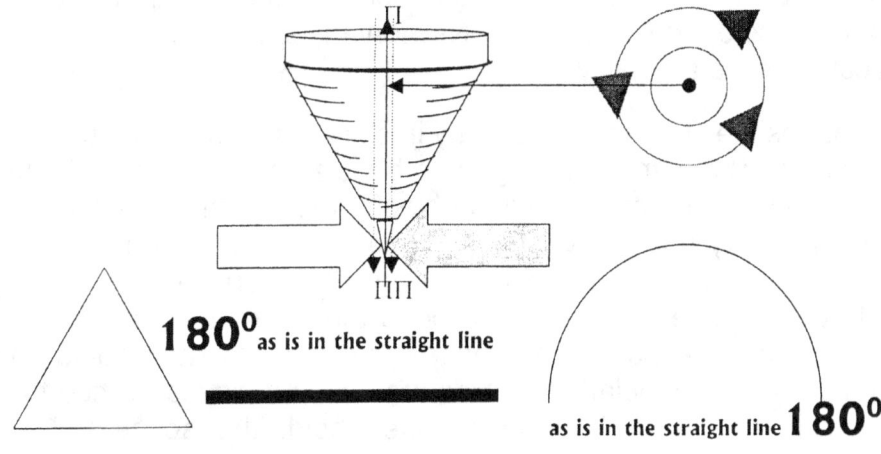

180⁰ as is in the straight line

180⁰ as is in the triangle

as is in the straight line **180⁰**

Even the motion of innumerable stars relate to a singularity in the centre that plays the part of the generated governing singularity and every faintest and slightest motion of every individual object plays a significant part in the generating of the governing singularity.

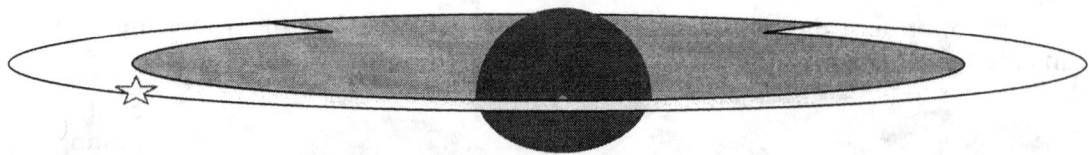

The Sun is on the outskirt of the Milky Way and the Sun is in an ova orbit around the Milky Way. The law of orbit is in principle that all orbiting structures follow an oval path.

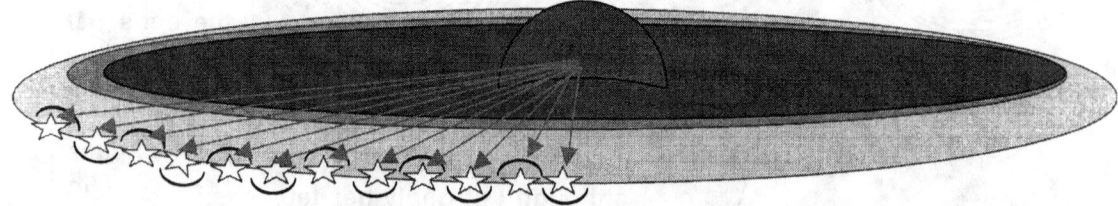

Exaggerated to a large extend the influence the Milky Way has to have on the Earth orbit comes to focus when a pattern comes in pace as the Earth follows not a circle but a wave around the Sun while the Sun sets its motion around the Milky Way. The fact that the planets orbit the Sun and the fact that the Sun orbits the Milky Way indicate an influence undeniable. The fact that the Sun is heading farther away from the influence should then lead to a variation in the planets orbiting wave. The Earth never, not once lands on the exact same spot by the completion of one more year cycle.

By not having a wheel rotate, the wheel becomes the factor of one, and the rotation becomes zero. The wheel does not disappear. In the cosmos, everything is rotating because nothing ever stands still. Therefore the mean equilibrium, the common factor there is to share, has to be one, eternity, the eternal Π, because all rotating objects has Π in singularity, and sharing singularity, gives every object in space a relation with all other objects in space. After trying for many years to bring them the candle, I concluded that Newtonians are incapable of realizing that mathematical principle as reality.

The comet rotates the sun, and the Sun by itself has a point of singularity where Π remains without r. The comet, holding the orbit, also has a point of singularity, but since there is space separating the two objects, they cannot share a mean point of singularity, the very point of existing. Since singularity means just that, being single, there cannot be two. The comet and the Sun have a mean point of singularity but the space they occupy divides their common singularity. That is why they orbit in an oval path, a path where the one structure holds on to more space from its point of singularity towards the space it claims. Since they do not claim equal space, BY THE DENSITY they hold, the space will not be in proportion.

They do share in the common fact of singularity a point away from their individual singularity proclaiming their cosmic individual reason to exist in the cosmos. That point of common singularity holds space between individual singularity and that point of mutual singularity saves and protects the points of individual singularity. Since the start of time at moment-Alfa where both found the space they occupy, in the space they hold, maintaining a time to that space in accordance to the singularity they hold that point will be their individual eccentricity from singularity. The two objects are holding eccentric space around their individual but common singularity. That point of singularity is Π the circle without the radius because the singularity removes all forms or values of r, discarding r to infinity and leaving Π to be singularity.

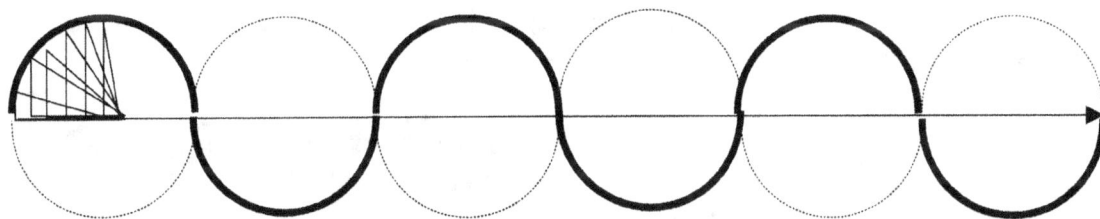

$a^3 = T^2\, k$ then $k^3 = k^2\, k$ and this is showing that the space k^3 is equal = to the motion $k^2\, k$ of the space k^3 seen form one specific point.

In the shown formula introduced by Kepler years before Newton changed it, the formula shows that a space, which is clearly defined by a property of being individual, has the ability of motion and that is what Isaac Newton saw and named as gravity. He saw the falling and initially named that falling gravity. However Newton was of the opinion that a force was in charge of the pulling the one object closer to another object. He put the emphasis on the material having the status of pulling each other. To my mind that is incorrect because I see the flowing of space that may or may not contain material and the space containing material or not is flowing towards singularity. I announced years ago there is no such thing as gravity after which most Academics thought me to be a loony and even my friends would have me institutionalised behind my back. I then changed my song somewhat not because I do now believe there is gravity because the gravity Newton referred and which Newtonians accept, that there is not. There is no contraction between particles pulling them closer.

That is why gravity is a fixation of Newton's mind making Newton bullshit, and his $F = G(M_1 M_2)/r^2$ is utter nonsense. The moment you say Newton or any of Newton's laws, the Newtonian brain stun. Not once did I find one Newtonian surprised at this, I could not once find one single Newtonian to see this. It always leads to an argument and the argument is about Newton being in use for centuries. One Professor even answered me by saying that I should realize Newton's formulas placedman on the moon, and if that is not proof of his correctness to me I will never obtain proof. That is besides the point. That is miles from the issue. If you say Newton is wrong, you commit the worst blasphemy possible. One may swear at God and all is understood but mention your not accepting Newton's gravity and they all fall on their knees, cover their eyes in the ground, start stuttering and moaning and you cannot make them see anything but Newton. Dare say there is no such a thing as gravity because Newton is wrong, they run outside and hide the woman and children from your rage of mental instability.

Because I have had unmentionable arguments that I in the end lost because the mental Newtonian block all Newtonians hold covering their senses, where I could not reach a single spot of healthy logic within their minds, I wish to run through the facts once more and find what is so incomprehensible about the issue.

This in fact, is the very same findings that brought Johannes Kepler his own everlasting fame when he declared that the planets stand to a value of $a^3 = T^2k$ as they orbit the sun. Never once did he mention the presence of a force or gravity. Newton came up with this bogus idea all by himself without the help of other "giants" as he called *Galileo and Kepler*. In a later stage I indicate that I might prove the possibility that Newton did not have enough information to draw conclusions about Kepler's work. Newton saw a circle in Kepler's formula and there is a Universe of information hiding in that formula because that formula depicts the key to science namely singularity.

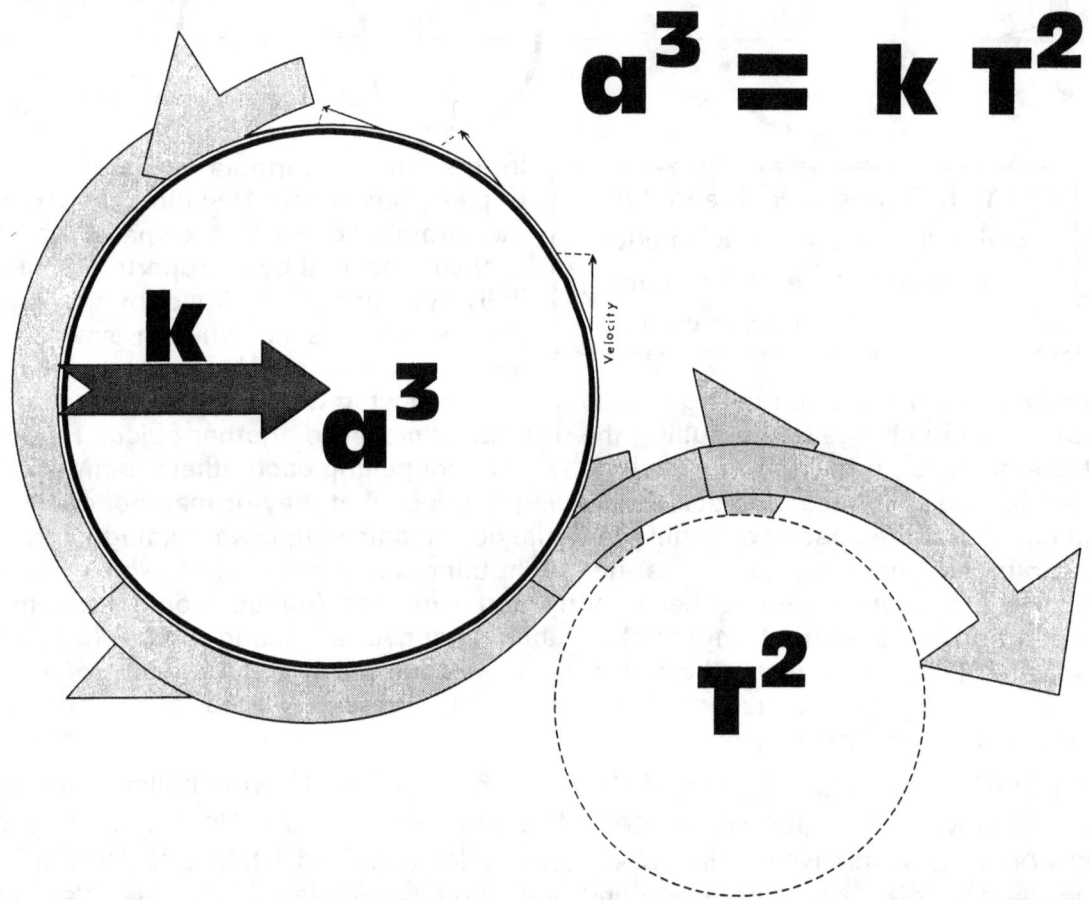

$$a^3 = k T^2$$

Newton made the formula one big blunder as far as the cosmos is concerned. Newton works perfectly well where there is equilibrium and unchanging in space and time as we find on Earth with the Earth forming the basis for space-time. Taking Newton to outer space is a blunder and Newton created the blunder by re-adapting his original formula of $F = r^2/ (m_1 X m_2)$ to fit Kepler's vision of $a^3 = T^2 k$. This very same bogus idea helped Einstein to ignore the space factor of R^3 and place a relative value of one to space. This he stated (without stating it) when Einstein put the Universe in a single dimension property at the point where gravity was stretched to the limit. Einstein put the Universe to a three dimensional value of matter, space and time and then out of the blue he places space at a factor of one when gravity supposedly destroys time. This notion stands totally unrelated and divorced to reality. I do admit that Einstein is absolutely accurate when saying this, but the space he refers to, as outer space and the space disappearing in time are as far apart as the cosmos is wide. Einstein was the one that said that space and time could never be separated because it was the very same thing, a point I agree with in all my findings. The difference between my point holding the

Universe and being the Universe is within every atom because it is there where singularity is. From singularity through the atom space has the relation between Π^0 as singularity and Π forming space inside singularity holding time $\mathbf{T^2} = \Pi^2$ in relation to the triple value of $\Pi\Pi\Pi$ forming Π^3. I am afraid that Einstein made much more sense when he was still an amateur, working as a clerk in the Swiss patent offices. Then he landed himself under the spell of the Newtonian disciples and all his initial ideas that were factual, became integrated and confused with delusions of the "Xepted scientific Newtonian High Priests" called "acclaimed scientists" and their mesmerizing fantasies about gravity.

There is a way to explain space-time by finding **space-time**. Space-time is not some force well and truly out of our reach, the one we may dream about and wonder why we have to adhere to it with so much respect. Einstein the master of physics was completely lost in his physics. He went looking for a flat Universe, he saw singularity in the dark of the night hiding as obscure fairytale characters behind Black Holes, where he saw gravity lingering around stars with nothing better to do than to wait for passing light just to bend seven variations of different types of shit out of each of them. Singularity makes every atom rotate that makes every cosmic object rotate that applies the overall rotation to the Universe. **THAT IS TIME**. Each time I try to share the idea of mine with the **"Accomplished Scientists",** I do not get farther than the phrase*: "Newton and Einstein are wrong."* After completing this sentence, I get treated as a raving lunatic with extremely dangerous hallucinations indicating a murderous tendency. Why would not one academic listen to the rest I wish to say before bluntly denouncing me?

Nobody even listens or pretend to listen to the rest of my case. Every time I see the light in their eyes go blank and they sit patiently and wait for the motor mechanic to finish his senseless rambling. It is so obvious they consider me as mindless person with arrogance and having a nerve to criticize the two highest-ranking Newtonians of all time! I can assure you I am not mentally disabled! It took me twenty-one years of research and another six years in compiling and writing this book.

That is the relation matter has outside singularity. $R^3 / T^2 = 1$.

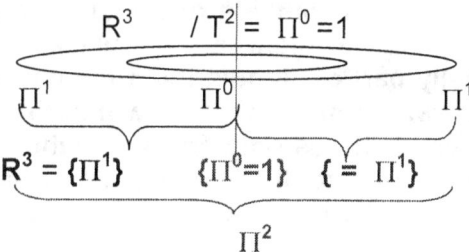

In any of the pictures on the next page one does not see space, because you see a space filled with particles. It is the atoms holding the space secured that forms the picture. Why on Earth would nobody realize Einstein was seeing the Universe from a wrong perspective? What Einstein saw was one hundred percent correct but Einstein saw what he saw in the space of the atom and not in space at large.

That is space-time holding every aspect the Universe hold to a specific relevancy The space outside singularity holds the time outside singularity because everything in the Universe is spinning. Science knows there has to be a difference because in space an object might be weightless, although it retains its mass, and no one can say the difference, except to put it down to "gravity". Us, the tax paying public, are letting these Master Minded Academics get off the hook so easily, because every one is so scared to ask "why and how". In the pages above, I pointed to the most basic mistakes about the "gravity" which science ignores, because the answers they do not know. Even Nobel Prize winning work is blatantly misguided. I challenge any person to prove how an atom can collapse on itself, by force, by weight, by pressure or by any other means. No atoms will ever touch one another let alone compress to diminishing space, and if they do, the result is a nuclear reaction

IF IT DID NOT SPIN, IT WAS NOT ROUND, AND NOT BEING ROUND THERE WILL NOT BE SINGULARITY.

According to Einstein, the speed of light is a constant throughout the Universe. The speed of light results from two factors, being distance (kilometres) and time (seconds). This speed is accepted at 3×10^6 kilometres per second. Scientists know that it takes Sunlight 10^6 years to reach the surface of the sun, and we know the Sun is not thousands of billions of kilometres in diameter. The "Xepted scientific Newton Mistaken" explanation about this fact is that the sunlight "bounces against matter" and this retards the Sunlight dramatically. When light hits matter, (except in the case of glass), it joins singularity immediately. Therefore, the Sun holding matter on the inside has to be all- glass, or the "Xepted scientific" explanation is not very scientific at all. It all comes down to the density of matter in space valuing the time in that space away from the point maintaining singularity. Why can nobody but me see that?

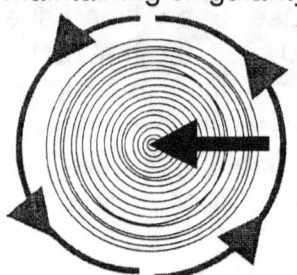

This is the dot that has no start. It is 1^1, the part that released from the spot 1^0 when motion parted singularity. It came apart when motion unleashed the dot 1^1 that has no start from the spot 1^0 that has no end. It is the Universe born from motion that was driven by heat. It is still there because once anything is part of the Universe and forms a principle within the Universe it has n where to go but to remain within the Universe. Walk outside at any time and you are a witness to the result.

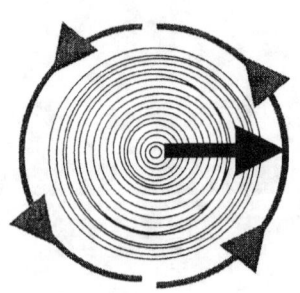

When you walk outside and look at the vastness of the blue sky or at night at the black night sky, you are physically part of singularity in the part of 1^0, the part that moved away from 1^1. You are within the part 1^0 that has no end because it has only one side, which is the inside. It is 1^0 going nowhere. It is the part that I named the spot that had the dot 1^1 moved away from

This shows that light uses pulsating circles to move and while the one side of the Universe is dark the light flows in the opposite direction to how it flowed in the previous half cycle. In that was it could travel forward, backwards and equal at the same time using the same pace.

The pendulum arm covers a specific distance per time unit, every instant it swings. This is because of Singularity in position $\mathbf{a^3}$ during time $\mathbf{T^2}$ in instant $\mathbf{k}$.

The space $\mathbf{a^3}$ holds precise accordance to the time $\mathbf{T^2}$ that it takes minus the compromise singularity claims from k by reducing space to the increase in heat.

Π^1 Π^0 Π^1

Space-time depends on the relevancy of matter occupying space change position in accordance to all other matter relating or relevant or even only influenced by the space $\mathbf{a^3}$ in the e duration of the time the matter changes position T^2 in the instant of changing. **Space-time is everything excluding singularity diverting from singularity** and that is what Galileo recognised without realising in his observation of the pendulum. Where Π^0 is singularity and Π^1 is the diversion from singularity forming $\Pi \times \Pi = \Pi^2$ being gravity or time.

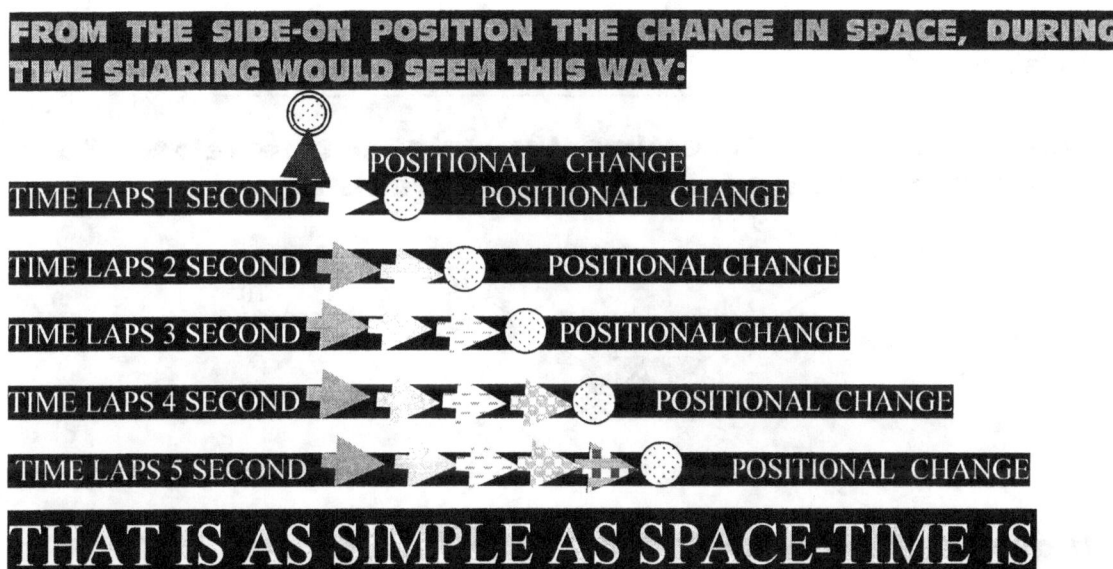

FROM THE SIDE-ON POSITION THE CHANGE IN SPACE, DURING TIME SHARING WOULD SEEM THIS WAY:

TIME LAPS 1 SECOND — POSITIONAL CHANGE POSITIONAL CHANGE

TIME LAPS 2 SECOND — POSITIONAL CHANGE

TIME LAPS 3 SECOND — POSITIONAL CHANGE

TIME LAPS 4 SECOND — POSITIONAL CHANGE

TIME LAPS 5 SECOND — POSITIONAL CHANGE

THAT IS AS SIMPLE AS SPACE-TIME IS

That is what Kepler (again I cannot say whether he wittingly or unwittingly) declared by using the formula $\mathbf{a^3 = T^2\,k}$ he announced space-time in a formula, the formula Newton raped to his advantage because in $\dfrac{M_s \times M_c}{r^2}\,G = F$ there can be no pointing to singularity in the cosmic sense. In his initial formula $F = r^2 /$ (Mm) singularity point at every aspect because as matter falls to Earth, matter continues down a precise path that singularity provide holding that specific position that leads the way. What it does point at is the motion caused by the Earth's singularity applying on much lesser objects holding or not holding singularity. I change a to R and T to T holding $\mathbf{k}$ to Π^0, which is singularity in the instant

FROM THAT POINT SINGULARITY IS IN EVERY PROTON HOLDING SPACE AS RELATIVE AS TIME. R^3 / T^2 = one

This in fact, is the very same findings that brought Johannes Kepler his own everlasting fame when he declared that the planets stand to a value of $R^3 = T^2$ as they orbit the sun.

Illustrated it would be represented as follows:

Illustrated it would be represented as follows:

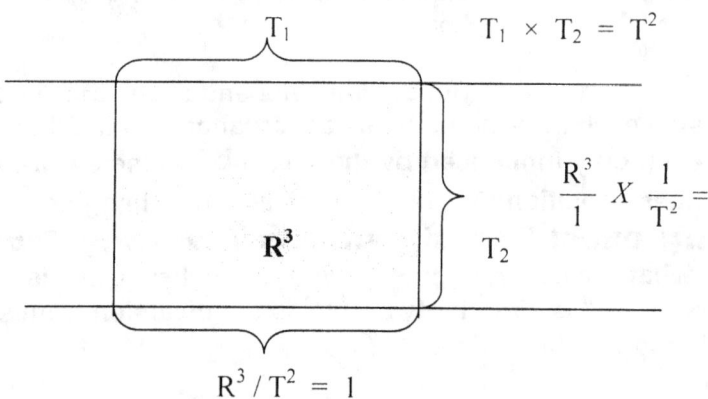

$$T_1$$

$$T_1 \times T_2 = T^2$$

$$\frac{R^3}{1} \; X \; \frac{1}{T^2} =$$

$$R^3 \qquad T_2$$

$$R^3 / T^2 = 1$$

That means when the time that a structure relates to, is effected, the space will be effected pro-rata.

That means when the time that a structure relates to, is effected, the space will be effected pro-rata.

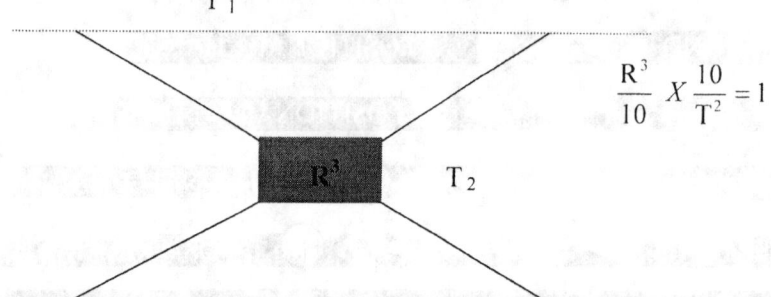

$$T_1$$

$$\frac{R^3}{10} \; X \; \frac{10}{T^2} = 1$$

$$R^3 \qquad T_2$$

If one can illustrate the universe and its relation with space-time, the following illustration would fit like a glove.

$$R^3 \qquad\qquad T^2$$

SSSSSSSSSSSSSSSSSSSSSSST
SSSSSSSSSSSSSSSSSSSSSTTT
SSSSSSSSSSSSSSSSSSSTTTTT
SSSSSSSSSSSSSSSSSTTTTTTT
SSSSSSSSSSSSSSSTTTTTTTTT
SSSSSSSSSSSSSTTTTTTTTTTT
SSSSSSSSSSSTTTTTTTTTTTTT
SSSSSSSSSTTTTTTTTTTTTTTT
SSSSSTTTTTTTTTTTTTTTTTTT
SSSTTTTTTTTTTTTTTTTTTTTT
STTTTTTTTTTTTTTTTTTTTTTT

$$R^3 \qquad\qquad T^2$$

In the search for time in space, the most obvious place to look for the factor, time as such, must be where it is excluded from the space factor and stands alone. Therefore, one should find the place where space is zero leaving time to be eternal. Such a point would be impossible to locate and to place a value on time. However, in the cosmos at large, there is no such a place, because there is no such a thing as zero time or zero space.

Before I start with the true purpose of this letter of introducing my new method of revising cosmology, I wish to say in my defence that I chose one aspect from a wide range of possibilities to explain the way the cosmos formed. However that would constitute to a book much larger than the one you are reading and therefore I limit the development only to where matter, space and time parted and then I immediately thereafter focus on the point where the Big bang came into place.

In the beginning, there was time Zero to moment Alpha. There has never been a Big Bang, as such and there were too many Bangs too numerous to count. Everything is a variation of time duration in space. During the period of time Zero to moment Alpha the value of 1 second was equal in duration to about 1 000 billion, billion, billion, billion years (I am only stopping with the billion part in order not to bore the readers), measured in geodesic space-time values that currently applies. It could be even billions times this duration because the value of time then, was measured far beyond the speed of light, since light did not yet exist. We have no way to calculate the duration of time. **The closer time is to singularity the longer the duration would be. The method time is expanding is by heat and only heat can expand while only by reducing heat can there be a demise of space.** That applied during that geodesic space-time era as much as it does today, and we must accept it as one equal to infinity shorter than eternal. It is heat in all its splendour because only heat can expand. Even boiling soup produces space that expands. When a bowl of soup is boiling, have you seen the bubbles of air rising from the soup? Has any Newtonian ever taken the time to explain that process in detail? I think not, because such explanations would be far too "everyday-like" to bother their mighty brains.

Well, that boiling soup tells the complete story about the creation. Creating is a fact of creation, however creation was not created and left on its own, the Universe is in creation being created every smallest fragmented split instant there can be. The Universe is generated as it moves and such generating of the creation is a process of creating what there is. We speak so lightly of creating and no one comes close to understanding the concept of creation. Poets and painters and writers always wishes to say how "they created their creation". That is rubbish; they created nothing. They brought nothing new to the cosmos, they only rearranged what was a small part of the cosmos into a new order, that one can detect a distinction from. Creating is producing what never was before. When looking at the boiling soup, there are bubbles rising from the soup at the top. In the soup's brew, there are only liquids and solids before the heat came. In such a manner the expanding of heat created space. No one placed air in before the event or during the event at any time. Yet from the brew of liquid and solid rises gas, or if you wish space. That space was not there previously. That SPACE WAS CREATED.

That space is energy and energy is the interaction between heat and space. As space becomes a part of the soup, a part not there before, with no room to be, it moves out. We refer to that process as boiling. That space creation is applying heat to time, and time in singularity will respond as space in singularity. The space created will vanish just as it came, back to singularity. By applying heat to time, brings forth space, and from the three components, only the heat factor is not in singularity. It removes space in singularity from time in singularity to establish room (space) for heat (time).

That is how creation started. Time in singularity overheated and the product of that was space. That is the 180 $^\circ$ of the straight line as much as it is the 180° of the half circle.

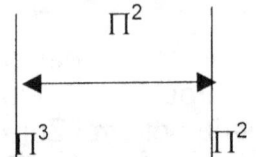

The Π^3 space from
The Π^2 is motion of liquid heat
The Π is time in space to singularity.

There is a time as a line that we find in the centre of the top, which is singularity. However there is another time, which offers material, the space in which material are able to duplicate. That too is time but it is the relevance between the holding time and the space-time where material is located.

Material uses the relevancy of time within, which developed as singularity and time without which was the expanding of singularity to commit to motion

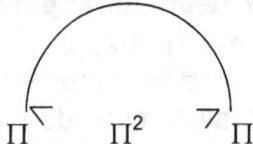

The half circle is 180° placing matter in a circle but because space only applies, to one half, 6/2 and matter holds space to value 6/2=3 only half the circle comes into effect. Half a circle is 180°. Because space has three parts in effect, it also becomes a triangle.

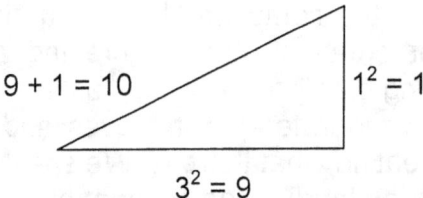

That means where space holds three and time is one, the heat within that space becomes another dimension, the fourth dimension holding space-time (3^2+1^2) = 10. That changes the matter inside space in singularity at ten and "gravity" at Π^2. This is why "gravity" Π^2 is space (10) losing one dimension (Π^2). "Gravity" is all about space (occupying matter and heat) losing one dimension back on a long journey to singularity.

As time is in singularity, and space is in singularity and both are the same thing ($\Pi^3 \rightarrow \Pi^2 \rightarrow \Pi$) the 10 of matter (heat) that affects space (10 Π) will also affect time (Π^3) and therefore time carrying heat will become 10 (Π^3) with space 10 Π. Anyone with a simple calculator can divide 10 Π^3 by 10 Π and see where Π^2 fits in. It is the doubling of matter in relation to time (7/10 + 7 /10) times the double factor of time in space (10) standing related to the line of time that refers to matter (10/7). That gives gravity its value of (Π^2).

Through the Coanda principle the motion of liquid Π^2 confines space Π^3, to what space Π^3 confines as the atom $\Pi^3 = \Pi^2 \Pi$) to the solid. In this containing of space by liquid in motion with the limiting or putting a border on space by liquid flowing, which confirms the space what establishes the Roche limit. In that there has to be a liquid (the neutron at Π^2 lies in the two components of space-time occupation or "gravity" manifested in the Roche limit. All objects spin and spinning is a circle Π^2 while all objects are moving in a direction $\Pi^2/2$. Again only, half of Π has any dimensional validity at any given time, therefore the dimension surrounding an object is Π. That is how gravity forms the atom as the surface of the cosmic object extends from Π^2 to Π but only half of the circle of Π (180°) can apply to time (Π^2) being in a straight line $\Pi^2 \rightarrow \Pi$, "gravity" will form at that point of $(\Pi/2)^2$ giving the Π in space the "gravity" to hold.

$$\Pi^6 (\Pi^2 + \Pi^2) / (6 \times 10)$$

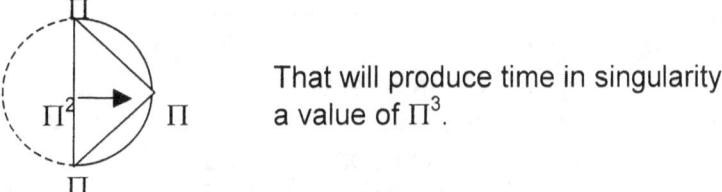

That places Π in a total of Π^6 with 6 sides in space (10) affecting the proton $(\Pi^2 + \Pi^2)$

That is why space will forever comply with 7 / 10 Π^6) / 60 = 112, (the Π^6 is (Π^2 +Π^2 + Π^2)) and time forming the line (180°) between the half circle (Π to $\Pi = \Pi^2$) at a 180° will form the triangle of space in half (180°). The matter component of the Titius Bode law effectively applies to the value of space, therefore 7/10 comes into the calculation. That places any atom with an existence in space at a premium of 7/10 (Π^6) (6/10). The reason why plutonium at $5(\Pi^2+\Pi^2)$ $(\Pi/2)^2(3/5)=244$ is at the element limit is obvious; when dissecting the relevancy in detail. The complete element holds the very edge of what an element in space and time can endure in this era, but two or three era ago it had the function cobalt has at present

That will produce time in singularity a value of Π^3.

Explaining the other five stages of gravity (Π^2) development is extremely complicated and for that there is no room in a book meant to introduce new ideas such as this. My motto in this book (part one) is "Keep it simple"

With time in singularity, time was eternal.

Π°

Time is the spin rate of heat in space. This translates to heat in spin (the atom sealing time off by the spin of the electron, which then produces a motion relevancy with the proton and time in space, which brings about the time line. As we can see when the top spins the top forms the spin of heat (the top spinning) in space. That means the way the movement changes where matter and heat relate to other matter and heat in space. All the movements are relating to a circle (Π^2) going somewhere (Π) in space 3. The Π will form the radius to the circle (Π^2). Any novice can see that the longer Π becomes, the wider Π will be and therefore the longer change in the repositioning of matter will be.

Any person wishing to uphold Einstein's view about the speed of light being the limit through which matter can apply velocity, then that person should first explain how the Black Hole works. It is very distinct that whatever is inside takes that which is inside, to exceed the velocity of light. In other words the Black Hole is able to force matter into speeds that goes way beyond the speed of light. The contraction produces a spiralling of particles that takes matter into a motion dimension far beyond the speed of light. The concept involves not the moving of the particles but the slowing of time to force a duplication cycle that goes beyond the capability a photon can withstand. Inside the Black Hole must be matter, because there is no space, yet time does apply because it takes the particles spiralling inwards to the centre time to move from point to point. Matter in motion is time. However, no light can return to the surface, therefore the light is slower than the moving particles within the star. The only thing about the star is that it maintains a higher relevancy than the relevancy the speed of light can apply. By accepting the existence of a Black Hole, any of Einstein's claims about the speed of light being the fastest that matter can travel becomes fictitious.

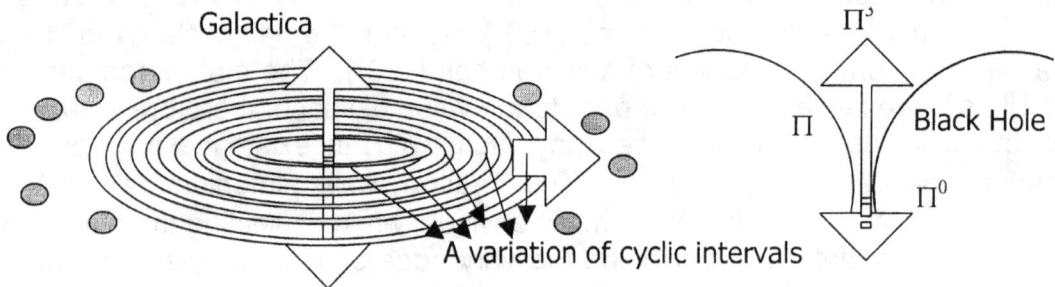

A variation of cyclic intervals

Another place where the speed of light becomes obsolete is within the centre of galactica, where the accumulative movement of matter exceeds the speed of light. That is where doctor Hawking saw a Black Hole that is not a Black Hole, but the precise opposite. Light, matter and heat, moves inward in an effort to maintain cooling as the group of proto-stars belonging to an era to the future) where they still claim their share of heat maintenance. Those particles in such close proximity, establish a time in motion well above that of the speed of light. Everything in the cosmos is all about relevancies. Particles in that phase are still very close to time eternal where motion of material took space into singularity. Time started at such a high velocity, it had to be eternal. Nothing that diverts from eternal can become more than eternal so it has to be less than eternal. It is fragmenting eternity into parts making eternity smaller.

Professor Hawking holds the opinion that there is a Black Hole centred within the centre of the star, which of course cannot be possibly true. The dynamics of a Black Hole is such that it is a star as massive as they come, that fused all the atoms into one structure. The nature and the essence of a star is to unify the singularity that was divided amongst all the atoms during the process of cosmic expanding. The star is a collection of atoms, which unite in motion that then through the unit generate motion to establish a controlling centre governing singularity. The rotation of every individual atom spinning is collected as a generated effort and the collective drive accumulates the effort to the centre of the star. The more the star develop the more is the drive of the star vested in the centre of the star and is it less concentrated in the material compiling the heat and therefore the drive. As the star becomes self secured, the maintaining of the star removes the duty of finding heat to secure the star from overheating from the atoms to the centre governing singularity. The singularity finally takes control of the motion of the star as the star evicts all space and drive time back to eternity.

Then a point arrives where the star abandon all motion. The star then achieved the main goal all stars have by producing a gravity that controls the motion of time. The spin has moved from within the star to time itself and be abolishing space all together the space became what singularity can offer.

By using Kepler's formula the relevancy placed infinity in control of contraction and $k^{-1} = T^2/a^3$ which means the space used by the star is infinitively small as the motion producing the gravity calls on the entirety of time to move and to establish such infinitive immobility. On the other hang the space that the star then control is eternally big $k = a^3 / T^2$ since the entirety of time establish by motion the outer limits set by the Coanda effect. That makes the controlling gravity the entirety of the time aspect because singularity being infinite commands time to motion where such command is stretching the ultimate. It is more complicated and I do explore the working of stars and the development of stars leading to Black holes much more in detail in another book I have being "*STARSTUFFIN*".

In the very opposite it is the motion of all the heat and all the stars proto or otherwise that forms the unit driving the galactica to improvise a Black Hole situation within the centre of the star. The Coanda effect that generate the singularity which control the star becomes generated as a result of all the heat and particle motion that turns about the star centre. Where the motion is valid enough to sustain a drive that would generate an equal gravity to that which the Black Hole demands, the totality of the liquid in motion in the galactica invests into a gravity that does form the drive equal to the drive of a Black Hole. But the drive forms what seems to be a black Hole. There is no real Black hole because if there was a Black Hole, then the galactica had met its destiny before any of the stars within such a galactica could journey onto a road of development. From a Black Hole nothing escape and every star is a future Black Hole on a journey of development to finally become the ultimate, the Black Hole. However there is one star more supreme than that, but there is no space to go into that explaining.

Gravity is motion. Motion is either the expanding or the contracting of material because of heat interacting with space. When material overheats, heat expands the space of the material and when heating diminishes the cooling reduces the space that material claims. In both instances it is motion applying. While moving the material overheat thus it expands. The expanding may be controlled and therefore the progress of expanding is controlled but motion has to be by way of expanding even when the expanding is under the auspices of contracting.

I would suggest we think of stars in the following terms. A star that generates and transmits a lot of light is weak on gravity because their progress started recently. They command a lot of space-time but the demand they have to keep their cooling acceptable is very low. In that they can generate a lot of light but with the demand on cooling low and the gravity in the centre not very developed, those stars cast a lot of light back into outer space. It is just because of the size the stars hold that tell the that the stars are still young and have a weak developed governing singularity. The stars will have very prominent hydrogen and helium layers, with the inner core not very prominent. The control of the star is still very much in the individual atoms and in that the motion the atoms have to produce in order to maintain their individual singularity will only come about through motion. The atom has to make contact with as much space-time through motion as possible since it has a very poor ability in contracting space –time in support of the cooling system.

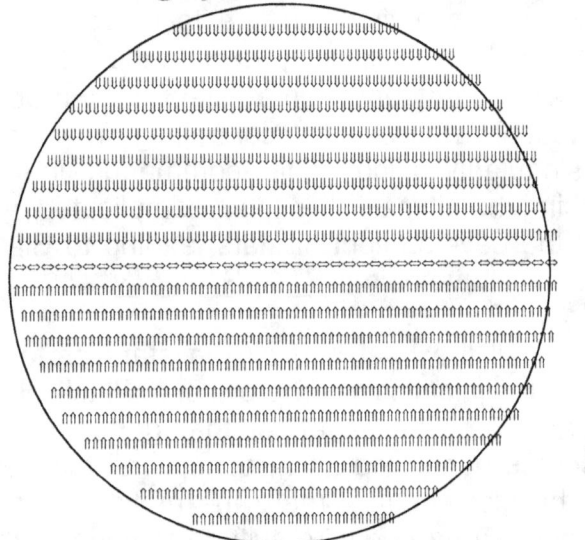

The entire motion and the entire contraction of every atom culminates as one effort and this produces a single combining effort which is then displaced to the center of the sphere of the star where singularity is normally nurtured as a result of the shape of the sphere

The contracting action is at present the only part of gravity that Newtonian science credit as gravity. There is a lot more to gravity than such simplicity. Every atom in a star is pushing the atom in front by filling the space the atom in front vacated. Every atom in front of every atom behind is pulling the atom behind as the atom behind is urged to fill the space that the atom in front vacated. That is motion, which is the most complex issue one can find in the Universe. Since every atom is driven by singularity and no singularity are able to move it bring about that every singularity must remove and rebuild the space every atom fills or vacate as the atom moves along. There is a building of an entire Universe going on in every split second and this split second is so fast we cannot name it. By naming it there will be so many time units gone by, by the time we said the name, the Universe might not even be recognisable. We might call it energy but I hate to call it energy because energy is a lot like Holy water. It can come from anywhere and you can use it for everything and in the end it does not even become something durable because its use eventually comes to nothing.

> The atom restricts dismissing of space by the containing structure to the atoms relevancy being Π^0 in singularity bringing on Π relating to $(\Pi^2+\Pi^2)(\Pi^2\Pi)3$. As the layers swap there aligns between duplicating and dismissing the atomic relevancy adapt to comply

Since the star performs as an accumulated atom where innumerable atoms inside the confinement of the star combine to select one centre spot forming singularity that represents the star, I have chosen the to use the same symbols that I found in atoms to describe the relations in space –time to singularity within the space-time of the star. I refer to a star as a cosmic atom in other books.

Early stars still in the envelope of heat within the centre of the Galactica have only space duplication and growth through the cover of such enormous heat. These class stars are not visible but are shrouded in a blanket of heat covered by light. The atoms forming the stars are small and under developed. They remain cool because they contrast with the heat surrounding the star where the star material supports the cool space and does not form part of the liquid heat forming the outer limit. I would like to draw your attention once again to the fact that the sun at one stage was a cool $18 \times 10^{6\ 0}$ on the inside and a freezing cold at $6500^{\ 0}$ on the outside while all the time outer space was a blistering $10^{34\ 0}$. This was considered the coldest place in the Universe because the sun was still part of the deep frozen space inside the blanket of heat. Look at any galactica and see in the centre there are stars surrounded by a blanket of heat with stars conversed by heat sitting like a duck frozen in this pond of liquid heat.

It is the duty of the star to contract that which the galactica expanded. The galactica expanded as a compromise for the overheating but in the expanding the galactica, such expanding also develop and control the progress of young stars into adulthood. The galactica expands, expanding the stars and the stars has the role to contract the Universe back to singularity. While the galactica expands it gives the stars the opportunity to place heat stored as space of heat frozen by spin in the atom. The atom generates a governing singularity that demolished the space as it accumulates all the heat back to singularity.

BACK THEN when the Universe was new

+   e⁻ negative space-time displacement field

positive space-time displacement field

negative space-time displacement field

Positive space-time displacement claims on space through motion.

If space outside the atom grew to where we now see the Universe the space within the atom also grew substantially and if a star demolishes the space it has, such a star must then also reduce the atomic space because after all that is what the star is all about, dismissing of all space including the space surrounding as well as the space inside the atom.

PRESENTLY we refer to the sizes we find space has in the sun as quantum meaning they are inexplicably big

+   positive space-time displacement field   e⁻   negative space-time displacement field

negative space-time displacement

Positive space-time displacement claims on space through motion.

IN FUTURE TO COME they are going to get a lot bigger than the quantum size now present.

+   positive space-time displacement field   e⁻

negative space-time displacement   negative verplasing field

Positive space-time displacement claims on space through motion.

The manner, in which the schematic layout presents it self as follows.

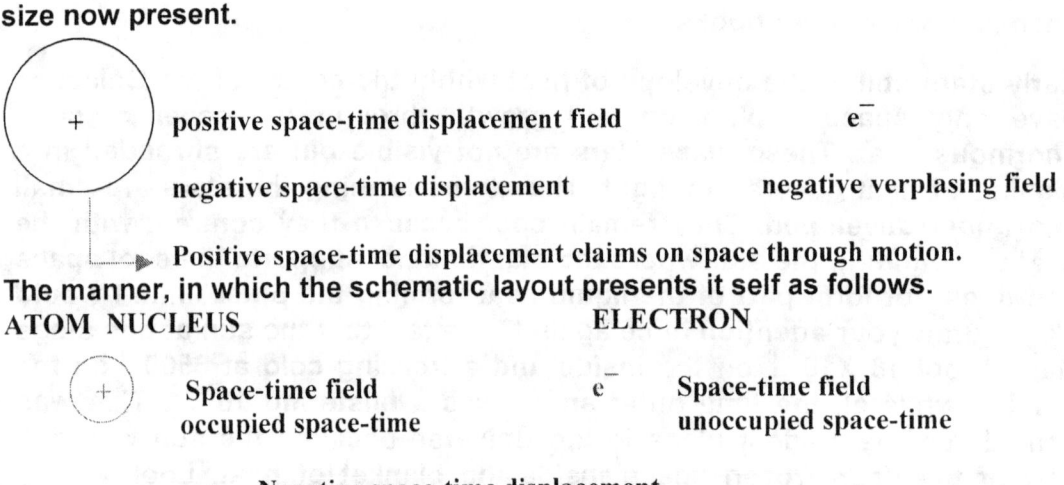

ATOM NUCLEUS   ELECTRON

+   Space-time field   e⁻   Space-time field
occupied space-time   unoccupied space-time

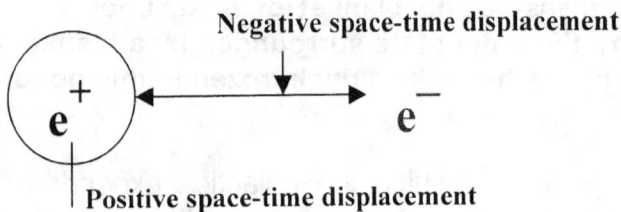

Negative space-time displacement

e⁺   e⁻

Positive space-time displacement

It   is   not only outer space that grows because $k = a^3 / T^2$ is as much the cosmic value as the value within the atom. That means that $k = a^3 / T^2$ is also in place within the atom and that shows the space within the atom grows as the

The Hawking Black Hole

Universe grows because the atom represents the Universe that is in growth. As gravity brings space-time reduction from the centre of the proton so must the growth come from the centre to the atomic proton cluster. As the atom expands in space-time, the proton can also grow dimensionally bigger through the neutron growing in stature.   It expands in captured space–time by pushing the electron walls to allow the atom more space to occupy. It is pushing the electron to achieve a distance every time in the same manner that the body lets nails and hair grow.   There are three factors of space-time where space-time is released.   Cosmic unity and space and heat parted as singularity released the space heat holds by forming motion which produces time to set boundaries and relevancies applying.

The Hawking Black Hole found to be in the centre of some large Galactica, which is generated by the total motion of all the spinning stars, material and mass surrounding the centre with the governing singularity. We also must consider that due to out position we have in the Iron period, we can only see half of the material since the larger stars using more advanced inner core would seem dark to us. What should enable us to form a concept or to form and idea of what constitute Black Hole we can imagine a star where all the mass within such a galactica was in one star. This was long before our individual Big Bang because at this period we find ourselves to be in there were seven Big Bang events going around that forms the Universe we now see. The star we see as a Black hole comes from a time before the iron period we now experience was due. The star then developed as all stars presently do and eventually fused all the atoms that formed the star into one gravity generating monster that absorbs space –time far beyond the speed of light. Having the ability to generate gravity beyond the speed of light puts the singularity core beyond what we think of as our personal Big Bang where our light became visible. In our little Universe we call Earth the atomic relevancy at present is $(\Pi^2+\Pi^2)(\Pi^2\Pi)3$ but in the Black Hole the relevancy already returned back to Π^0. In all aspects of the Universe there are the motionless which forms the role of the solid. We may think of the stars in the galactica to be solid but since those stars form motion around the centre of the galactica all space-time in motion including such stars are liquid as far as that centre is concerned. In comparison to this state of affairs there is liquid and only liquid can move. All stars spinning around the centre holding the Hawking Black Hole is therefore liquid and the motion by principle using the Coanda affect generates a gravitational centre that has the velocity to generate the gravity we will find in the Black Hole.

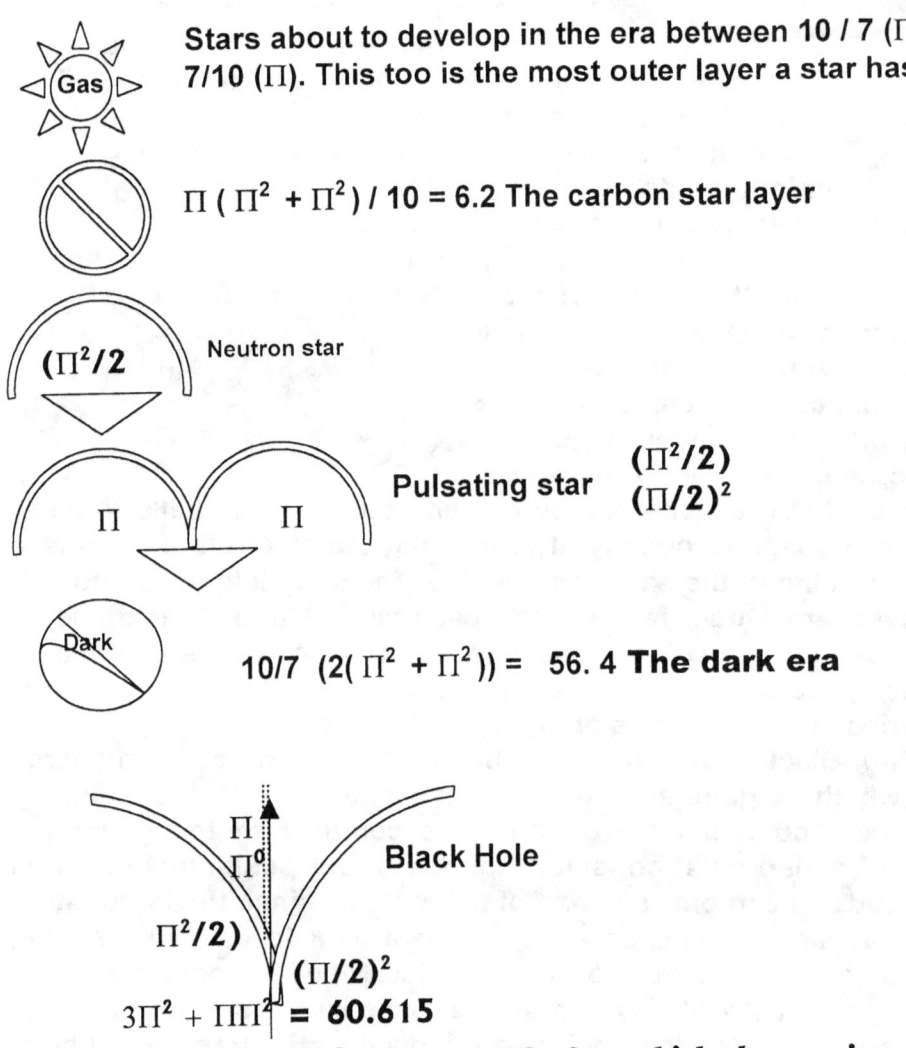

Stars about to develop in the era between 10 / 7 (Π) and 7/10 (Π). This too is the most outer layer a star has

$\Pi (\Pi^2 + \Pi^2) / 10 = 6.2$ The carbon star layer

$(\Pi^2/2$ Neutron star

Π Π Pulsating star $(\Pi^2/2)$
$(\Pi/2)^2$

Dark $10/7 \ (2(\Pi^2 + \Pi^2)) = 56.4$ **The dark era**

Π
Π^0 **Black Hole**

$\Pi^2/2)$
$(\Pi/2)^2$
$3\Pi^2 + \Pi\Pi^2 = 60.615$
$4\{(\Pi/2)^2 \ X \ 2\Pi\} = 62$ after which the star is a Black Hole

Every star is on the inside many different stars because every layer holds a different (k) or relevance making the space in the star very different from every other layer in the star. This is because every layer has different motion in relation to the governing singularity and therefore has a different gravity confining space. The layer is the result of the gravity effort of all the atoms in such a layer and therefore the space in that layer will bring about the time factor that produces the proton cluster relevancy

When contracting, gravity takes place by means of lying down newly acquired heat to maintain the cooling of the structure.

However by contracting it is accumulating material that produces a build up of material in order to enlarge the existing heat surface. In that manner it spreads the heat in a wider area than what the area was the instant before and in that manner it duplicates slightly more than it did duplicate the instant prior. That means even by contracting the measure is still expanding the material by relocating the material from an uncontrolled zone to a controlled zone (the atom.) Still this accumulation by contraction is expanding by motion. When saying this please be sure of one thing: it is not the Universe that is expanding but factors within the Universe that relocate that which takes up space and provides space in the Universe. The Universe remains unaffected by all this by never increasing or decreasing. Two of the three factors swap ends and that places the third factor at a different relevancy. The Universe can expand as little as it can reduce. The Universe expands by the curvature of space –time as Einstein proved but my solution proves much simpler than the way Einstein went about.

I will in a short while indicate how Einstein is correct about the curvature of space time in his theory about "The curvature of space-time" because the curvature of space –time is the form of Π, which is the value of singularity and that forms the Universe. It is a building of what there is by the dynamics that singularity provides in relation to the accountability space-time has relating to singularity. There is the space-time complying with singularity and filling the space-time in singularity is heat and matter valuing space-time. Space-time (Π^3 to Π) cannot bend, cannot curve, forms a straight line, but what fills space-forming time is matter in motion (Π^2) and heat (3) in time in space. That part changes. The atom cannot be gas, or liquid, but is a solid, because the atom is densified in occupation of space-time. The atom is space with heat under control of directed motion. It is the heat in unoccupied space-time that produces the gas, and a liquid, is closely connected as much as part of the solid that all substances form however it is not within the enclosure of the atom.

It is THE HEAT in SPACE that produces TIME, that can and does curve, bend or whatever. That HEAT in SPACE forming TIME that forms the relevancy of space-time does bend because it can flow and flowing is changing direction or constructing by altering flow directions in space wherein matter flow but that then is part of being part of time. If, by applying the forming of gas, or liquid to the element where it is the space between the elements, of course you will get the incorrect vision of Π, where the space-time (matter holding singularity forming singularity) is doing all the bending that applies to the curvature of space (validating time) and time in singularity (a straight line) will be solid. Einstein placed the relevancy incorrectly on singularity, instead of heat.

I do admit, IT IS A LOT MORE COMPLICATED THAN WHAT I ALLOW IT TO BE AT THIS POINT, but the motto is, Keep it simple. If you wish to keep time in space constant, everything in the Universe will be oblong. That is why the Newtonians have an absurd view of the cosmos, and they present facts in the cosmos in a way nobody (least of all the Newtonians) can understand. Please allow me to explain this part first.

When material in space in time first appeared there was a displacement relevancy that was in place then that had 139 protons in relation to 138 protons relating to 136 protons. It was an atom that had many atom variations gathered and that held the construction of one atomic atom worth on the outside 139 protons, in the centre 138 protons and in the gravity zone 136 protons. There was no electron at the time because the electron came about at $10 / 7 \ (4(\Pi^2 + \Pi^2)) = 112$. This was the phase where the neutron was established as part of the Universe. The first time I can detect space / time and matter is when the proton came to a relevancy of $\$T = 7(\Pi^2 + \Pi^2) = 138$

That produces space at the square in relation to matter at 7.

In that this relativity indicated that at such a point the proton was already demanding space ($\Pi^2 + \Pi^2$) in time set to the proton's conditions, and controlling Space-time not yet established, but creating the correct environment for controlled space-time to be. Time was slow, time became faster and faster because by extending the position of Π, Π^2 will produce speed.

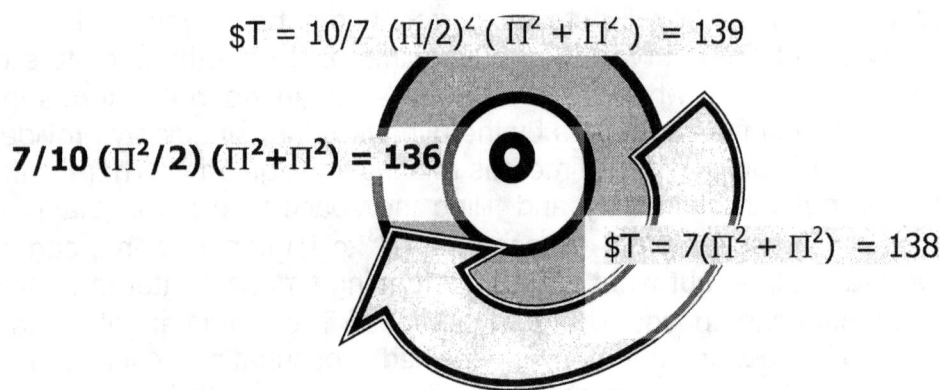

$$\$T = 10/7 \ (\Pi/2)^2 \ (\overline{\Pi^2} + \Pi^2) \ = 139$$

$$7/10 \ (\Pi^2/2) \ (\Pi^2+\Pi^2) = 136$$

$$\$T = 7(\Pi^2 + \Pi^2) \ = 138$$

Explaining is as follows:

$\$T$ is space-time ($\$$) in the time sector (T)

7   refer to the 7° of any sphere.

($\Pi^2 + \Pi^2$)   refer to the double proton.

At the same time space formed as a consequence to the Roche limit and so did matter. Forming the Aanplasings-Atomic-Epitome (the point where matter breaks in singularity)

$\$T = 10/7\pi^2/2(\pi^2+\pi^2)=139$

Keeping these factors in mind it is clear that Π^2 are the choice of gravity and not r^2.

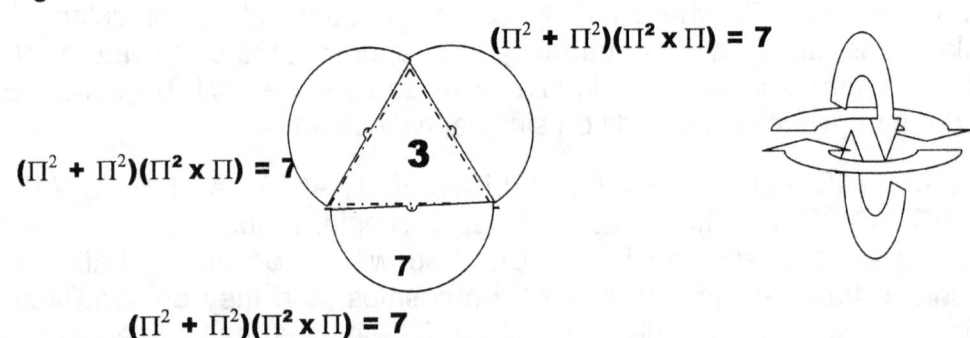

$$(\Pi^2 + \Pi^2)(\Pi^2 \times \Pi) = 7$$

$$(\Pi^2 + \Pi^2)(\Pi^2 \times \Pi) = 7$$

3

7

$$(\Pi^2 + \Pi^2)(\Pi^2 \times \Pi) = 7$$

$\$T = 7(\Pi^2 + \Pi^2) \ = \ 138$

Explaining as follows:

10/7   Forming the space factor and

7/10   Forming the matter factor in the application of the Titius Bode principle.

$(\Pi/2)^2$ the interaction that matter forms in sharing space and time by applying the Roche principle relating to matter-to-matter.

($\Pi^2 + \Pi^2$) the double proton

This is what the Universe consisted of, everything that is today, was then, in a dimension that only holds "gravity" or gravity- motion.

The only way new space can form is by unleashing the forming of heat. Some particles still overheated which evidently forced more space. From the overheating and the consequential space that came about, particles broke down allowing matter to dissolve to matter (neutrons) and space (electrons). It must be very clearly stated that neutrons and electrons did not yet form but only a suitable set of rules formed that later would apply, where matter could supply matter to initiate a sufficient heat supply as heat, locked in time, converted to space. The process that happened at this point lies far beyond that of a nuclear reaction, it lies far beyond that of a "Black Hole" or as I prefer to call is a proton star. All of this still fall under the same aspects and parameters we find energy or the application of energy to be at present.

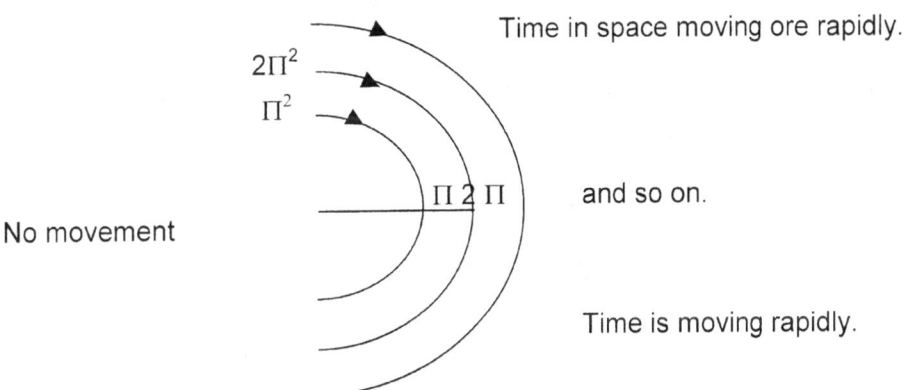

Time started at what ever point you wish for it to start, and we can take time back indefinitely. That will be pointless.

Heat converts time in forming space by expanding through overheating that creates space and that was energy back then, at present and to the end of the future. In order to keep matters short, I shall proceed to the "Big-Bang" explanation, the period of nuclear heat forming space. Space formed for the very first time an identity separated from matter, apart from time, becoming the fourth dimension. In other words, the atom as we see it came into practice. Matter solidified at $\$T = 7(\Pi^2 + \Pi^2) = 138$ which enclose a formation by the Coanda effect spin which brought the atom into place, but this was relevant to a displacement flow of liquids. However the contrasting between expanding by overheating (10/7) and contracting by cooling (7/10) did not end because the displacement of space –time in time in space was in progress. That brought about that the Universe found a means in resolving the issue at the constituting of the six sides in the three dimensional Universe $((\Pi^2+\Pi^2)+(\Pi^2+\Pi+3)=(35.75 \times \Pi)= 112.313)$ and that capped the limit to whatever a compliment of protons in the six sided Universe can withstand. The time factor was at an expanding limit of $10/7(4(\Pi^2+\Pi^2))= 112$. This was when the three dimensional Universe first announced its presence with an apparent Big Bang.

The last element formed and had one proton, but still the temperature was out of control. Then the final element came into place, the cosmos. This was now possible because although the Coanda effect was in place from the first instant the flowing of liquid that was no longer a part of time made diversions of what materials would finally become more possible. The Coanda principle can sustain gravity under the guidance of motion $a^3 = T^2 k$ where liquid flow T^2 adheres to the form singularity (Π) dictates by securing space a^3 in relation to the relevancy **k.** For the first time then there were liquids that could flow T^2 and by the measure of the flow T^2 compacted space a^3 in expanding **k** the realm of space a^3. There too was solids a^3 that was contracting by means of cooling k^{-1} gravity in applying the liquid in motion T^2 to the space a^3. There was a dimension of motion by space 10/7 that allowed motion of space 7 / 10 in space. This went on until our Universe in the form of the atom arrived at 112 protons displacement relevancy.

$T = 7/10 \ (\Pi^6) / (6 \times 10) = 112$

7/10   is the matter has the dominant value.

Π^6 matter has the six sides it holds in the fourth dimension.

6   are the six sides to space occupying matter.

10   are the value or dimension in which space holds a ratio to time.

The cosmos began, not to a specific space, because all the space that was there, initially is still there at present. Any atom above 112 cannot apply to the fourth dimension not then and not now.

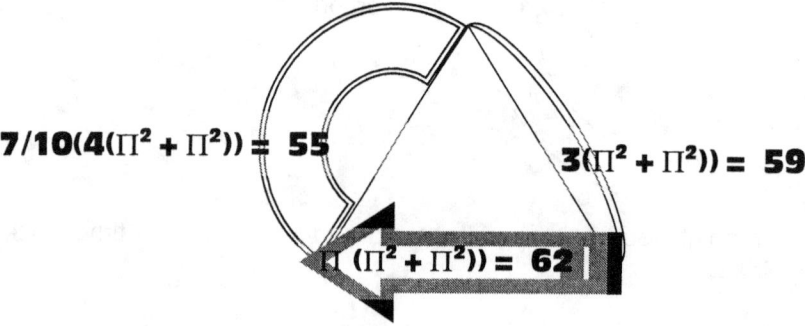

$7/10(4(\Pi^2 + \Pi^2)) = 55$ $3(\Pi^2 + \Pi^2)) = 59$

$\Pi \ (\Pi^2 + \Pi^2)) = 62$

FROM THIS SPACE HEAT AND MATTER DEVELOPED

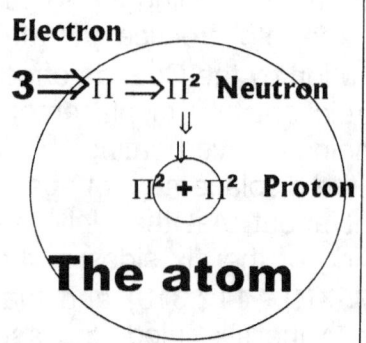

Electron

$3 \Rightarrow \Pi \Rightarrow \Pi^2$ **Neutron**

$\Pi^2 + \Pi^2$ **Proton**

The atom

From the value that outer space can support being the sum total of the particles forming the atom (($\Pi^2 + \Pi^2$) **the proton** $+(\Pi^2 + \Pi)$ **the neutron** $+ (+ 3)$ **space** = **35.75 X** Π **singularity** = **112.313** the star deliberately reduce space as the star intensify motion and that reconstructing of space-time changes the qualities of the atom from what we presume the atom to be to suspending the atom beyond the boundaries of **7 /10** (Π^6) / 6 = **112.16.** The converting of space- time from outer space through gravity to the star centre is the same route electricity follow

At this point I feel advised to remind you that mass does not produce gravity as is alleged by our respected Newtonians but is the restriction of the flow of space-time, which then is gravity. The restricting is at the top $10/7(4(\Pi^2+\Pi^2)) = 112$ and at gravity level it is $7/10(4(\Pi^2+\Pi^2)) = 55$ while in the end it is at the bottom of $3(\Pi^2+\Pi^2)) = 59$ with space collapsing into eternity at $3(\Pi^2+\Pi^2)) = 59$. At a value of $\Pi \ (\Pi^2+\Pi^2)) = 62$ the atom collapses and the neutron is banished from the atom and that is where the Neutron star is in development. That means the element blessed with the number of protons to the value of 62 is the lucky participant that has the honour to collapse space-time while the runner up can achieve the generating of gravity has a number of 55.6 protons in its cluster.

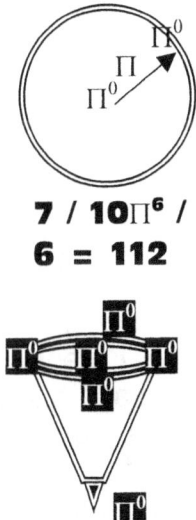

7 / 10Π^6 / 6 = 112

It is motion that started space because when motion stops space collapses. Motion brings focus to time by establishing time in relation and as a factor of space but not as space without time as a factor. The motion presents space-time. The spinning of Π^0 around the centre Π^0 establishes Π and Π is what produces the form gravity has. Still it is the relation or relevancy there is between the centre Π^0 and the spinning Π^0 that gives status to the form that Π represents. In out Universe we are accustomed to and are familiar to the rules we want to place seven points holding singularity to the centre holding singularity in a relation of 7/10 Π^6 / 6 = 112. In that Universe everything less that a duplication ability to the value of 112 protons fit but only atoms to a maximum of 112 protons fit.

What that says is that the square of the proton diverted from singularity by the square as the proton claimed the square from singularity and hold an overall control of space in a sphere (7) and that was the form matter from that point on would always be. Space is the condition set by the neutron, and without a neutron there is no space. To understand this one must firstly understand the principle behind the theory I introduce. At the most inner point one finds time or if we can supply it with a completely fictional name: "The gravity - motion". The gravity - motion carries the value of Π^3. This value determines time in eternity a position matter has no space, but is occupied in singularity. Taking the neutron position to that of the proton we find the value created when the three dimensions (six sides) came about $\Pi^2 + \Pi^2 + \Pi^2$, which carried to the fourth dimension in cosmic or geodesic space-time became Π^6. When relating Π^6 to singularity it becomes Π^1 (space) x Π^1 (time) x Π^1 (matter).

I do realize this explanation does not suit normal mathematical principles but we are working in dimensions and Π^1 (time) in a straight line is 180° and Π^1 (matter), which is half a circle is 180° and Π^1 (space) which is a triangle is 180° as each Π^1 represents one dimension establishing another dimension and providing that dimension's existence.

In relation to the " gravity - motion" space through a straight line will be Π and through the half circle matter with heat positioning space (Π = 1) to 3 sides 3. Relating the gravity - motion Π^3 three will always be a Π^2 and a Π combining 3. The half circle will be Π^2 and the straight line Π. Because there are three directions of flow of space-time two of the three is in opposition at 10. 7 going away into overheating and expanding while 7 / 10 are contracting where cooling reduces space.

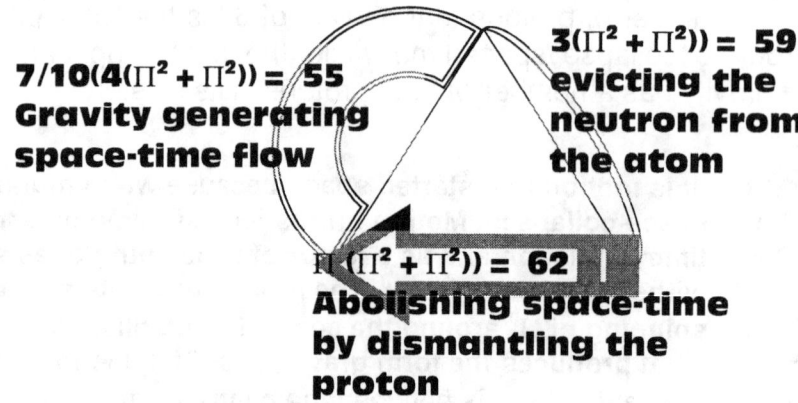

$7/10(4(\Pi^2 + \Pi^2)) = 55$
**Gravity generating
space-time flow**

$3(\Pi^2 + \Pi^2)) = 59$
**evicting the
neutron from
the atom**

$\Pi(\Pi^2 + \Pi^2)) = 62$
**Abolishing space-time
by dismantling the
proton**

Behind all of this explanation is one obvious rule, the one subatomic particle positions in such a way as to displace space-time in the form of heat breaking down the value of space in order to meet the requirement of time. It is again all a relation between space and time and the less space has heat, the less the value of space becomes, because space has no value. Without heat in space and without matter there is no such a thing as space, therefore space does not exist but for matter and heat valuing space to form time.

Π^0 singularity diverted to $(\Pi^2 + \Pi^2)$ diverted further by $(\Pi/2)^2$ in forming neutron space.

When saying this, I wish to include the following explanation for those that may have an interest however they are more likely to condemn by being sceptical.

Energy is a term for a power or a force, an effort to get work done. Energy relies on movement. The influence one object holds relating its position in accordance to another position. Energy is the flow of space-time. It is gravity as much as it is electricity as much as it is nuclear force and every other force Newtonians keep in their back yard as home broken pets. I was thus far unable to distinguish between electricity and gravity. One might go as far as calling the flow of liquids energy (10/7) and the flow of gravity (7/10) but this might be recognised in the light that it is merely a directional distinction there is between the same things.

When an object remains in one position relating to the rest of the surrounding objects, the time it remains in that position is unchangeable. Therefore, for that duration it will remain eternal. To reflect this relation to a formula will be $R^3 = 0$ and $T^2 = \Omega$. Once the movement starts the position changes, therefore the space relation changes. The movement relates to time because any movement takes time, even with an uncontrolled explosion. Changing space always takes time.

The misconception claims its incorrectness in the way Einstein placed the speed of light. The speed of light is not time, but merely particles in time occupying space. The Universe is not in singularity: the Universe is the way matter divert from singularity, claiming space and time within the constraints the Universe in singularity will allow. Singularity is time, holding time at an eternal value and allowing space to bring about the infinity that interrupts eternity. The time aspect is in the matter that keeps space in singularity apart from time in singularity. In fact the only Universe there is, is the Universe outside what we think of the Universe locked in singularity with all other factors being generated by the Universe in singularity.

THERE WAS SINGULARITY

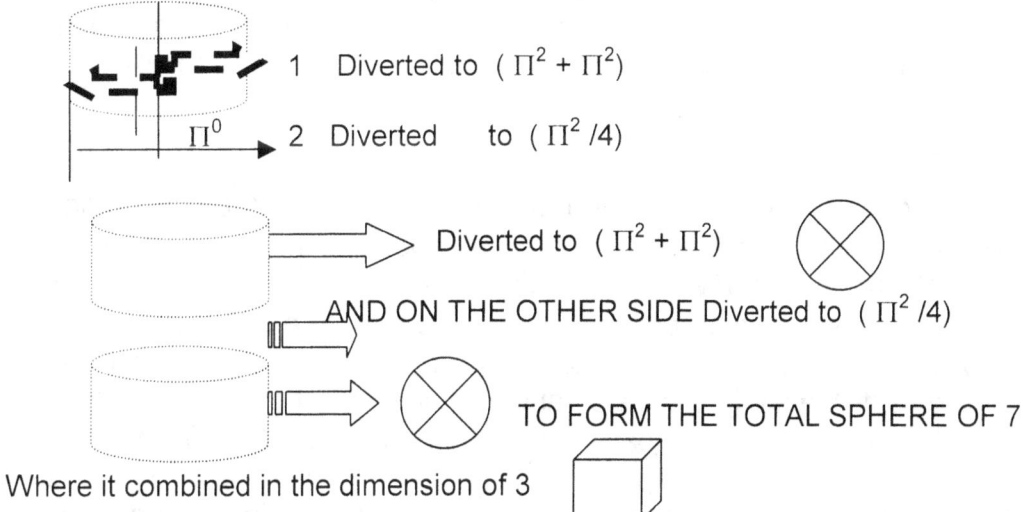

1   Diverted to $(\Pi^2 + \Pi^2)$

Π^0 2   Diverted   to $(\Pi^2/4)$

Diverted to $(\Pi^2 + \Pi^2)$

AND ON THE OTHER SIDE Diverted to $(\Pi^2/4)$

TO FORM THE TOTAL SPHERE OF 7

Where it combined in the dimension of 3

TIME IN ITS ORIGINAL VALUE DID NOT DISAPPEAR. TO THIS DAY AND UNTIL THE COMPLETION OF THE COSMIC ETERNITIES, TIME WILL RELAY ITS GROWING VALUE TO SPACE DUPLICATED IN TIME PERFORMING AS THE DISTORTION OF TIME.

Time started ROLLING AND ROLLING IT REMAINS. **TIME CANNOT STOP; THIS IS CRITICAL IN UNDERSTANDING THE COSMOS. THERE CAN BE NO MEASURING OF T, ONLY OF T^2 AS T^2 INVERSELY RELATES TO R^3.** **Therefore the formula to determine time t $=\sqrt{(1 - (C^2 - V^2))}$ is nonsense**

The first question that one can ask is why would there be the value of $(\Pi/2)^2$ between orbiting structures positioning themselves in a time relation to space. Humans have a connection between human historical occurrences done by humans onto humans utterly confused with time. Those are eventualities that occurred and if it was not historically documented it would be lost as waste. History is not time but merely a colander, reminding whom ever has interest in such reminding of historical re-occurring of events relating to human behaviour in memory of human perceptions, which is far too short lived to have any bearing on time as such. Time is cyclic reoccurring events repeating events that can never again repeat as it did before and is totally one off. Time is a circle running on a straight line of cosmic progress and cosmic development where everything is happening at least once but also only once ever and will never precisely repeat after occurring all detail afterwards changed for ever. Most important, time can never stop because if it does and where it does, space will collapse onto time and disappear into time.

Every person knows about the entry restriction an orbiting spacecraft finds that forces the craft to comply with the entry maximum of 21,991 and the minimum entry of 7. This is without doubt, the number of Π (21,99/7). The Earth holds its value to $4\Pi^2$ and when an object is not part of the surface of the Earth, even say a mountain; it becomes a holding value of 7. Later in this part I explain the sound barrier in more detail and the 7 will then become better understood.

At this point one must see the Earth $(\Pi^2+\Pi^2)$ where all other particles will be in relation to each other by be either Π^2 or Π. When water is in a vapour form, it will have a value of 3Π, having heat separating the water to the factor of 3. By dislodging the thunderbolt, the 3 receive a square value and displaces to the Earth in the linear light to time stance of 3^2. With heat (3) grouping by initial spin value, it will remove from space leaving the water to the value (Π to Π) and this will then give the water a relevancy of Π^2. The factor of Π^2 places the water no longer amongst heat as gas, but heat as a liquid (rain) or solid (hail).

The relevancy of the water will change from 3Π to Π^2 placing the water's position from space (3) to liquid or solid (Π^2). Where does the Π that one find in the Roche limit $(\Pi/2)^2$ and the vapour(3Π) finds its relevancy to gravity? Every particle that enjoys space-time outside the Earth's structure $(\Pi^2 + \Pi^2)$ will hold a neutron position of $(\Pi^2\Pi)$. The Π^2 end will be at the point where heat passes through the object directly to the Earth and this position of space-time relates to the neutron time link of Π^2. The space link of the neutron will then form the Π link. The value of the Π link we find to be $(\Pi/2)^2$, but the explaining to why it is $(\Pi/2)^2$ is rather more complicated and is therefore left to another book for another day.

In nature there is the Roche limit placing a limit to the reduction of space and the inflow of heat to sustain proton cooling. At a point of $(\Pi/2)^2$ the reduction of space disallows any object the cosmic object cannot reduce, an entry to its area of reducing space.

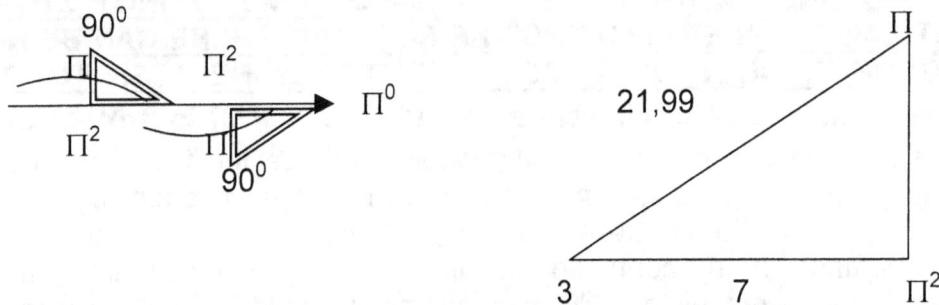

Time is a line with space in time and time in space on either side and that is 90 ^{0} to either side of the line being 180^0. It is Pythagoras in every sense there can be.

At the end of the space relevancy 3 where matter occupies space (21,9 / 7) is a border Π. That border is the exact point where space reforms to a square of time placing all matter (occupied heat) and heat (unoccupied matter) to a value of the square of time. That specific point is in relation to the square of the diminishing shield around the Earth. However it takes matter (R^3) from the 3 dimensional position to the square (Π^2) in relevancy to time in singularity. With time holding space in singularity the 4 sides of Π truly relates to half of the total square value. Let us take the "Sound Barrier" from a point we see phenomena apply laws that matter complies with because the sound barrier occurs when the 90^0 square breaks. The "gravity" factor of Π^2 becomes one and only holds the square to the Π position as it holds space to singularity at a square (time dissolving space at a square) and the time value (Π) remains dimensionless in singularity at (1). It then is $(1)^2$ where the one becomes the space position ($\Pi/2$) representing time (1) at that point. This makes the position time (normally Π^2) but now directly links to Π^3) relating to the singularity position of space diminished from the three dimensional to times single dimension in the square. That makes the Roche limit holding the position of $(\Pi/2)^2$ when the neutron position of time (Π^2) links directly to singularity (1). This may only represent figures, something to accept through intellect but lying far outside the reality surrounding our everyday understanding. To the best of my ability will try to convey my comprehension on the matter can, bring across how I can see it.

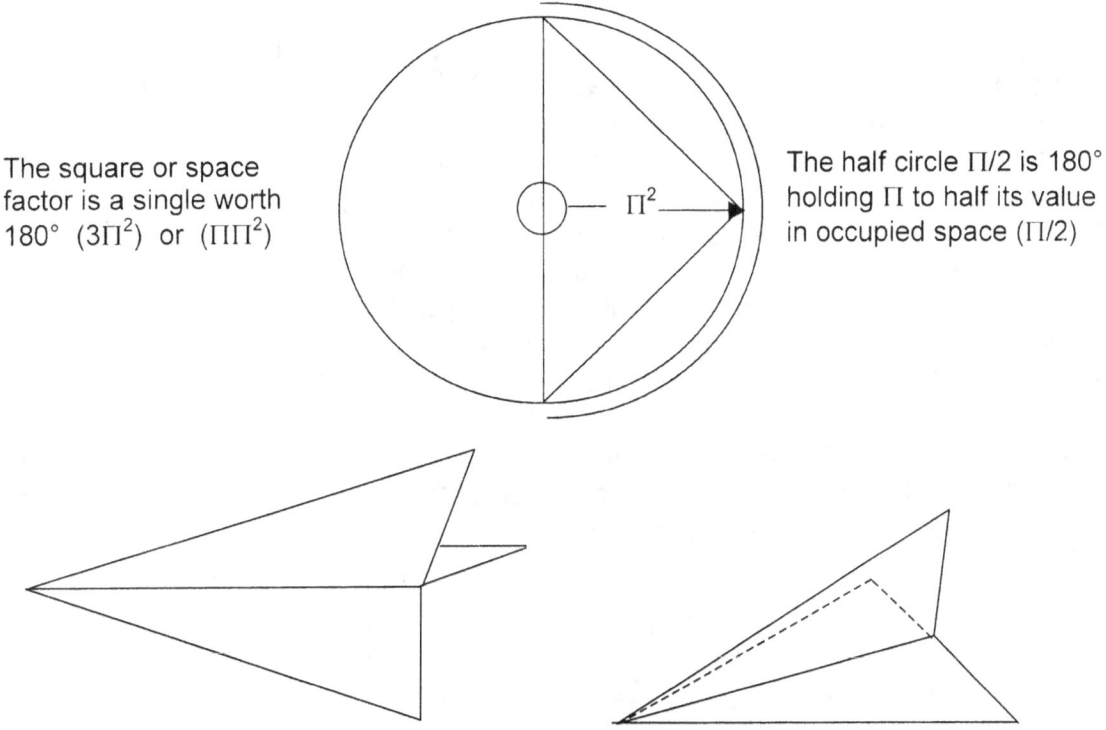

The square or space factor is a single worth 180° $(3\Pi^2)$ or $(\Pi\Pi^2)$

Π^2

The half circle $\Pi/2$ is 180° holding Π to half its value in occupied space $(\Pi/2)$

That is the shape needed to fly in the surrounding atmosphere of the Earth that we presented with a name THE ATMOSPHERE. In outer space a flying object can hold any shape because any shape fits the three -dimensional form. It could be round, oval, flat or any combination of all shapes and forms. The dimension is 6 with a linear value of 3 applying in one direction. This implicates the fact that wherever an object directs a direction in time revaluating positional change in space, it will forever be heading in three directions at the same time. The only control in direction is the concentration of heat at a point opposing the direction of travel. By applying the release of heat, one is applying "anti-gravity". One is producing space that never existed before, because the release of heat will produce space. To the object in travel another dimension brings about influence to his three directional space application. This application of the heat brings about a fourth dimension that applies directly to time. I shall come back to this point duly.

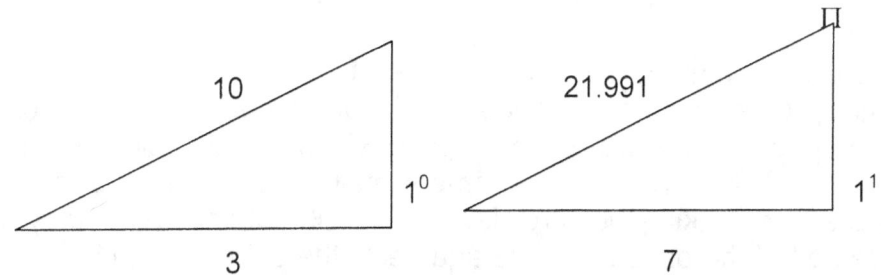

Without the application of specific heat, the object remains in the three directional moving of six possible directions. The value of space unoccupied therefore remains $\Pi\,\Pi^2$, as it was before the "Big Bang" event, whichever "Big Bang" you wish to refer to, because there were many. However, space unoccupied holds time to the value of 10 to 1, and is in relation to time following the law of Pythagoras.

As Pythagoras also indicates, the triangle being 180^0 in line with time is 90^0 in relation to the flow of time by the measure of half a circle. The total formed holds space to Π. Therefore unoccupied heat holds the relation to space in applying 3 directions of influence $(10)^3 = T^2$ $(R^3 = T^2)$. Always part of this equation is the dual function of space in $R^2 = T$ while at that very instant one have R=T. Therefore in space in time you have $10(10^2) = T^2$. Applying the fourth dimension does not bring about another part in the six dimensions, but actually cancel the influence of the one dimension by favouring a specific dimension. Bringing about the fourth dimension will lead to halving one dimension because of favouring the direct opposing side. Then the equation of influence becomes $(10/2)\,(10^2)$ and this means the implication of any heat, be it heat, be it light, will apply under the factor of $5(10^2)\,\Pi$. The implication of this may not dawn on one the very instant of realization, but to scientists, there is no greater shock than just that. To any application of movement, the factor will be

$3\Pi^2 / 5\,(10^2)$
$2(5)(10^2) = 100$

$49+1=50$

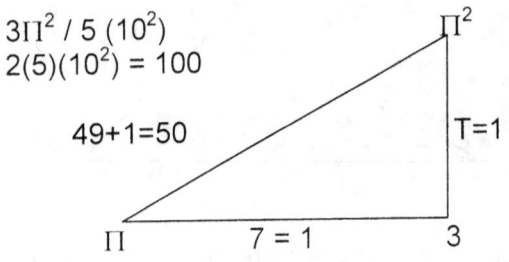

Π $7 = 1$ 3

$50 = 5\,(10^2)$ where the complete is Pythagoras be comes a factor
$\sqrt{100} = 10$ the value of space.

The factor of space comes about as $(7^2 = 49)$ plus $\Pi^0=1^0$ adds to become 50 and that is in the three dimensions of space R^2/T where T holds the relation to one and R/T again where T relates to one. At this point it is most important to remember that Pythagoras works on the application of the sum of the square of the two sides. When seven has a direction in the fourth dimension applied to it, the opposing dimension will be one and this applies in time relevancy, therefore the interchanging in time between infinity in time and eternity in time will place matter at $7^2 \times 1$ relating to circular and $7^2/1$ with $7^2 \times 1$.

The square in matter holding 49 plus one (singularity) always being a factor of one. Space in time however, never can be a cube, it will always be a square with one side pointing the direction of time from time to the past (1) to time to the present (1) to time to the future (1). This means with space in singularity in relation to time in singularity there will always be $R^3=1^3 / T^2 = 1^2$.

It is the value of matter, be it surrounded by heat solid, liquid or gas that differentiates between time in space and space-time. Without applying heat release the space in time would be $10(10^2)$, which is the circle we think we see space to hold. This is where astronomers make their biggest judgement error, they look at space thinking they are viewing the cube of space, but they are in fact viewing the half of the base times the square of the cube $(10/2)\,(10^2)$. What they think they see is 1 000 instead they are seeing 500, but as I will later explain, they are not viewing 5 (10^2) they are not viewing light setting the time factor $3\Pi^2$ but they are in view of light in space (Π) being in time ($\Pi^2=1$) the connection to space is $\Pi/500$. That means whatever the radius between the objects the light reaching it will have a comparable velocity of the distance of travel divided by 500. This will be discussed later in more detail as well. From this there is a measured point where the dimension of space in heat forming gas ends and space in "liquid" forming the neutron begins. The dimensional time application in space diminishing starts at that point. We think of it as the Earth's atmosphere. This no person, no force, no money can recreate. I have read articles about some Newtonian Wizard that plans on building some ship where this ship will "create artificial gravity by centrifugal spin."

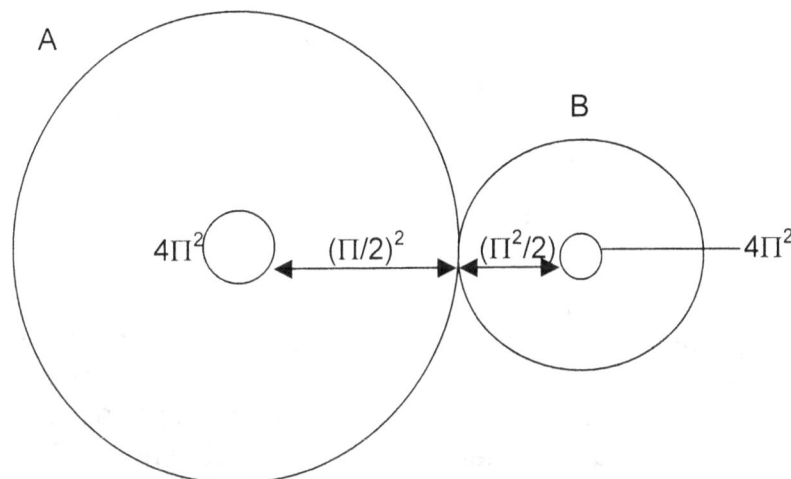

Again and again Newtonians show just how little they truly know science. The depletion of space through increased spin contributes to the view that the Earth is acting as the nucleus of an overgrown atom. The point Π indicates where 3 gain one side of its dimension becoming Π. In that space all objects are at the mercy, not of a FORCE called GRAVITY but of a change in dimension   At that point a dimensional change takes place where the Universe sacrifices one dimension from 3 to Π (in reality it is 6 toΠ.). In that space objects which wish to fly, need a replacement for the dimension sacrificed. The aircraft needs wings to fly. Hot air balloons and the Zeppelin Hydrogen flying cylinder use the natural tendency that hydrogen provide by enlisting concentrating heat to maintain at the value of Π and not reducing to Π^2. An aircraft of any kind, that wants to maintain a vertical flight would have to apply three directional wings for stabilizing because of the loss of one dimension. There may be the possibility that with enough application of heat at various points, the introduction of the fourth dimension (heat application allowing space creation) may even bring about that stabilizing may occur in such a manner. Although the prospect seems light, however with no research to widen the view on the possibilities available, it is hard to tell.

What is overall important though is that the point the Roche limit indicates, being $(\Pi/2)^2$, the dimensional change takes place. Any object holding space-time in own value, will either reduce space occupation, or enlist time duration enhancing, or most probably a little of both as it will cover itself with heat. This is the result from the atoms on the outer edge of the atom, repositioning space-time occupation where the normal application will be in the dimension of the Titius

Bode law developed to a certain degree. This change will bring about that the atom will revalue its position to incorporate the Roche limit in protecting its own space it holds in the time of that specific space. The value of the space-time of the atom of the aircraft therefore will reduce its electron position, as it locks itself in a position under the cover of the heat shield. The point of the Roche limit is as much part of the Macro Cosmos as it is of the Micro Cosmos.

Both structures hold eternity to Π^0 in singularity as time $\Pi.^2$ and space Π diverts from singularity at Π. The deviation of Π^2 and Π will be in accordance with the measure of deviation that heat occupied and heat unoccupied hold in relation to both objects claiming space from singularity.

$$S_1$$

At this point of $(\Pi/2)^2$ the reduction of space through positive space-time displacement will start. Any other cosmic solid that will not comply with the flow of heat to the object will remain outside that area. This means in more suitable language, that the second object will block the flow of heat to the first object (A). To object A everything outside its proton sphere of $(\Pi^2+\Pi^2)$ must either be Π^2 or Π. If the object is in a neutron time position, it will hold the value of Π^2 and if it is in a space position, it will hold a value of Π. Should the object B maintain a density where it will become a singular value to the proton value of object (A) in as much as maintaining its own proton value of $(\Pi^2+\Pi^2)$, it will cause overheating in the proton heat flow of object A.

The demand in heat supply of object A will either remove the heat object B needs for cooling, thus increasing object B's chances of overheating, or it will start overheating itself. By overheating the space that object A occupies, will increase because the overheating will demand more space occupied. By increasing object A's space-time occupation, the demand on heat will increase as the point of space reduction shifts further away. The increase to the value of the $4\Pi^2$ factor, will increase the $(\Pi/2)^2$ linear factor.   As the demand for heat rises, the competition for heat sustaining will also rise in the occupied space of both structures' time zones. That is the Roche limit and that is the purpose of the Roche limit. At a point where the space reduction and demand on heat supply starts in all earnest $(\Pi/2)^2$ seen from the view of the object in question $(4\Pi^2)$ it will either reduce the other object to a suitable value of (Π^2) or Π, but it will not allow any object to compete with it's demand $(\Pi^2+\Pi^2)$ on the supply of heat. It will reduce the competing object's value of $(\Pi^2+\Pi^2)$ as it relates to its own value down to a mere manageable (Π^2) or (Π). The reducing method object A will provide, is removing as much heat flow from object B in order to get object B to overheat. Thus object B will increase its space-time value from a solid $(\Pi^2+\Pi^2)$ to that of a jelly (Π^2) or even better still to a liquid (Π) by removing all gas (3) from object B in its demand for heat. If object B can comply with its own demand for heat, it will remain in space of it's own, providing it's own time within the cosmos $(\Pi^2+\Pi^2)$.

The fact that it takes the space-time back to what the Universe was when it was in total development $(\Pi/2)^2$ $(\Pi^2+\Pi^2)$ = 48,7. At present that value supersedes the space value of 10Π completely, so it holds a relative position far above the geodesic value (10Π). That is not all, because the one object acts as a neutron to the other object, both claim a proton flow of heat from space, holding a time of (48,7 + 48,7) = 97,4. That places the active space-time development with the combined structure to an atomic era that was relevant just after the "Big Bang". Heat became a gas at the atom value of $7/10(\Pi^6)$ / (6 x 10) and darkness parted from light at the atomic value of 3^2 $(\Pi^2+\Pi^2)$ = 98. As you can see, the two objects pushes each other back in space to a time that applied three eras ago.

Therefore one can observe that the Lagrangian point of S_1 is in fact the electron position because the two objects hold the status of a compound, sharing space-time and heat flow. In other words, the two objects become two super sized atoms forming a compound in space. By the same measure a Lagrangian system becomes an enormous five-electron atom with all the cosmic objects working in conjunction as a unit with one mean goal and that is self-preservation through group preservation. Since this falls outside the spectrum I allowed this book to have, I mention that only to fill in a picture that may otherwise seem empty.

If the superior object does not take the minor object as a threat blocking its supply of heat flow, it will consider the minor object as a neutron that can help it grow. Even if the two objects maintain a relation of $(\Pi^2+\Pi^2)$ it will hold the other object to a value of $(\Pi^2\Pi)$ and outside the combined effort they will jointly relate to the rest of the Universe as 3. Therefore seen from the two separate views there is the following atomic order:

$(\Pi^2+\Pi^2)$ $(\Pi^2\Pi)$ S_1 $(\Pi^2\Pi^2)$ $(\Pi^2+\Pi^2)$
Proton   Neutron   ▼   Neutron   Proton
Electron

But placing the two in a battle for survival the relation becomes super strained where the one will not give any way to the other object's demand for singularity dominance and then the relation becomes that of

$(\Pi^2+\Pi^2)$ $(\Pi^2/2)^2$ $(\Pi/2)^2$ $(\Pi^2+\Pi^2)$
Proton

Space-time establishing

This will bring about space-time development that applied when the Universe was fresh and new, hot and lively and full of spinning power. As the objects cannot destroy one another, there is the Hubble Constant effect that comes into place. Each structure has to build a Π value, to maintain and secure its relevant position until the end when the final eternity arrives. In using other words, each object will secure its place in future as a Black Hole or Proton Star. By producing a value of Π separating the Π^2 from three and thus ensuring itself an own atmosphere it will have to revert to space-time development that even preceded the "Big Bang". I will produce once again the value of Π from its position of Π^2 (time) and matter.

Matter is seven and time is Π^2. Both hold claim to the same space diverting from individual singularity, therefore the development of matter and time will produce Π and through that the Titius Bode Principle.

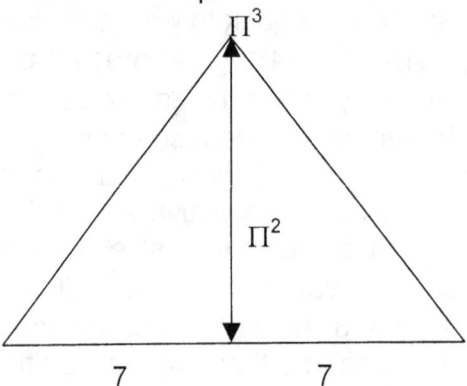

They join space-time therefore the matter factor is the same. This is where one can visually see the one object, filling the space of the other object's atmosphere.

$$7 \times 7 = 7^2 = 49$$

That is matter Π^2 (time) times matter (49) = 483,61.

As this is all under the law of Pythagoras the law will evidently place a square root to that value of 483,61 and therefore $\sqrt{483,61} = 21,991$. This leaves the space value of the Roche-limit, as it develops into the Titius Bode law giving them a shared value of 7 (matter) and 21,91 (space) the value of 21,991 / 7 = Π.

Then the relation becomes

$(\Pi^2+\Pi^2)$ $(\Pi^2\Pi)$ $(\Pi\Pi^2)$ $(\Pi^2\Pi^3)$ holding space (3) still outside. They therefore will share space and that sharing will continue till times end. We know by now that matter is 7, and space is 3 holding time to a relevancy in singularity of 1.

Sharing the space means that 21,9 will become (10) space to the one side
1   to the instant position of time (k^0)
,99   lost to space depletion $\Pi^2/10$
7   the relation to matter.

Through that the Titius Bode law comes into affect of 10/7 or 7/10, depending on whether space or matter holds a superior position to time. From that stance, all objects will relate to one another by the value of $\Pi^2\Pi$ and seen in a whole sale total 7/10 or 10/7.

We on Earth take our measurements per second, minute, hour and so on. This time relation we derive from the ever-changing position the Earth holds to the Sun first, and thereafter to the Universe. That is time in space and more precise space-time.

Time is the spin of matter in space. **It is the changing relevancy that particles have in the eternity Π^0 to matter relating to the outside but also relating to matter holding the Earth Π^1.**

This leads to a point where I would love to show how this cosmic code can be interpreted and from there one may make some deductions from where one can come to some conclusions about how the solar system came into place. The evidence provided in this letter is far from convincing but please see it as a much-abbreviated version of the much more detailed and much more complex book ***Seven days of Creation***. In that book I am proud to announce that I do bring proof and not only by accordance to scientific facts but I prove to the letter that the Biblical version of events occurring during the Creation of heaven and Earth stands proven scientifically beyond doubt. This letter however I present as an example as to indicate to you as the reader that there are other possibilities of finding solutions and there are other methods which we can use to translate what is out there within the limits of the cosmic code. This we can achieve when we use the cosmic code to the correct effect. Now I will repeat some of what is said to show how one may implement the code and use the facts to interpret what is already said in the letter.

All atoms hold an electron to the very outside of the space which that atom occupies. All electrons are negatively charged. No two sub atomic particles that have equal charge can touch; because of the way they will repent one another. Should there be a way to force the particles to touch, say with "matter" versus "anti-matter" a nuclear discharge will result from such a manoeuvre. Therefore, short of causing a nuclear discharge, no matter can come into direct contact with matter because of the negative relation the electron holds to other electrons. NO ATOM CAN EVER COME INTO DIRECT CONTACT (TOUCHING) WITH ANOTHER ATOM.

This is what the Universe consisted of, everything that is today, was then, in a dimension that only holds "gravity" towards and away from singularity controlling.

This piece of logic shows that matter places matter in relation to the heat separating matter. There is never any empty space keeping matter apart. I hope that this bit of minor logic will end all theories about the fact that "Gravity" has the ability to contract or draw matter closer. If you wish to stick to the terminology using "gravity" as a term, then the best way in explaining the way "gravity" works is the ability of the element to reduce the heat between the particles and as such reduce the space. Explaining the factual working of "Gravity" would be rather complex to use in this book and the intentions in publishing this book does not cover the most intimate technical details. Time and space were still parting in relation to the Titius Bode law growing from the Roche limit.

$10/7 \ (\Pi/2)^2 \ (\ \Pi^2 + \Pi^2) \ $ and
$7/10 \ (\Pi/2)^2 \ (\ \Pi^2 + \Pi^2)$

Explaining as follows:

10/7   Forming the space factor and
7/10   Forming the matter factor.
$(\Pi/2)^2$ the interaction that matter forms in sharing space and time.
$(\ \Pi^2 + \Pi^2)$ the double proton

The only way new space can form is by unleashing the forming of heat that expands. Some particles during the development in that era still overheated which evidently forced more space. From the overheating and the consequential space that came about, particles broke down allowing matter to dissolve to matter (neutrons) and space (electrons). It must be very clearly stated that neutrons and electrons did not yet form but only a suitable set of rules formed that later would apply, where matter could supply matter to initiate a sufficient heat supply as heat, locked in time, converted to space. The process that happened at this point, lies far beyond that of a nuclear reaction, it lies far beyond that of a "Black Hole" or as I prefer to call is a proton star. All of this still fall under the same aspects and parameters we find energy or the application of energy to be at present.

Even before making the following statement I find myself in a position where I have to repudiate myself from making the following statement. The Universe is neither big nor small as the Universe is beyond size. However this is a letter and the letter can't change human culture in one go. If I stick to what I say, the letter will be too short just to explain everything I say every time I say it. I shall quickly name one instance to prove my case. I say the Sun is cold and that is true. However by my addressing the sun and referring to it being hot I am not denouncing myself, but merely sticking to human form. The Universe cannot be small because it fits everything within and yet the Universe packs what ever it has into what is a dot smaller than what may be found in the Universe. The dot that came from the spot is the size of something not fitting into the Universe at present and yet it does hold whatever is in one package. The Universe that started was 1^0 and it grew to 1^1.

The Universe holding the smallest moving fragment is the very same size being 1^0 going to 1^1. The Universe is in relevancies but only man works in size because of man's incompetence to realise what is in the Universe. What is in the Universe is relevant to singularity and all points confirming singularity is equal at $\Pi^0=1$. That proves the Universe is in a state of equilibrium. The rest we give size and measure is generated by singularity and as far as the Universe is concerned in the matter, it is locked in the relevancy between two points of equality holding singularity. However my writing is with words and speech demands laws abiding culture and for that reason of not getting all-bohemian I will in this letter for your benefit stick to what is presumed as being the normal and the norm and confine myself to what culture insists on in order to make myself understandable. When I refer to a small or a large Universe or a hot Sun I am not in conflict with myself and neither can I repeat this explanation every time I am forced to contradict my discipline in the parameters of this letter.

At first, the Universe was small dominated by the density of unoccupied space-time; the value to unoccupied space-time was little less than the value to densified space-time, leaving no chance for occupied space-time. Ever so slowly time in space evolved from a continuous eternity, flowing in undistinguishable periods of eternity in duration forever weakening the strangle hold of time on matter. Then the age of the atom arrived, and light, which brought fore ward the atom encircled by electrons that is at this stage quite unknown to man. By the time, the atom was a recognizable unit, more eternities passed than we have sells in our brain.

The cosmic dynamics are all part of proportions and relevancies changing as relations change the applying dynamics. One must see the cosmos starting from the size where what ever was in the cosmos at the time would currently fit on top of a needlepoint. However that does not make what was applying back then, small in comparing to what is now valid because what was small is at present large, because what ever were in the cosmos, is still in the cosmos, and grew with the cosmos so what ever was small had grown with the cosmos and is in practise as large as it was. The dinosaurs are a vivid reminder of that and in **_Seven Days Of Creation_** I explain this extensively.

These first atoms were small producing very insignificant positive space-time displacement in the sea of negative displacement. The atoms were small, but to our great fortune growing as they extended their size and influence. By the time, time was a distinguishable factor the nucleus was presumably less than the size of the electron. The electron's orbit was right next to the small nucleus. Because of the size of this tiny atom, it could participate in the geodesic space-time value back then, which was a little less than the photon's velocity value back then. I have to emphasize that by explaining in this manner, I put the relevancy on the human aspect in view that it concerns the Bible, and therefore mainly the perception that life holds on events.

Heat converts time in forming space, that is energy back then, at present and to the end of the future.

In order to keep matter short, I shall proceed to the "Big-Bang" explanation, the period of nuclear heat forming space.

The cosmic cube we live in is **7 /10 (Π^6) / 6 = 112.16**

In motion applying gravity 7/10)

The six-sided cube (Π^6) / 6

Formina the star

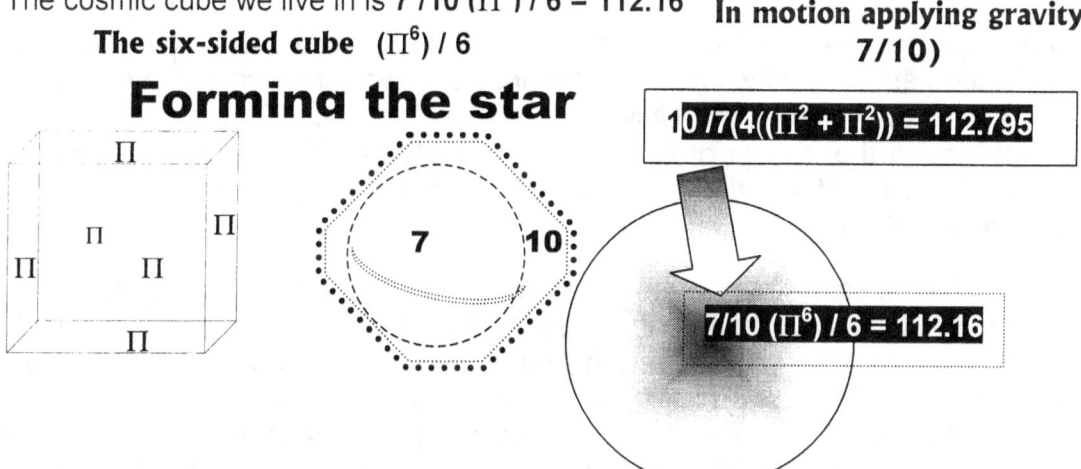

$$10 /7(4((\Pi^2 + \Pi^2)) = 112.795$$

$$7/10 (\Pi^6) / 6 = 112.16$$

Space formed for the very first time an identity separated from matter, apart from time, becoming the fourth dimension. In other words, the atom as we see it came into practice.   The last element formed and had one proton, but still the temperature was out of control.   Then the final element came into place, the cosmos.

$T = 7/10 ($\Pi^6$) / (6 x 10) = 112

7/10   is the matter has the dominant value.
Π^6 matter has the six sides it holds in the fourth dimension.
6   are the six sides to space occupying matter.
10   are the value or dimension in which space holds a ratio to time.

The cosmos began, not to a specific space, because all the space that was there, initially is still there at present. Any atom above 112 cannot apply to the fourth dimension not then and not now.

The Universe had two options. The matter which is densified space-time could join the geodesic space-time and spin itself into another eternal oblivion, or its positive space-time displacement, could withstand the geodesic space-time displacement could withstand the geodesic space-time negative space-time displacement. Fortunately equilibrium prevailed.

Obviously, our luck was in and the positive space-time displacement within the atom stood its ground against enormous odds. With this effort, the expanding of the Universe began, as space-time transferred from the universal value to the densified value. The more time transferred this way, the less time had a value in the geodesic sense and the more time had a value in the densified sense. In this, the space value grew in a geodesic sense as the time value in a geodesic sense declined. This brings about the eternal expansion of the Universe that will only stop once the geodesic time value becomes zero and the space value becomes eternal. This then will be the second eternity, to which the Bible also refers.

This very process was the inspiration, which lead the Universe to be what it now is. In the chapter, designated to moment-Alfa, an elaborate explanation shows how the value of Π has changed four times already, and will change another three times. The development through all the different era brought about that the Universe had to endure because of this very principle. I do not wish to go back to the old cliché about gravity and all the theories that were thought up to explain the means which will enable the Universe to retract again. The retraction of the Universe lies in the growth of time as it is transferred to matter. This process consists of the transformation of unoccupied space-time to occupied space-time to densified space-time.

In this lies the value that all stars have in their role they have to fulfil to bring about a successful conclusion to the life story of the Universe. Objects are growing more massive as time transfers from one value (Verplasing) to the other value (aanplasing). As R^3 is showing an increase in both the occupied as well as the unoccupied space, T^2 reduces from the unoccupied value to the densified value. In the star R^3 reduces as T^2 transfers to the densified space-time. This is what stars are made of, when they grow up to be stars.

With time in eternity, space in zero and matter being time and space, what would ever bring about that this situation changed? No Official Policy Protectors ever came forward to explain this. Why did the "Big Bang" start and what brought the "Big Bang" about? Only the Bible produces any logic to this question. Time, matter and space froze in one; there was no reason in nature for things to change, since this situation lasted eternally. Nature with all nature's laws did not apply; therefore one cannot say that nature started it. Nature was frozen. Nature was not even solid; it was in a state beyond being solid. Nature was nowhere!!

If the spoken language can refer to the Earth as Mother Earth then surely it is all the right in the world to address the Sun as Father Sun, for that is precisely what the Sun is to us Earthlings. If we have the Sun as a father, then it is a very young father. We as humans have received an infant star in the sun and a juvenile as far as stars go, where it is on the edge of becoming just a toddler. A fully-grown star that has not yet reached its middle age development, will have discarded its hydrogen and most probably its helium layers. As it ripens in age, it will discard its carbon and all helium related layers to have only silicon, an iron and an inner core that has outgrown the element value altogether. This process leads the star to a route that takes time back to its origin of time = eternity; space = infinity.

The spin in the Universe slowed down, up to a point where the spin was equal to that of the speed of light. At the point where the Universe spin equalled that of the speed of light, the Universe was still in total darkness. The light (photon) was there, but did not yet produce light. Light only came about as the spin reduced to below the speed of light, and only then light became obvious. The Universe grew away from darkness as this event lasted many eternities, during the period where the light separated from darkness.

How did this "Big Bang" take place? The best way to examine the reason is to see why anything in the Universe expands. To get anything to expand one has to heat it. All matter expands when overheating. Science may come up with whatever brilliant theory, the fact of the matter is that when matter overheats it expands. The bigger the overheating, the bigger will the expansion be, it is as simple as that.

This means whatever leads to the forming of the "Big Bang", whatever preceded it, it had to come about from matter that overheated. With the event of the nuclear age, the proof came about that matter is heat in some frozen form. Unleashing heat from its frozen form brought about a jolt of heat, never yet experienced by man. By breaking matter from the frozen state, of which it is in, within the atom, heat produces light and heat. Where this process clearly shows how new space-time forms is where the releasing of heat caused winds that stun man's logic.

The nuclear explosion shows quite clearly what the "Big Bang" was, with the nuclear explosion being a very minute form. Yes, we have all heard the rubbish about matter and anti-matter. What can anti-matter then be, since matter is heat and that we know from our nuclear experience? The Coanda spin defined the heat to a certain space that then heat occupies for that time while it is confined to the space. With matter being frozen heat, what would form anti-matter. Anti-matter means the opposite to matter, and if matter is frozen heat, anti-matter must then be overheating heat. This in itself is quite ridiculous. Anti-matter can only be matter with an opposite spin in relation to what we regard to be matter.

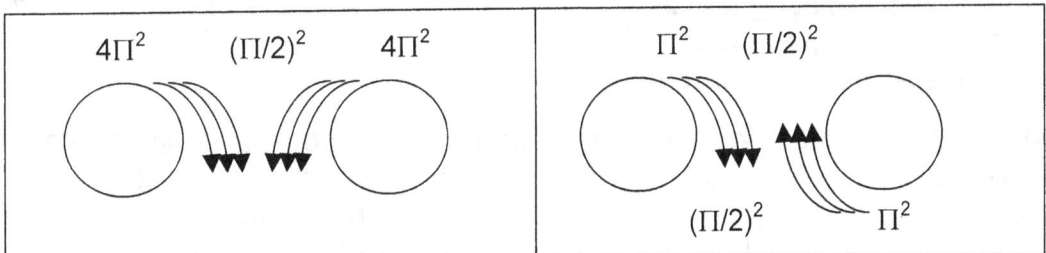

In the second sketch Π^2 (Π/2) / (Π/2)^2 Π^2 the Roche limit cancels the border that is set by the equally negative electron and with that the space of the electron effectively disappear. The electron is a direction of flow of space-time and when two opposing flows do not reject each other they will cross the parameter separating both and both will dissolve on another through spin. The two protons touch destroying each other and the neutron as well as the proton demolish and become heat 3^3.

The process where matter then touches matter, will bring about a reduction in the feeding process of heat, where all matter in that space will overheat and expand, producing unfrozen heat. This means there was heat occupying space, and matter with both in relation to time.

The proton with a positive space-time displacement less than 136 placed its displacing properties in negative space-time displacement. In short, to substitute for mass shortcoming of less than 136 grams/molecule and still finding sufficient cooling properties for the proton to survive the overheating deficiency, it has to spin more rapidly, therefore by spinning it makes contact with more heat than it would otherwise do. One may consider this as "breathing". If the proton does not find an adequate supply of heat and does become motionless, the atom has to substitute the movement through motion. Forming an object that has an increase in heat supply through work commonly uses this fact. In nature, just the opposite is true because of the motion by movement, as one finds in the case of wind.

There is no "force" in the cosmic because everything is a "force" in one way or another. The proton takes heat from space in an effort to maintain temperature and stability. This flow of heat brings about the reduction of space by increasing the heat in that space. The flow of the heat, through the electron by means of the neutron to the proton is time. The amount of heat taken by a proton is a constant throughout the Universe but relative to the space reducing effort of all the protons influencing that specific space. That is "gravity" (a term I denounce and reject). "Gravity" is nothing else but additional application of time. The higher the gravity is, the slower the time will become by prolonging the duration of time, NOT TIME ITSELF. TIME REMAINS A CONSTANT, BUT THE DURATION DEPENDS ON THE SPACE THE HEAT IS CONFOUND TO.

By taking this statement and introducing that to a Galactica, the shining luminous middle part, holds time to eternity through movement. The atoms in the middle admits light because it holds time duration to the speed of light, the longest duration that time can be and still remain in the fourth dimension. The centre of all cosmic structures determines the time that applies to the structure itself. As the cluster of protons supply the density, that influence the space of occupation, the density is a collective reducing of space with the increase of heat. This can apply through an object relating to space through movement by the object and by the object reduction of space through the density of the accumulative effort of the cluster of protons we named elements.

As we Earthlings had a double advantage from the process of binary stars, I would like to ponder on the binary stars, as they are the key to time in the cosmos. The first influence of a binary system we benefit from is the Sun and some other minor star, which by fragmenting formed the non-gaseous planets. The second encounter was the binary to which the Earth and moon developed, and from which LIFE then was able to take full advantage.

The two stars develop in the galactica in close proximity as they help each other in transforming negative space-time displacement to positive space-time displacement. With the combined effort, they can grow extensively in supporting each other. In spite of all our Official Policy Protectors teachings, stars do not form as result of cosmic dust storms. It is just not possible to form that way, because even in the present era we find ourselves the iron era, only elements with a space-time occupation value exceeding 17,5 can reduce space sufficiently to apply "gravity".

Planets never form naturally because a planet has no role to play in the Universe. Cosmic objects do not form in any way or means through the application of "gravity". Matter occupies space presenting that specific space a time value in accordance with the time in space the matter holds. I realize this is rather a mouth full to understand. To us humans, planets are a necessity, which one cannot do without. We humans have a measurably small place in the Universe, so small the Universe do not recognize life as a substance.

We see around us a planet overflowing with life. Life is so abundant, we can hardly distinguish between nature and life because to us nature is life. We have to think of nature without life because cosmology is without life. An easy way to distinguish is thinking of a mountain burning. The mountain cannot burn, it is plants that burn and the plants are not nature. The plants are life, therefore the mountain cannot burn and even less is the possibility the mountain can move independently without only being part of the Earth's motion. It is only life than can manipulate space-time. Other than that all other objects may extend the space-time they occupy by natural converting heat from a gas to a solid through proton growth.

Whenever you wish to distinguish between life and the cosmos, the rule of thumb is: Can it move of its own accord? No mountain can fly to the next continent for whatever reason. That action belongs to life, and it is the manipulation abilities of life that renders a piece of metal such as an airplane the mobility of motion. If not for life, the aircraft could have an ability to commit motion. The actions we see must accompany life, and therefore be associated with life and the qualities attached to life. It is not in common use wherever we might go. In the rest of the cosmos heat drives objects in relation to the flow of time and life has no ability in those locations to intervene. The aircraft will fly, but the principles belong to the cosmos and by life's manipulative abilities the craft will only fly as an extension life brings about. We can make life manipulate cosmic law on Earth and on the planets (well, some of them but not even most of them). It is where life can be, on the inner planets and their moons, but life will not function even on the micro stars such as Jupiter, Uranus, Neptune etc. On the little fragmented pieces of a star gone by, yes, that is where Newton's physics apply and apply they do. The comet obeys rules applied by the cosmos, in the terms the cosmos provide. Newton's physics do not apply to the cosmos however which way Newtonians try to explain it by applying Newton's laws. A comet cannot leave the Earth and journey around Pluto, and come back with a detailed report. This action belongs to an object that is acting on behalf of life as an extension of life. I am very adamant about this. If Science wish to understand the cosmos, do not make the cosmos one big Earth. Up to now, such an application only brought miserable results that apply to nothing and helps science establish the joke it became. Life is the extension of God, the Universe is the Creation of God, do not confuse the two.

The Sun is not an Earth on steroids. Life cannot be on the Sun because the Sun holds completely different relevancies. The relevancy has a comparing formula but the end result will be totally different in every aspect one wishes to apply.

By referring to "gravity" only one aspect of the space-time relation of any elements apply. There is no mention of the second and crucial part of "Gravity" where the motion of the object brings about the space-time relation, or if you wish, providing the cooling aspect. At first, the proton cluster's total positive space-time displacement has an insufficient "gravity" to secure a stable cooling effort. This spinning of the element clusters is inherent from an event, even predating the "Big Bang". To find proof of this statement I just made, look at the photograph of any galactica. In the centre part heat is a liquid flowing like a river and taking matter for the ride. This is one part of the relevancy. The other part is that matter is spinning at such a tempo, it is capturing space in the form of pure liquid heat. The proof of this is obvious in every galactica.

Galactica is living proof of cosmic development and what is much more is the fact that it proves cosmic development in stages. In the centre of a galactica one can see that the Big Bang is presently arriving for what ever matter holds space-time within. Then as the circle expands, it is clear that the widening brings about less heat occupying more space producing darker regions holding bigger bodies of matter. Every galactica proves the Big Bang theory in precise detail.

The spinning motion of the element clusters (or proto stars or if one wishes to use the name of future stars, it will be just as applicable) hold their relation to heat secured by maintaining motion. As the time value in the cluster's space occupation (mass) secures an era related value, the structures that were spinning, reposition in such a fashion as to apply a new linear displacing value. At first the motion is such that the linear position is negligible, but as the mass grows, the linear distance grows accordingly, placing the revolving structures further apart and at the same time, "pushing" the rotation of the objects in a wider revolving orbit. By widening the rotation circle, the objects rotate at a "lesser" pace and this pace coincide with the space-time occupation ("mass") of the totality in the effort of all the protons put together. In this one will not find a "force" but it will be a complete balance between matter, space and time. By securing an ever-increasing space-time occupation (mass) the future star will reduce its negative space-time displacement (motion) and increase its positive space-time displacement (gravity). The higher the positive space-time displacement (Gravity) becomes, the lesser the negative space-time displacement (motion) will be. At present only stars holding an iron$_{56}$ inner core can maintain a star status, and any object with a lesser element in the inner core will not bring about fusion, or in fact, any form of luminosity.

For instance by the time the "Big Bang" arrived, only elements with a "mass" of 112 had the ability to release from the Galactica luminous core and during the Era of the Quarks, the releasing mass of the time determining elements carried a combined proton-cluster "mass" of 88. As the single proton's time holding value increased (molecular mass) the time grew less and the Universe "grew bigger".

I uncovered a mistake in science. From the onset, the mistake seems as insignificant as it is small. Because the rest of the book is about the mistake, I do not intend on elaborating about the mistake itself. The mistake came about with the culture of education and the mistake in itself seems harmless. When admitting that, one must also admit that any pilgrim that got lost and died of starvation through an incorrect travelling direction, made the very first part of his ultimate mistake by looking in the wrong direction. How harmful does looking in a specific direction seem, and yet such a mistake leads to his ultimate mortality. The traveller could when taking his first directional flaw with that the first incorrect step, only put his foot skew in avoiding a rock. Or he could have turned his face to avoid a branch and that move pointed him in a direction that lead to his fatality.

It is not the mistake that becomes the penalty and it is not the origin of such a mistake that leads to the penalising, but the ignoring of accepting signs telling the wonderer of an impending error and his stubborn ignoring of such telling sign that makes the lost party pay the ultimate price. By ignoring the mistake, for whatever reason, the ignoring of such a mistake is his undoing because the price due comes from the inability in recognising the sign indicating the presence of the mistake forming the reason for his final demise. The sooner such a person sees and admits the wrong, the less will be the consequences of his final price to pay.

The Newton mistake is one born in culture and the penalty from this mistake is bred by arrogance. At school minds are young and accepting, although developing. Through many tens of millennia humans came to a habit in surviving as a specie where culture taught them that accepting the advice from their elders is the same as to ensure survival of the following generations, and by such doing is also following the quickest way to an adult mind. By accepting the advice from their elders the accumulated of generated knowledge went to benefit those that protected their culture and they gained experience without question proves the dominance of the tribe in relation to other tribes of the same race. This is culture we cannot do without and still maintain progress. It is an inheriting method humans grew on and is the corner stone of all civilization. We cannot abandon it.

This culture of accepting the teachings and advice inherited from the elders is the foundation of civilization. As far as any students mind can tell the news is as actual and accurately tested beyond a crack of doubt as anything notwithstanding the fact that such news may be with the human mind for thousands of years while it was actually never tested. Whatever the teacher tell students will generally be a first time experience to the student with the information the student receives being so new he has no time to digest the information. The teacher will not allow any pondering on the matter and an in-depth analysing because of a program the teacher has to complete in time. Taking into account his youthful ways (which we all had), he has little stomach to scrutinize it because learning is a painful process to all. Without pain and perseverance there can be no education of any sort. He does very willingly accept the facts as tested and correct without flaws of any kind. The scholar has to because in any education system time will not allow students to ponder about detailing information and securing a prognoses to all learnt every day. What ever the Master tells the scholar is taken as Biblical correct without any thought about testing the results. Where there are cases of scholars having doubt and subsequent questions, the Master takes such behaviour as being obstinate and being a reflection of the student on his (the tutor's) personal integrity and knowledge. The young mind will very soon discover that his behaviour is not tolerated by the system, and the truth is the system cannot tolerate such behaviour for the good of the rest. Time must be spent on learning and accumulating as much information as that which the young mind can accept.

No information could be more affected by such a culture than that of Newton. You better understand Newton and are then considered as being smart, or you do not understand and therefore accept Newton however that has a price tag attached. If you are not readily accepting Newton as the word of God, then with that comes your accepting that your stay on campus will be very little short lived which excludes you from any future you might have had in the world of science. Newton is science. No Newton understanding automatically becomes "no science" education or learning. When in the past I went around explaining the degree of incorrectness I saw, the only attitude I would draw was sympathy because to their mind I did not have the mental capacity to understand Newton. What is there to understand anyway? The bloody comet misses the sun by a country mile and using Newton's formula it should hit the sun bull's eye in the centre. Without Newton, there is no other and science will be a faculty filled with vacuum of containing nothing. This is very unfortunate but is the ultimate of truth. It is either Newton's way or no way at all. With this culture also brought along the stigma that only the minds of the sharp and the sighted can accept and understand Newton and when not understanding Newton one tends to fail your personal I.Q. test. It is a sure sign of the slow witted when the student fails to recognise what Newton said. The only way to advance in science is to understand Newton and indicate to all your pears how brilliant your mind is in accepting information.

All students have little understanding about Newton, and that I can and will prove through the next two hundred or so pages. The mistake Newton made and which I discovered is laughable small, yet it took me (not being that bright I may add) almost one lifetime to recognise the mistake whereas it took mankind three hundred and fifty years of research without recognising the flaw. Others in the past may have come to see what I saw, but if there were such persons they never saw what I saw because if they saw what I saw they should also see that behind such an almost invisible puncture hole in the tube, is a reason for science to deflate and not accumulate. When "understanding" Newton it becomes the very same as learning Newton from the heart and accept that what you memorise is what you know. The memorised knowledge is beyond question, as it has to form part of the identity of the student having secured the knowledge.

It is not the hole forming the puncture and preventing inflating being as such the obstacle that is of importance but what that hole does to the tire and the car and the travelling with the car that becomes a menace. In the extreme effort to keep the journey on course, the Academic Masters are spending an all out effort in inflating the deflating tire faster than the deflating tire can deflate. The recourses attached to this effort is enormous and without cause. It will lead to nowhere and that is where science is heading in their attempt not to head in that direction.

When students become masters the Masters seek to break new academic ground. Masters do not ponder the ways on which they lead their students, but have an all out effort in establishing their own support of new territory that will distinct them from the rest in future to come. My uncovering of such a mistake as I did could only be from a person as ill-educated as I. I did not go through the learning process where the learning process is the very same as a brain washing and mind controlling process. This remark may seem harsh and is not intended to be such, because reality demands no other way in education. The mistake and the carrying of the mistake with the support in refusing to recognise such a mistake is part of human training and there is very little to do but to admit that such may occur whereby to follow in acting to correct the mistake after the discovering, rectify what ever can be salvaged and build from there.

Friend and foe think of me alike as being slow of mind and not quite in faculty to understand Newton. This was what I was told on more numerous occasions than I care to remember. Mind you, it was never done in direct accusation and was always handled with great sympathy as to show those in ability has a passion for us with lesser developed minds, after all I am only a motor mechanic with little schooling and I go around admitting that. Pointing at the mistake I see, I am told by the wise as well as the not-that-wise, such as I that through my lack in education I cannot dream to understand Newton and therefore are committed to the position of the ILL-EDUCATED, a position I learned to accept with grace.

Thus my position renders me the place to shut up and do as I am told because my brain is to under developed to form a comprehension about the complexity of astro-physics. I must say that except for two or three occasions those of the Highly Schooled part of life did treat me and therefore my diseased mind with the utmost sympathy their positions could allow them to muster on that particular occasion. In all cases there are more sides than one and therefore I think of myself as poorly educated, the contras to my position must be those fortunate to be SUPER-EDUCATED. In this letter there will be the addressing to you holding the position of the reader and (I hope) the un-bias judge, with me presenting my case to the un-bias (you) in concerning the third party, the SUPER-EDUCATED. Referring to the party in opposing me as the SUPER-EDUCATED is by no means in disrespect and much to the contrary holds my whole-hearted admiration, (at time somewhat limited I admit.)

Quoted directly from the Oxford dictionary of Astronomy the following:

The definition of space-time is as follows:
Space-time is a four dimensional position of the Universe where the position of an object is specified by three coordinates in space and one position in time. According to the theory of special relativity there is no absolute time, which can be measured independently of the observer, so events that are simultaneous as seen from one observer occur at different times when seen from a different place. Time must therefore be measured in a relative manner as are positions in three-dimensional Euclidean space, and this is achieved through the concept of space-time. The trajectory of an object in space-time is called world line. General

relativity relates to curvature of space-time to the positions and motions of particles of matter.

The definition of space-time is as follows:
Space-time is a four dimensional position of the Universe where the position of an object is specified by three coordinates in space and one position in time. According to the theory of special relativity there is no absolute time, which can be measured independently of the observer, so events that are simultaneous as seen from one observer occur at different times when seen from a different place. Time must therefore be measured in a relative manner as are positions in three-dimensional Euclidean space, and this is achieved through the concept of space-time. The trajectory of an object in space-time is called world line. General relativity relates to curvature of space-time to the positions and motions of particles of matter.

The definition of singularity is as follows:
Singularity: a mathematical point at which certain physical quantities reach infinite values for example, according to the general relativity the curvature of space-time becomes infinite in a black hole. In the big bang theory the Universe was born from singularity in which the density and temperature of matter were infinite.

While it probably is the greatest mind to walk the Earth that produced the spectacular in the above, a much more simple mind as the one I have noticed much more simple aspects of nature that only one with a simple mind as I have could recognise because my mind does not have the capacity for the greatness of the great minds.

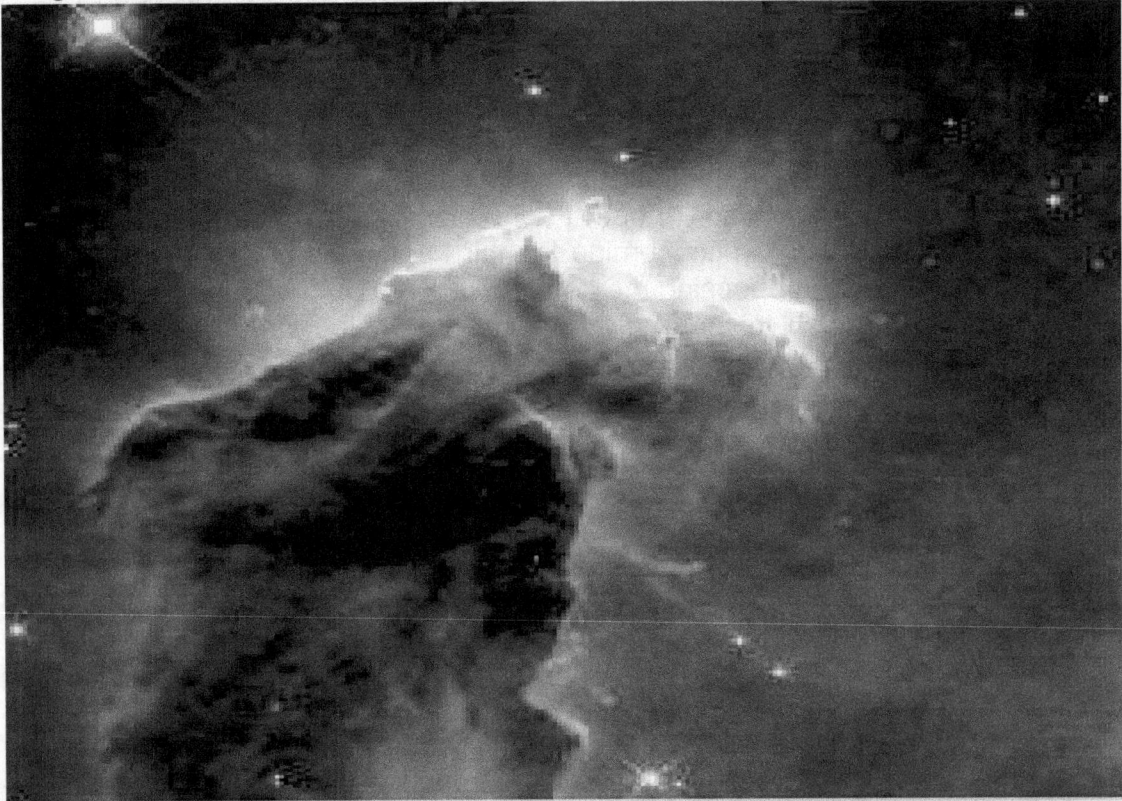

If the Universe did start from one single point and time matter and space flowed from that point, then that point must have a relative connecting base because such a point holding singularity must be eternal as space matter and time link eternal. There then therefore must be one point linking the entire Universe when regarding the fact of singularity. Then according to the theory off relativity there has to be one exact point holding time in a relevance notwithstanding the fact that time depart from that position and relate differently to all space-time away from such a point.

Every person I have discussed facts about creation recollects images in the trend depicted in a presentation as one may find shown with massive clouds of unbounded material That depict chaos and in chaos I would have no ability to use mathematics. The fact that I can use mathematics presents a Universe of order and that is just what gravity is. Where there is gravity chaos is prohibited. That would be the most unlikely way Creation came in place. The recalling of pictures representing images about creation must have form, but to mathematics it had no form. From this thought the very opposite arise where Creation came from nothing but such an idea is mathematically simply not possible. The mathematical presence we have in the distribution and organisation of the Universe and even the calculation where able to present where chaos is abundant, tells a story of organised growth and not a blob of material with no centre from where gravity control and no even distribution of material. No wonder those Newtonians will fill a Universe with nothing and then stand back to have a view of the entire nothing they can see. Please keep in mind I am the under-educated zombie that has no brain function, which would enable me to understand Newton while they can have their view at night on what they fill the Universe with and call it nothing. How on Earth can one say one can see nothing and still point at something you see?

The thought of nothing is just what it is, a thought of nothing and although it is in the human mind common nature to present nothing as a value in the recalling of something, nothing is a presentation of the figment in the human mind. There can be no number such as nothing and that was (possibly) Newton's biggest error. Nothing represent non-existing and that is just what nothing is, it is non-existing.

In order to prove my point I wish to ask the reader to define the shortest line there can theoretically be. If he should answer anything but that the shortest line will be at a point where the beginning and is the very same spot he will be wrong. The shortest line that can ever be anywhere must have a start and finish holding the exact same spot. The line will be humanly impossible to create but we humans are capable of very little.

When the line has a beginning and an end at the very same spot and it wishes to extend the position as to further the possibility it has, which direction should it favour. Humans in the west would naturedly think of extending from left to right while in the east humans may want to go from right to left. Some persons will tend to go up or down, but all of the options are about human preference and not mathematical conclusions. Extending the line in any one direction will favour one direction without a conclusion about not extending in other directions. Such a conclusion has no sound mathematical foundation. The only option about extending will be in all directions equally in order to give a meaningful non-bias flow of mathematical equilibrium

The shortest line in the realm of possibilities must have a start and finish holding one spot and such a line will also be a dot or a circle. Not favouring one direction puts all directions at equilibrium meaning that any form what ever may be can develop from such a spot with the end and the start being the same. This reasoning prompted me to look for singularity in such a spot because if the prime spot from which all came was a spot, then the spot must hold the shortest line but more prominent it will hold the smallest form including the smallest circle.

One possibility that the shortest spot can never have is having a starting point on the zero mark. If the mark of zero holds the start it must also hold the end because the end and the beginning has the same position. If the position of zero then is the beginning, the end will also be zero leaving the line without an end as well as without a beginning.

The conclusion from this is that no line can start at zero because that will be a mathematical impossibility. A line or spot starting at zero would therefore be shorter than the shortest line possible. A line growing or extending from zero can never leave zero because of the influence of being zero disqualifies any possibility of growth. If the line then had to grow in all directions at the same pace the line must therefore be a circle. The value of the circle is Π, and that is where creation started.

That gave me the clue where to start looking for singularity. One would find singularity in the value Π and the value Π will be in all things rotating in a circle. To start my explanation about my cosmic theory I wish to firstly start with some nostalgic and the relevancy will become apparent later on. Such is the importance however that I wish to place this at the very start of the prologue.

When we were boys we played with a top we called the spinning top. I cannot imagine that there is one boy in the western world that did not hold such a devise in his hand. Tying a string securely around the tapered cone started the operation and then with a jerking or pulling throw the devise is launched in a projectile manner and the big knack to success was getting the nail end firmly on the ground and with a releasing jerk the top was rotating. The champion was always the one boy that could throw his top to spin the fastest and that would create a humming sound. The louder the sound produced the bigger the champion

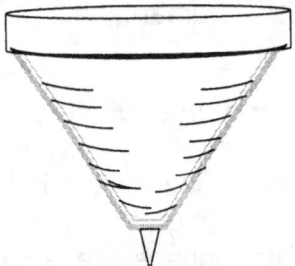

When a back braking effort produced a throw of enormity the spinning top would not only produce sound varying in pitch but also create a spin that would seem to have some instability. There are very many limitations about the spin, parameters that determine the slowest and the highest sin rate and spinning is within the parameters of such settings. The question arising is why such parameters are there in the first place?

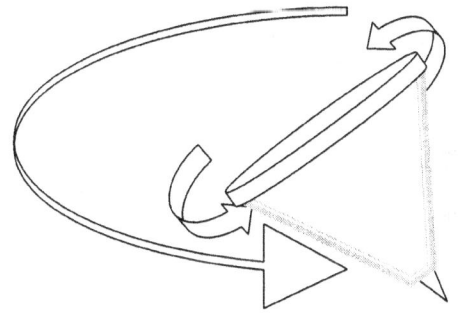 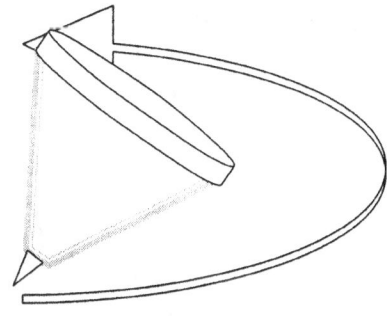

An enormous effort will have the top going oblong while spinning violently and as the pace reduced the top will stabilize by coming to an upright position. In the upright position it wall then spin for the remainder of the period where it will in the end start tilting to the side and in a last effort throw a few wild oblong turns and fall over.

Boys playing games will never realize scientific breakthrough explaining and grown ups do not play with toys. In this little toy played everywhere everyday by almost every one is the answer most brilliant of human Brainpower seeks answers about all the cosmic riddles no one seems to understand. In the spin as such one may find two vital boundaries in the motion and the boundaries are marked by a wobble coming about as if the top is fighting some other influence. Spinning too fast pulls the centre off centre and so does spinning too slow. It is the same influence coming about at both ends of the limitation in the spin. There are influences at work, but force…no; it cannot be forces setting such boundaries. From that I started per cuing what sets such limitations because that limitation must be universal as all matter is spinning in one way or the other.

I MAINTAINED DURING ALL MY CORRESPONDENCE TO SO MANY I HAVE CONTACTED, I STILL MAINTAIN AND I PROVE MY VIEW POINT THAT:

1) There is no gravity and therefore GRAVITY DOES NOT EXIST.
2) With no gravity it stands to reason that I also maintain that NEWTON AS WELL AS EINSTEIN IS ALL TOGETHER WRONG!
3) With no gravity NEWTONIAN VIEWS ON THE WAY CREATION CAME ABOUT IS ALTOGETHER INCORRECT.
4) THE BIBLE IS ALL TOGETHER CORRECT ABOUT CREATION AND FOR THOSE SCEPTICS OUT THERE I PROVE IT AND THIS TIME THE ATHIEST MUST BRING THEIR PROOF IN DOUBT. For instance that darkness filled with nothing being the night sky is light and that was the very first command "Let there be light". That is light out there! Now it is the atheist turn to prove that which is filling the night sky is nothing. Go on… prove it mathematically that $149 \times 10^6 \times 0$ = the distance there is between the Earth and the sun.

In the past these remarks made me the clown in the courtyard and no friends came to my aid because no friends were in support of my statements. A description that would be closer to is that no friend wanted to admit any friendship because such admitting may also reflect on his or her sanity.

When looking at the cosmos from whichever angle indicates the fact that the cosmos is moving. It is forever spinning and it is going to as much as it is coming from. Everything is on the move and always encircling something of greater importance. A top can spin but the parameters of its spin are limiting the motion it can apply. By not spinning the top is still spinning as the Earth are doing the spinning on its behalf.

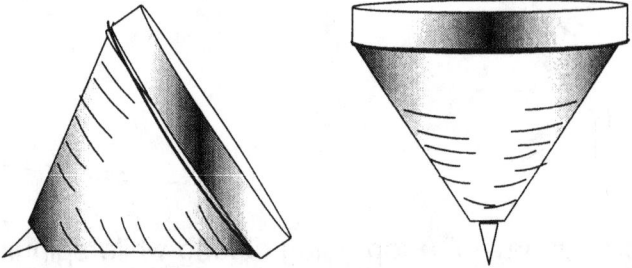

When spinning too fast the top fights something because the alignment keeping it upright starts to tarnish. The same apply when spinning too slowly but that makes sense. It is the fact that the same affect comes about when spinning too slow that triggers the questions.

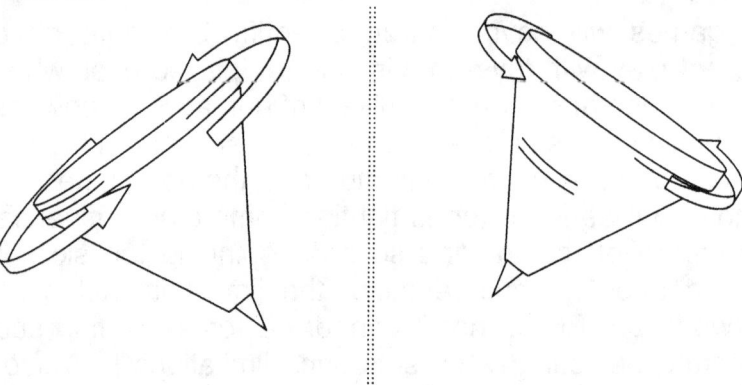

The spinning top is all the evidence any one needs to come to such a conclusion. By saying that I first have to admit (no not my mental stability), that I have no academic background and I do not enjoy any link to any university.

Without trying to not to be too presumptuous I'd say a fair guess would be that I know probably as much as any graduate about cosmology but lack certificates to prove my knowledge. I am not part of established science. In my developing of knowledge accumulation I came to some conclusions about cosmology that are unique and divert somewhat too drastic from the accepted norm. Most of the work I see the same way as the norm does but in a reverse. Allow me a short explanation

When looking at a red flower we say the flower is red. Nothing can be further from the truth. The flower is every colour in the spectrum, except the colour we attach to it. It is screaming with all might to its disposal that that specific colour it cannot accept. Yet, we maintain that that colour is the colour we associate with the object, ignoring the objects rejection of that colour. Only when looking at the cosmos from this stance, can the cosmos make sense? By recognizing a disassociation in spite of our cultural recognizing the association, can we understand the cosmos? We maintain the Sun is burning, while the fact of the matter is thjat the Sun is freezing. From our perspective on the outside we see the Sun burning as we see the red flower. What we see is not what is the truth. Only by applying the correct view to the cosmos can the four principles I introduce, make any sense and find any proof... and I do prove them. Only by telling the complete story as I do in the complete six parts of "*Matter's Time in Space – The Thesis*", can the explanation surface to a point of understanding. One cannot draw any conclusion from the outside; one has to be inside the star to see what is going on. To get such proof I had to do extensive research on cosmology. The proof lies in unrecognised and misunderstood laws and principles science know. These laws fall outside the parameters of applied physics.

I defined gravity; I defined energy, but before that I had to prove the existence of time and time's control over the Universe, time's role in the Universe and what time is. This was up till now not yet been achieved. I had to prove what space is, that time and space is sides of the same coin, with matter forming the separation. The main conclusion that brought about such conclusions was my different view of science. It's not the explanations science at first that made me question the validity of Newton, but the things Newton cannot explain but is factors in the cosmos nevertheless. As a school going youngster I was fascinated by astronomy and in particular the cosmology aspect. In a long and strenuous process of self-education I was completely stunned by the behaviour pattern that the comet had in its relation as it orbits the sun. Please forgive my boyish way of presenting the following but it is important that I bring it across as I saw it as a boy and as a matter of fact still see it today as a middle -aged adult.

We may start by determining the influence of gravity on planets as we find them in the solar system. First, let us concern ourselves with a comet. **It is common knowledge how the comet relates to the sun's gravity. Firstly, picture the comet at its farthest point, away from the sun. The gravity of the Sun pulls** the comet straight towards the sun... **this we all know. Gravity always pulls an object directly towards** the **centre of a cosmic body**: **that too is common knowledge. Therefore, the comet is drawn directly towards the centre of the Sun and throughout its journey the comet is picking up momentum directly related to the gravity that is cantered in the middle of the sun, (gravity is always cantered in the middle of a cosmic body)**. **As the comet is increasing its speed, the comet comes closer to the Sun and therefore the sun's**

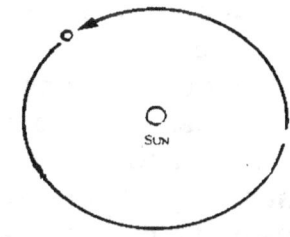

gravity pull is simultaneously increasing as the distance between the two cosmic bodies is reducing. Each instance the comet is drawn towards the sun, the gravity that the Sun applies to the comet becomes larger progressively.   When the comet is at its <u>closest point to the sun</u>, <u>something odd happens which cannot be explained by Newton's gravity at all! Remember gravity should now be at its strongest point because of the proximity of the two objects.</u>
The comet remains at an even distance encircling the sun.

No longer does the gravity of the Sun pull the comet towards the centre of the sun.

At this very point the gravity that the Sun applies on the comet does not pull the comet towards the centre of the Sun any longer, in fact, it seems as if the effect of the gravity has been neutralized.

1.   The comet stays at an even space from the Sumas it goes around to complete a half circle's orbit around the sun.

2.   No longer does the gravity of the sun pull the comet towards the centre of the sun.

3.   At this very point the gravity that the sun applies on the comet does not pull the comet towards the centre of the sun any longer, in fact, it seems as if the effect of the gravity has been neutralized.

4.   The comet stays at an even space from the sun as it goes around to complete a half circle's orbit around the sun. It only completes a part of its rotation around the sun.

5.   After this, an even more peculiar event takes place. **The sun, at the point where gravity should be at its most dominant, suddenly loses its complete grip on the comet.**

6.   The comet brakes free from the sun's pull of gravity and speeds off towards its destiny into the vastness of the cosmic space, undeterred by the gravity of the sun.

Then after a pre-determinate and pre-calculated time the Sun starts applying its gravity on the comet once more. At a point where the comet is at its farthest point, the gravity of the Sun becomes strong enough to bring about a complete turn around to the comet's direction of travel. **However, the gravity between the Sun and the comet is at this point, at its weakest point of influence.**

However, this is not all. When we regard the planets as they stand related to the sun, the effect is the same, but not as obvious. All he planets follow an oval orbit around the sun and therefore the same factors concerning gravity apply to the letter as it 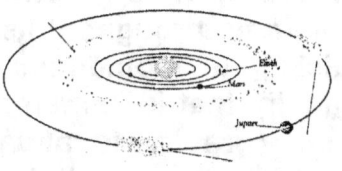 does in the case of the comet. Let us investigate the one planet we relate the best to, which of course is the Earth.

So, when the sun's gravity is at its strongest, the comet manages to brake loose and neutralize the sun's gravity pull in order to avoid its fatal collision with the Sun and when the sun's gravity is at its weakest, the comet cannot escape the pull of gravity. There is definitely something very wrong, either with the comets or in this case the Earth's circling behaviour or the laws made up by Newton.

Well, this is the part Newtonians are so able to understand and because I am a mechanic I am not able to Understand. The Earth makes a circle and Newton in all his mathematical splendour never provided for this circle. He even went further by introducing Π to what is the eternal circle! The Newtonians sympathise with my poor understanding because of my low intellect and education! Who should be sympathising with whom should be a better question?

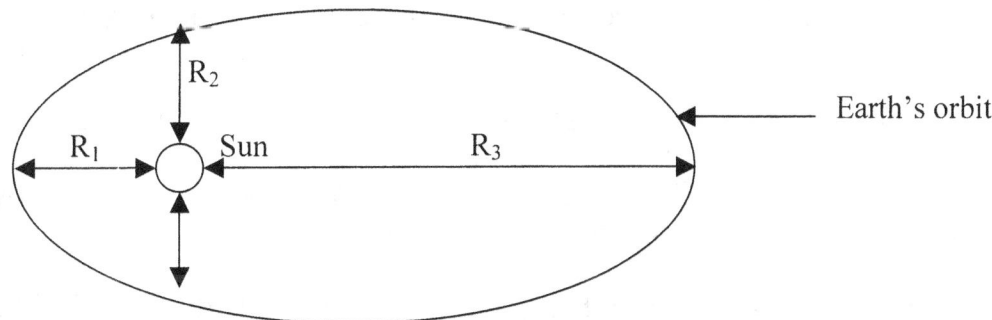

This illustration does exaggerate the radius of the Earth's orbit around the sun, but since it has taken place 4 500 000 000 times, it has no real effect on the validity of the next statement.

At one point (R₁) the distance between the Sun and the Earth **is less than** at another point we call R_3. Let us put a value of R_1 = one and R_3 = three. This means that each year, for the past 4 500 000 000 years the effect of the common gravity between the Earth and the Sun has a greater effect than at another point six months later. **At one point the Earth should be drawn or pulled closer to the Sun** and **after another six months** interval **the Earth should stand less effected by the sun's gravity**, therefore it should move away from the sun. Each cycle of twelve months would have one point where the gravity pulls the Earth closer 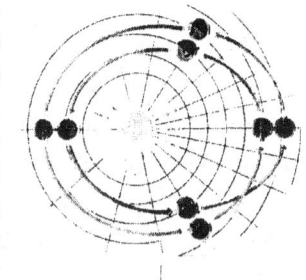 and exactly the opposite must apply six months later when the gravity is at its least. So, for the past 4 500 000 000 years the Earth has been re-establishing its seasonal swing towards the Sun and away from the sun, which means by now the Earth has to collide with the Sun in midsummer or escape from the Sun in midwinter, as it may then drift away into the unknown.

For the more mathematical minded person the argument is as follows. May I remind you, THAT NEWTON'S OWN LAWS ARE IMPLIED, and again the planets disobey these laws completely!

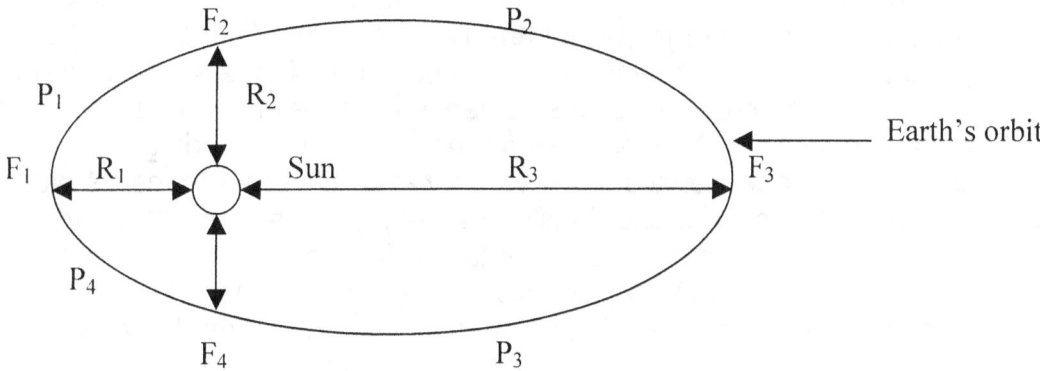

$$F = \frac{M_1 M}{r^2} G$$

We know that $F_1 \neq F_2 \neq F_3 \neq F_4 \neq F_1$
because that is what seasons are all about.

Even if $F_1 \neq F_2 \neq F_3 \neq F_4$, we know that $P_1 = P_2 = P_3 = P_4$.

Because r is at different values F could not be to the same value.
Therefore, the value of F has to be unrelated to force its value on to P.
Nevertheless, Kepler has proven that $P^2 = a^3$, although $a_1 \neq a_2 \neq a_3 \neq a_4$.

If $a_1 = a_2 = a_3 = a_4$, we would not have had season and climate changes on Earth. That means that to proclaim F $= \frac{M_1 M}{r^2} G$ is nonsense. The truth of the matter is that Newton actually proclaimed that in an ellipse, which has an uneven circle (Kepler's findings) the value of $F_1 = F_2 = F_3 = F_4 = F_1$, but because an ellipse has no constant radius, it actually means that $r_1 \neq r_2 \neq r_3 \neq r_4$ and thereby anybody can see that Newton's calculations are wrong. $F_1 \neq F_2 \neq F_3 \neq F_4$.

In this book, I dare to prove that **there is a difference** between findings of *Galileo and Kepler* on the one side and the work of **Newton and Einstein on the other hand**. Only one of these two group's findings can be right, because there is an unmatchable difference in the concept of these two groups' opinions. Newton considers that a force exists between two bodies in space: the mass of the two bodies' product is being brought into context with the gravity constant (G). This value is then divided by the distance r calculated as a square (r^2) value. I have to admit that I have not once seemed to bring across the importance I see in the arguments above when translating it to academics of stature. Every time I introduce the behaviour of the comet I get the impression that academics either will not or cannot see any truth in my arguments. In every incident I became the accused of not having the brainpower to understand and from my perspective there is little to understand.

I stood accused by many academics I crossed paths with in the past that I am not familiar with Newton and because of my poor academic background that I am not capable of understanding Newton. That is not the case and I have to be very adamant about that. I would accept such accusastions if Newton's science explained all of science. That is hardly the case because I can state four very prominent cosmic principles that no one can explain by applying Newton's claims. The Roche-Lobe, the Titius Bode principal, and the Lagrangian five-point position and the the Coanda gravity contraction is what Newton's gravity formula cannot explain at all. Neither can the Big Bang theory be the starting point if cosmology insist on using the application of Newton. I am aware of the Critical density theory and black matter, but those arguments has not found proof in the slightest and in truth serves as an escape corridor because science is at the end of its tether with the phenomena contradicting Newton. I admit that I am lost at finding a starting point introducing the book because the issue remains comprehensive when dealing with issues of cosmic proportions I shall explain the four unrecognised phenomena I use in proving my statements. With my introduction of the phenomena, which I named the four cosmic pillars you will find it obvious why science do not accept them even if it is documented throughout the Universe and is quite commonly found. It totally annihilates Newton's formula of F = G $(M_1 M_2)/r^2$.

From these cosmic phenomena I produce a path of cosmic development, preceding the Big Bang. The problem that comes from this, is that I take the reader from a point and lead the reader through the explanation of the existing principles, pointing out how they are flawed and introducing my explanations and proof and substantiate my argument. This is a path one has to follow. There is no point where one can drop in, or out and in again and maintain the golden thread of understanding. To conclude, only this: As I bring proof of existing evidence in cosmology about phenomena and of which science acknowledges the existence but science is failing to understand or explain the correctness thereof.

The Bible says: "Do not think of the heavens as Earth." Whether the Newtonians take exception in my quoting from the Bible means little, because that verse holds all the mistakes and misjudgements that science have about the Universe. Science holds the attitude that from the Earth, they can judge the Universe and by applying standards maintained on Earth, it is a "fit all bar a few minor adjustments." Science accepts the Hubble Constant and even uses it as a barometer. This Hubble Constant indicates how the Universe is expanding, growing in size. That means it is the measurement of how matter drifts apart. If the Hubble Constant applies to the Universe as a barometer, then that application should affect the Milky Way as much. Remember I am using Newtonian logic, not my own, so please do not misquote me. If there is the shift of matter, away from a centre point outwards, the circle in which the orbiting structures hold their position should then enlarge, because of such a shift. By increasing the distance that the orbiting structure travels around the centre, the time it takes to travel should also increase. If one wishes to form a concept of the implications in hand the most prudent manner is to return to the star of it all and then try and progress from that position.

The Big bang was never bigger than at the time the procession of time started. The Biggest Bang was when light or heat started the process we call the Universe. Heat expands and cold contracts and that is gravity. As heat expands it leaves an area by halving the heat in that area. The relevancy changing makes that the duplicating brings about the material two half by doubling the space it moves through. By the halving the heat, the motion is enforcing the area to cool off by half and that produces contraction. By expanding there is contraction because overheating expands while it then cools bringing about contraction. As the light came into the darkness by heat coming onto the eternal line of time the line was interrupted or broken by infinity releasing the line in eternity of its eternal procession. Infinity released from eternity, heat broke from cold singularity in the dot released from singularity in the spot, Π^0 relieved Π from eternity, 10 broke from 11 and all the Universe coming about fell into place in space in time in space-time an in time space. The line dotted on a spot the line was in progress one eternity long without progressing because it was in an eternal progress that had no beacon, no marker and no comparing to, which put the eternal running line eternally in one spot. The heat interrupted eternity by infinity parting from eternity.

Now please listen to my explanation and compare my explanation on how creation began with the nonsense of Newtonian science and that which according to them I am too simple minded to understand. Simple minded I might be but so was creation at the start never very complicated. It started from a spot that became a dot and by becoming something eternity began to flow, as time broke free from monotony, although at first by the instant of infinity.
The line had flow. The line did grow albeit from infinity to infinity relieving eternity from infinity. The Universe made the biggest leap it never could repeat afterwards and it was so small it can never be seen inside this Universe we are within.

The line diverted from Π^0 forming Π but there was so little established that although Π was Π also Π = 1 because 1^0 moved to 1^1. That put the figure of two in the Universe. Then from two the square of two diverted from the line of time by 9^0 0 which established a line of 180 ^{0} at a relevancy of 90 ^{0} to the timeline at 180^0

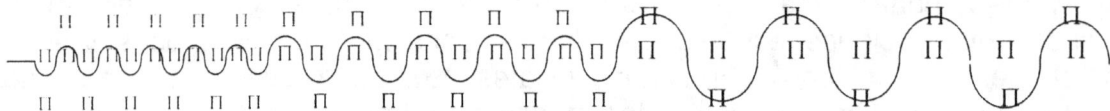

There was the line $\Pi^0 = 1^0$ that formed $\Pi = 1^1$ and $\Pi = 1^1$ ninety degrees to the line in time. Through the motion that came about that motion brought about a square in the line of time meaning that Pythagoras established mathematics.

As time moved on a lagging behind started to creep into the flow where the heat began to distort the flow of time by not repeating the flow in the same cycle

This lagging of heat formed the basis for space holding time end very much later it became even further behind and in such a lagging that the time solidified to form heat dragged behind as material. There at that point formed matter. But at first every point formed another value in the line of figures coming from one and ending at ten. The line established the forming of every dynamic present in the Universe that established numbers in relation to Pythagoras except there was no zero formed. The one number formed the next number in conjunction with Pythagoras and in that way mathematics formed the cosmos while the cosmos formed mathematics. However it is evident that Newtonians are under the impression that while mathematics was used to form the cosmos mathematics in principle never formed.

$$\Pi^0 = 1^0$$
$$\Pi = 1^1$$
+
$$\Pi = 1^1$$
$$\Pi = 1^1$$
$$\Pi = 1^1$$

$2\Pi + \Pi^0 = 1^0 = 3$
2Π = another r
dimension forming the
square $\Pi \times \Pi = \Pi^2 =$
motion of space in time

Then the Universe was 3 with two divisions on both sides of the divide.
It is clear that the adding was the first dimension because to get from one to two by multiplying brings one back to one. However by adding it does bring about three in the triangle while two is part of the half circle as well as the line of time. By forming a relation with the line and on the one side with a half circle while being relevant to a point on the other side of the divide clearly shows why 1800 fits all three forms and put the here on an equality that otherwise does not make any sense to us. However one can see that in the next development the square came about by multiplication of two points traveling equal along a centre line.

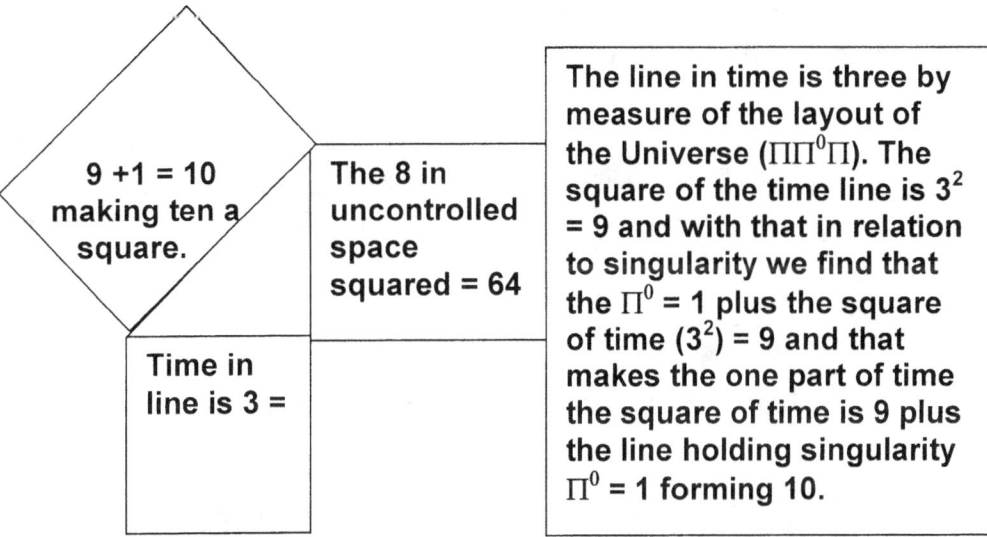

9 +1 = 10 making ten a square.

The 8 in uncontrolled space squared = 64

Time in line is 3 =

The line in time is three by measure of the layout of the Universe ($\Pi\Pi^0\Pi$). The square of the time line is 3^2 = 9 and with that in relation to singularity we find that the Π^0 = 1 plus the square of time (3^2) = 9 and that makes the one part of time the square of time is 9 plus the line holding singularity Π^0 = 1 forming 10.

That makes ten the square of as well as the root of and that is why we may be able to use ten as a decimal factor forming the basis of all numbers.

In the motion space found a relevancy by the ordering of gravity.

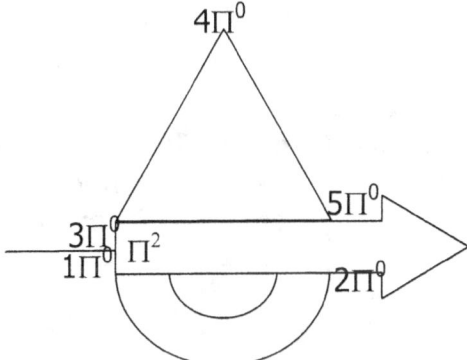

Motion is to the values of gravity, which is the value of Π^2. Crossing the division of singularity brings about a wave half circle connecting to a triangle on the other side of the divide. That places 5 points in singularity Π^0 in relation to the motion of time Π^2. The 5 points in singularity is 5 x the motion of singularity is 9.8 X 5 = 49. Put that in relation to Pythagoras

Three   Four   Three

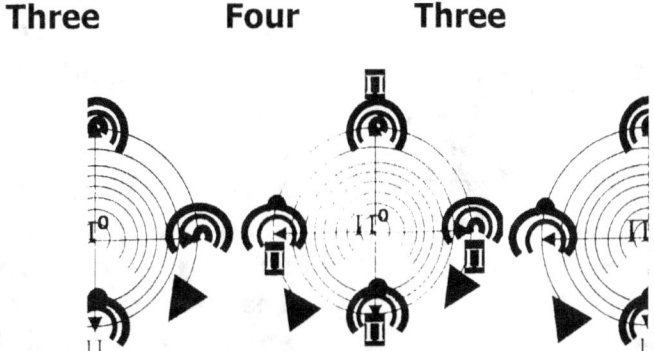

The atom formed the Universe long before there was an atom in the Universe. The universe became the atom long before the Universe compiled the atom

There was the line $\Pi^0 = 1^0$ that formed $\Pi = 1^1$ and $\Pi = 1^1$ ninety degrees to the line in time.

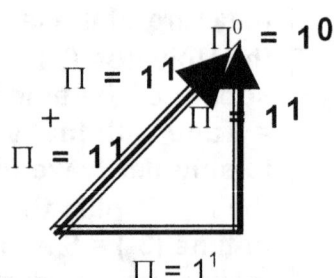

$2\Pi + \Pi^0 = 1^0 = 3$
2Π = another r dimension forming the square $\Pi \times \Pi = \Pi^2$ = motion of space in time

Then the Universe was 3 with two divisions on both sides of the divide.

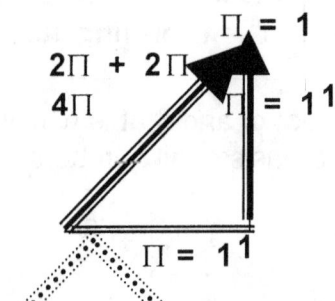

The motion of space- time in space in time brought about the compliment of time being four.

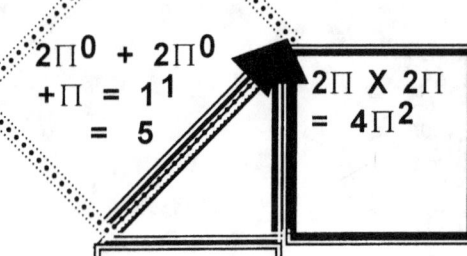

In this 5Π became valid in the cosmos forming a point one separated from time using 4

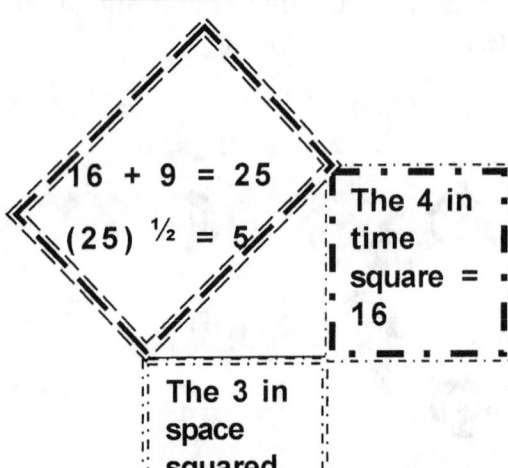

Time is holding 16 + material is holding 36 = 52. The compliment 52 – the 3 of the time line gives 49. The square of 49 is the end of filled space, which are 7. The sphere runs by seven points. That means compromised space starts at point 8.

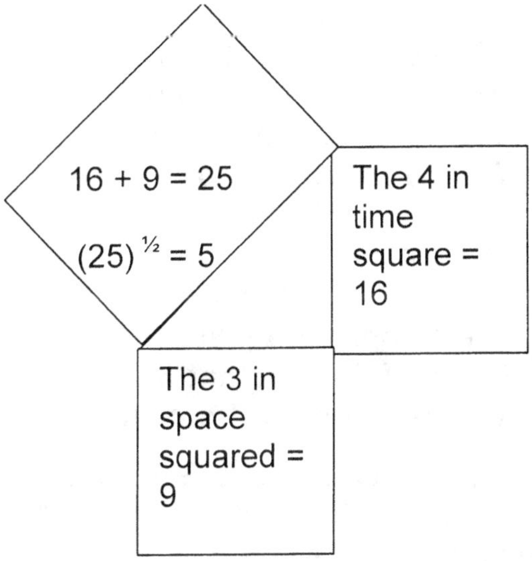

16 + 9 = 25

$(25)^{1/2} = 5$

The 4 in time square = 16

The 3 in space squared = 9

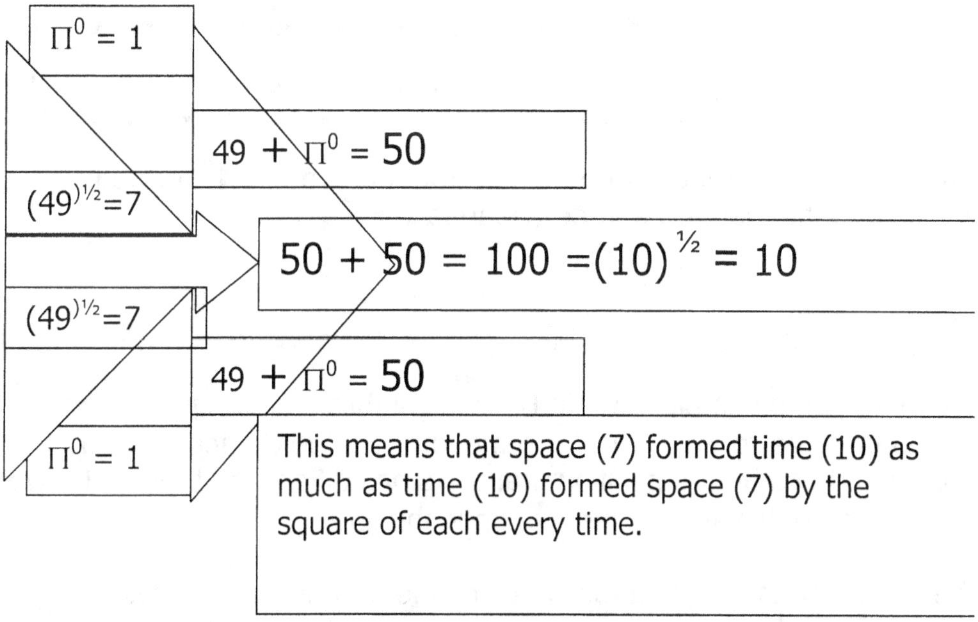

$\Pi^0 = 1$

$49 + \Pi^0 = 50$

$(49)^{1/2} = 7$

$50 + 50 = 100 = (10)^{1/2} = 10$

$(49)^{1/2} = 7$

$49 + \Pi^0 = 50$

$\Pi^0 = 1$

This means that space (7) formed time (10) as much as time (10) formed space (7) by the square of each every time.

The square of space is 6 X 6 (6 by dimensions) in relation to Pythagoras

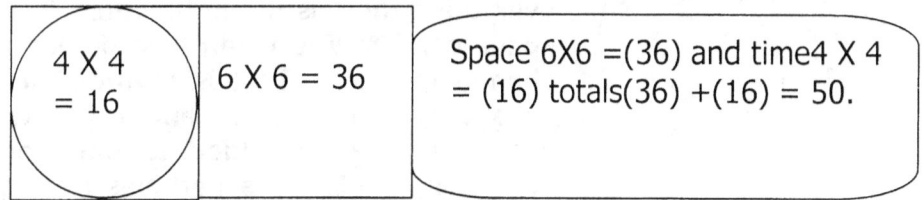

4 X 4 = 16

6 X 6 = 36

Space 6X6 =(36) and time4 X 4 = (16) totals(36) +(16) = 50.

The square of time is 4 X 4 (6 by dimensions) in relation to Pythagoras

$\Pi^0 = 1$

$50 - \Pi^0 = 49$

$(49)^{1/2} = 7$

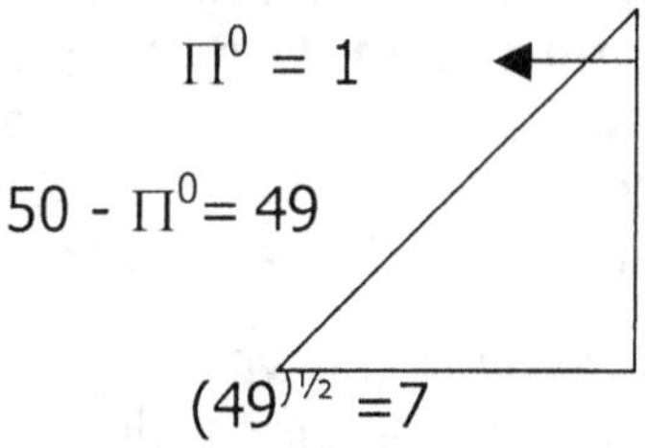

That way space in time formed the curve of space by 7^0.

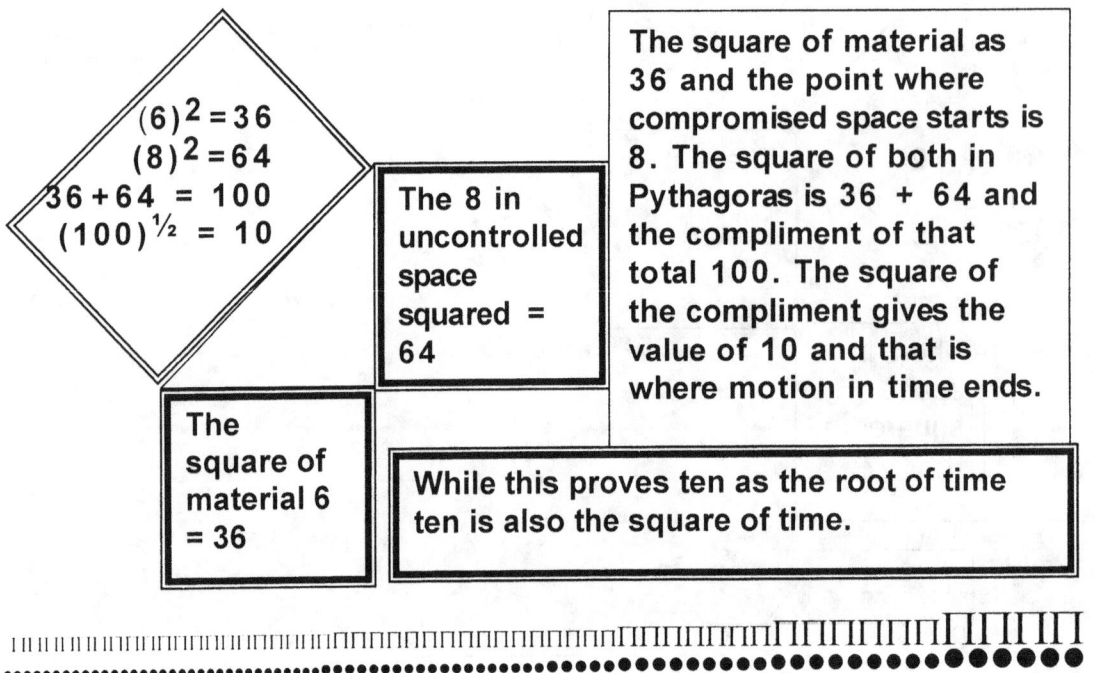

$(6)^2 = 36$
$(8)^2 = 64$
$36 + 64 = 100$
$(100)^{1/2} = 10$

The 8 in uncontrolled space squared = 64

The square of material 6 = 36

The square of material as 36 and the point where compromised space starts is 8. The square of both in Pythagoras is 36 + 64 and the compliment of that total 100. The square of the compliment gives the value of 10 and that is where motion in time ends.

While this proves ten as the root of time ten is also the square of time.

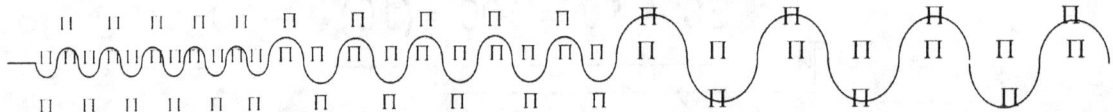

There was the line filling with ones that formed twos as 1^0 formed 1^1. That made two lines and that had to be three because 1 + 2 = 3.

Since three came into place while Π^0 was establishing Π I guess (I cannot prove it but it does come to mind) that the motion had took place while Π became the factor and 3 became the position that the small difference there is between 3 and Π was somehow initiated by the motion in the line.

In the motion space found a relevancy by the ordering of gravity.

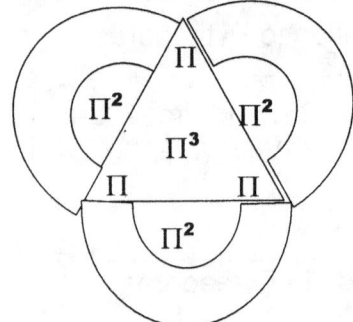

The Universe grew from relevancies but not out of the relevancies. The relevancies still dictate everything there is in the Universe. The only way space can grow is when heat comes about and change the form. That is nature. The only way expanding can come about is when heat becomes a factor added to particles. That is nature. The only ways particles reform or when heat remove some or all the solidness in the form. Heat coming into or removing from determine form. It is still part of the cosmos as it is still part of the cosmos that friction accumulates heat and heat results from a lack of usable space. That is nature.

Motion is to the values of gravity, which is the value of Π^2. Crossing the division of singularity brings about a wave half circle connecting to a triangle on the other side of the divide. That places 5 points in singularity Π^0 in relation to the motion of time Π^2. The 5 points in singularity is 5 x the motion of singularity is 9.8 X 5 = 49. Put that in relation to Pythagoras

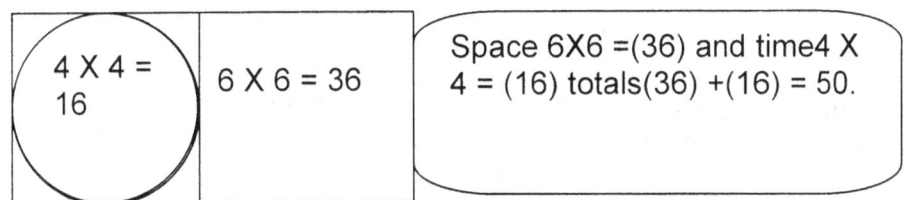

One point away from the divide $\Pi^0 = 1$

$50 - \Pi^0 = 49$

$(49)^{\frac{1}{2}} = 7$

That way space in time formed the curve of space by 7^0.

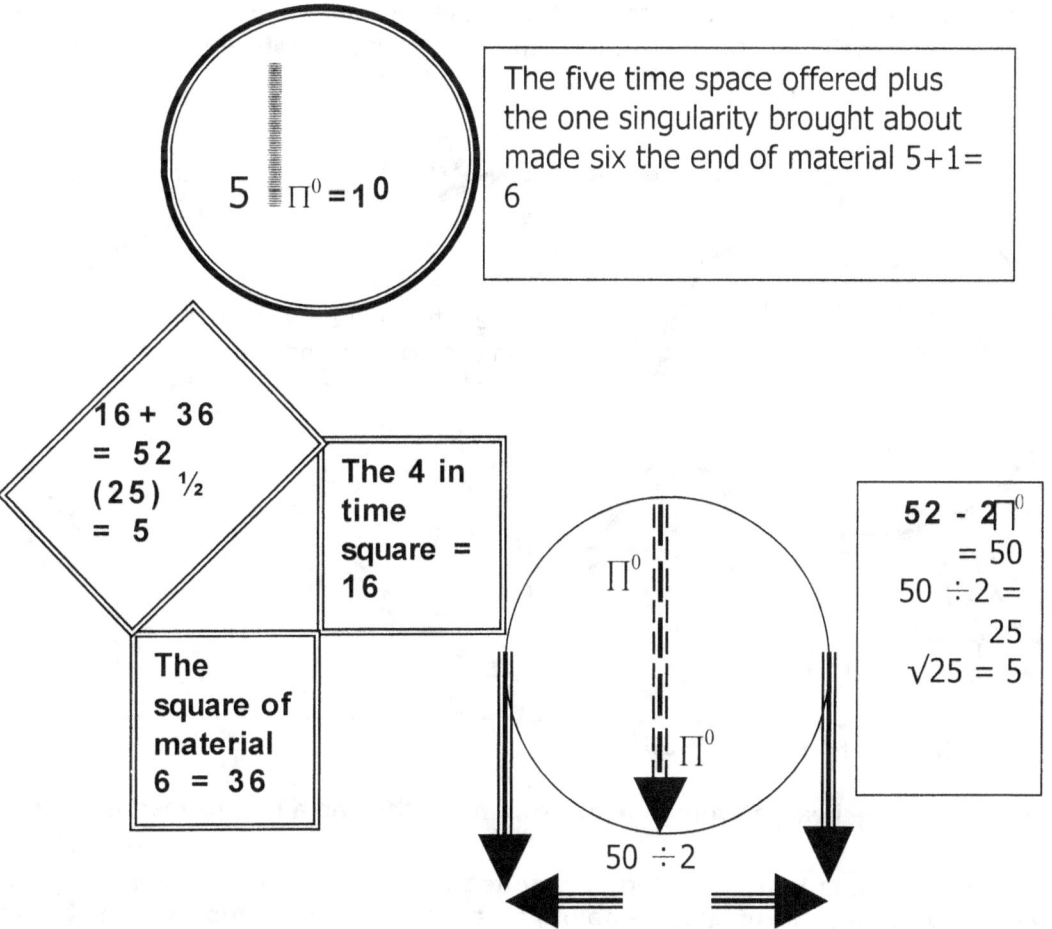

5 | $\Pi^0 = 1^0$

The five time space offered plus the one singularity brought about made six the end of material 5+1= 6

16 + 36 = 52 (25) $^{\frac{1}{2}}$ = 5

The 4 in time square = 16

The square of material 6 = 36

Π^0

Π^0

$50 \div 2$

$52 - 2\Pi^0$ = 50 $50 \div 2 =$ 25 $\sqrt{25} = 5$

That makes ten the square of as well as the root of and that is why we may be able to use ten as a decimal factor forming the basis of all numbers. Lets run through the process again in order to pick up some loose ends.

Every time Π formed singularity brought about motion by gravity in Π^2, which still maintained a line. Gaps formed and that defined and became the atom in space.

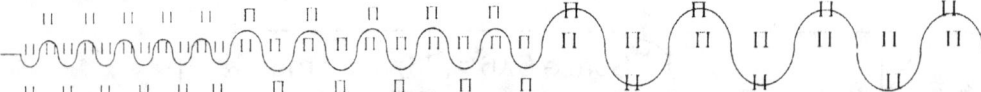

The line formed singularity by forming from the spot going onto become a dot. The heat surging moved as heat still does today. The heat expands as it moves into more space. That which formed was Π and the motion in the surge was Π^2.

Every time Π formed singularity brought about motion by gravity in Π^2, which still maintained a line. Gaps formed and that defined and became the atom in space. The heat moving became the motion that became the gravity. The moving was a relation between $((7/10 + 7/10) \times 10)/10/7$ and the moving formed Π^2.

$((7/10 + 7/10)$. Material forming as space becomes available provided by heat surging. It was material moving in time in space that was becoming time.

(10) The three in time-space or the space in which material has room to move.

10/7 heat cooling by surging back as the contraction retarded the responding growth.

In this was a line, which I now am unable to sketch since sketching firmly relies on a three-dimensional drawing, and at the time what was three-dimensional is very flat today.

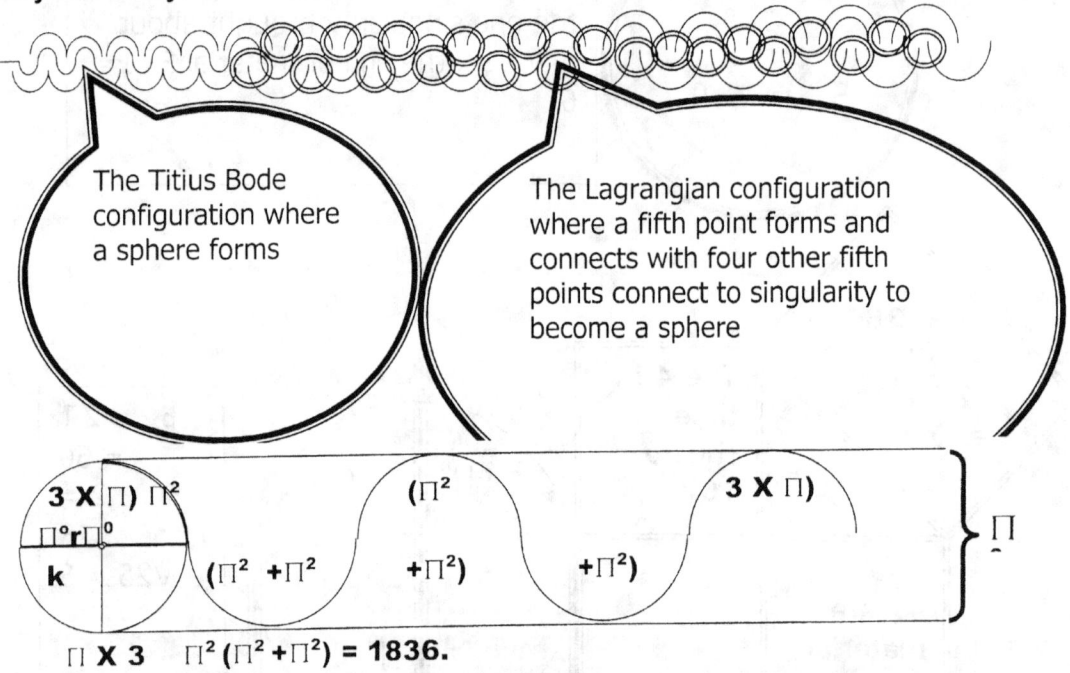

The Titius Bode configuration where a sphere forms

The Lagrangian configuration where a fifth point forms and connects with four other fifth points connect to singularity to become a sphere

$$(3 \times \Pi)\ \Pi^2 \qquad (\Pi^2 \qquad 3 \times \Pi)$$
$$\Pi°r\Pi^0$$
$$k \qquad (\Pi^2 + \Pi^2) \qquad +\Pi^2) \qquad +\Pi^2) \qquad \Bigg\} \Pi$$

$$\Pi \times 3 \qquad \Pi^2(\Pi^2 + \Pi^2) = 1836.$$

Finally with all the heat retarding an interrupting of the line a gap formed in time.

$$\blacktriangleright \Pi^0$$

All the while time is just a spotted and dotted line running along time as space duplicated with heat surging and cooling as cold contracted much similar to the actions of stars in the process of pulsating known by what ever name one wish to use. The star takes time back so slow we can see the pulsation of gravity cycles.

This was what lead to the process through which at a later time became the method that the atoms of various significant formed by the motion that was prevailing at the time. We find the evidence in the characteristics the atoms show in relation to heat. In the book Starstuffin I do explain this in length.

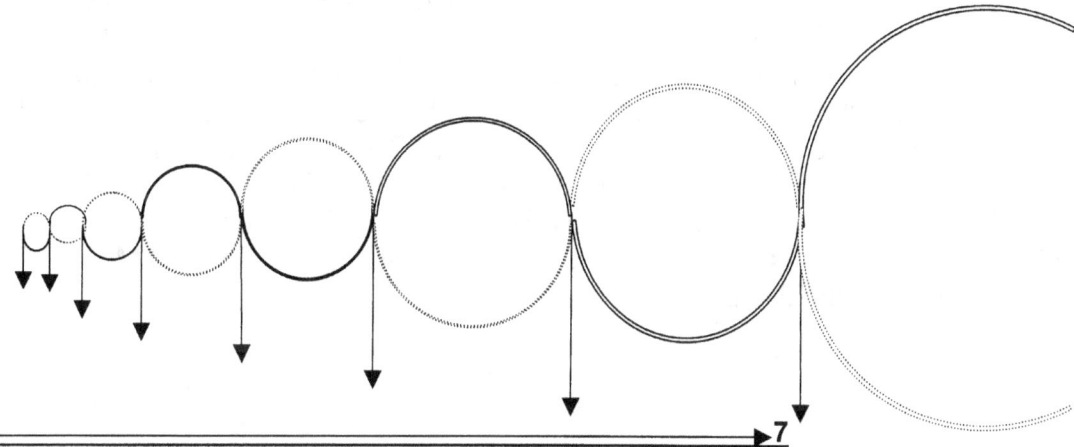

Newtonians put the Big Bang, as the start while the Big Bang they refer too was the seventh Big Bang and ever bang was considerably smaller than the previous one.

Amount of radius increase provided by the Hubble after the first:
2) 1×10^9 YEARS AFTER THE BIG BANG EVENT
3) 3×10^9 YEARS AFTER THE BIG BANG EVENT
4) 6×10^9 YEARS AFTER THE BIG BANG EVENT
5) 9×10^9 YEARS AFTER THE BIG BANG EVENT
6) 12×10^9 YEARS AFTER THE BIG BANG EVENT
7) 15×10^9 YEARS AFTER THE BIG BANG EVENT

They apply an age of 13,5 x 10^9 years in the case of the universe and 4,5 x 10^9 years in the case of the earth. This allows the earth's establishing date to fall into a position where the universe was 2/3 of what it is now. IF THERE IS A HUBBLE SHIFT, THERE ALSO HAS TO BE A HUBBLE SHIFT IN OUR PART OF THE UNIVERSE AND THEREFORE THE SOLAR SYSTEM ALSO HAS TO ADHERE TO THAT. THE SAME HUBBLE SHIFT MUST THEREFORE AFFECT THE SPACE BETWEEN THE EARTH AND THE PLANETS AT THE SAME AS THE RATE OF THE UNIVERSE.

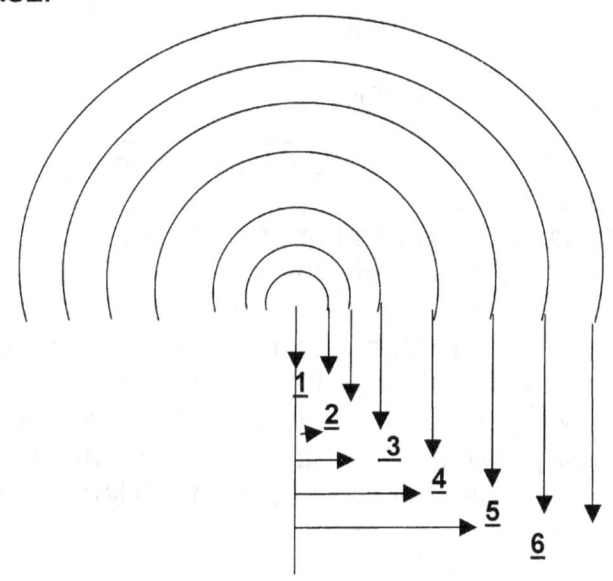

How long did the first instant in time last? It lasted one eternity minus one infinity. Then the next duration was one eternity minus three infinity and then the Titius Bode rule started affecting time by the measure of $(7 + 7) /10 /(10/7) \times 10 = \Pi^2$ They apply an age of $13,5 \times 10^9$ years in the case of the Universe and $4,5 \times 10^9$ years in the case of the Earth. This allows the Earth's establishing date to fall into a position where the Universe was 2/3 of what it is now. IF THERE IS A HUBBLE SHIFT, THERE ALSO HAS TO BE A HUBBLE SHIFT IN OUR PART OF THE UNIVERSE AND THEREFORE THE SOLAR SYSTEM ALSO HAS TO ADHERE TO THAT. THE SAME HUBBLE SHIFT MUST THEREFORE AFFECT THE SPACE BETWEEN THE EARTH AND THE PLANEYS TO THE SAME AS THE REAT OF THE UNIVERSE.

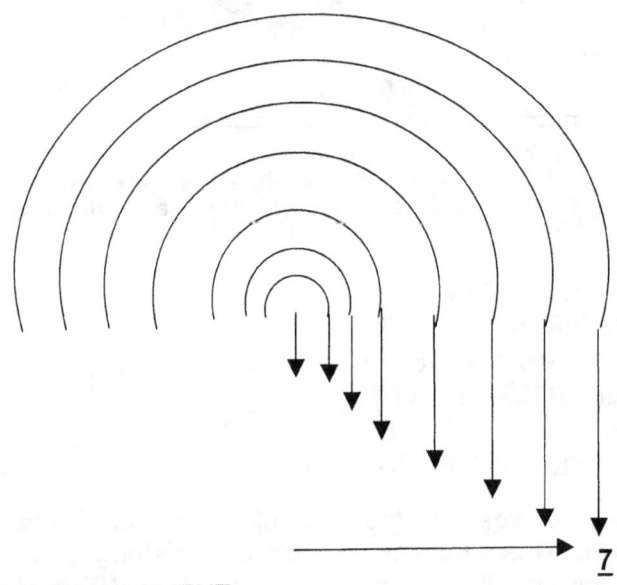

7

1) THE BIG BANG EVENT

AMOUNT OF RADIUS INCREASE PROVIDED BY THE HUBBLE SHIFT AFTER FIST:
2) 1×10^9 YEARS AFTER THE BIG BANG EVENT
3) 3×10^9 YEARS AFTER THE BIG BANG EVENT
4) 6×10^9 YEARS AFTER THE BIG BANG EVENT
5) 9×10^9 YEARS AFTER THE BIG BANG EVENT
6) 12×10^9 YEARS AFTER THE BIG BANG EVENT
7) 15×10^9 YEARS AFTER THE BIG BANG EVENT

Any person that holds a clear view and is not adamant in defending Newton has to be aware of many misgivings his formula brings to the Big bang, If he saw mass compacting mass by mass accumulation then the Big Bang was a missing event and we were no more here than we are there. There should be no star structures that formed because how easy was it for mass to pull mass when there is no separation between particles at all. Many such discrepancies go unnoticed or are either blatantly ignored by a sure lack of interest on the part on science. One cannot advocate a withdrawing Universe and claiming that very instant there is a contracting Universe the way Newton's law lays claim to that. While we all know that that is the case Newton does not prove it and to defend Newton is to become the devils advocate then.

With the Universe, expanding as it is, taking the Milky way along for the ride, and taking into effect the Universe square law where if the radius doubles, the circle grows four times, a year back when the Earth supposedly started, had to have been quite a measured mile shorter than it is today. Back then it had to take the Earth a few days to complete a year, since the orbit path was so much shorter. (Again, I wish to remind the readers that I am not applying my logic but that of the Super Educated.)

The Hubble's expansion brings about the increase in the radius of the distance between the Sun and the Earth. Every time the radius doubles, the circle grows four times by distance. This is a mathematical fact, beyond reasoning, to be accepted by one and all! Holding this mathematical law of value and positioning, the age of the Earth happening some 2/3 down the road the Universe took to develop, and judging the size the Universe got to where it is at present, the circle the Earth uses to orbit the Sun had to quadruple many times over. What is a year? The distance the Earth takes to orbit the Sun at present, or the distance it took the Earth back then. I am not the one to maintain a second back then is the second now.

Our Super-Educated are the ones that insist on maintaining to stick to the second minute and hour as it is today. By reducing the radius between the Sun and the Earth, and then multiply that figure with the number of four that the year was back some $4,5 \times 10^9$ years ago, one had to use some stopwatch to measure one year. If scientists took less time in studying far off galactica, and took the Hubble shift they regard with such prominence closer to home (say implementing it in our solar system) and work the Hubble shift from that angle backwards, a great surprise would await them. They would find that not one single application of the Hubble shift would nearly fit our solar system.

My being ill educated brought so much to resolving of issues in favour of academics in the past that I am putting science in two categories: them and me, and if I stand accused every time not understanding Newton for the lack of education then I am prepared to be the un-educated as long as the other party accepts the title of the SUPER- EDUCATED. I think it is only fair to bring distinction to both parties from both ends. Saying this I also do not wish to offend you in person so I wish to make the issue a three party affair with you in a refereeing stance judging me versus them (the SUPER-EDUCATED-ONES.)

In the first part of the cosmic development since the Big Bang arrival, there was a lot of activity that brought stars of many sorts and flavours to the cosmos from dust, but since there is no evidence of that happening in view of so many astronomers dissecting the sky every night, one has to conclude events slowed down the past 4.5 billion years or so with not even one star in a half built state to show. It must have been great turmoil as everything happened in a short period of time because the development produced all matter relative to the Universe, all space there is in thee Universe, positioned all particles in place to one another, set standards for stars and galactica, reduced galactica, produced dust by the truck loads and in specific semi oval, not to distant and precisely aligned positions enabling stars to form and in some cases even form and disintegrated several times. What an era to live through. Much different from the placid one we find ourselves in at present!

Lets give the Newtonians some leniency and say that the Hubble shift started by having a space of about fifty millimetres in diameter. Two thirds down the road of cosmic development the solar system received its day of birth. Gauge from the development where that should leave the Milky Way, and particularly our solar system. Some scientists are even of the opinion that a massive star developed, destroyed and from that debris and rubble the solar system came about. That means the last third of the cosmic development was a relative dreary and boring era, compared too the first and second era. Think about the billions upon billions of galactica that formed in the first event, followed by the massive stars that formed and demolished during the second era, and during the last third not one single planet formed, that we can witness to prove all the claims. The Sun and the planets formed during the second era; therefore, by the start of the third era, they were in the position and place they are at present. The working tempo slowed down considerably the last five billion years or so I guess some one somewhere formed a workers union and brought in human working conditions to ease the workers demands for overtime compensation.

Push this double standard applied back to before the Sun took its position and there was no Earth to indicate the year. How small was the year circle at that point in time and space. Take this right down to the:" Big Bang" where "the whole Universe were the size of a man's fist" (they go even further by putting us into a neutron), how far did the circle goes to indicate a year then? The year was immeasurably smaller, shorter and faster than at present. This is logic even the Newtonians must accept. There is no space outside insanity to apply time to the past at the value it is at present and far worse, to use something so extremely insignificant as the Earth to measure it by.

Using such logic makes science appear foolish. There is just no rational in the way Newtonians suggest that time lapsed verses how events occurred that can explain facts without. Since the time of Newton, the arguments tarnished from being brilliant to clever to fair, to poor and a hundred years ago to the point of being stupid.   That is what Kepler's formula is all about! That is what Kepler indicated with his formula $a^3 = T^2 k$. The space of an object (a^3) is equal to the time (T^2), which it is in, in every given instant (k). If the space becomes smaller, the time duration becomes longer in every instant that time flow progresses.

Singularity is a mathematical reality. Einstein may be the first to name it and Galileo (unwittingly) may have been the first to define it as Kepler was the first to formulate singularity, but in mathematical terms singularity is the most basic principle.

When science calculated the value of gravity and the gravitational constant, as well as the speed of light, they never considered the moon to play a distracting factor. It is quite understandable because they did not know about the Roche-factor influencing their calculations.

With Π^0 little more than a figment of the imagination there is actually two values of Π^1 facing each other in a relation combining Π^1 to hold the value of $\Pi^{1+1=2} = \Pi^2$ and with two sides being the very same but opposing each other there will therefore also be Π^2 to every side that holds Π^1.

From the above I can conclude that gravity is not 9,81 Nm/s, it is $\Pi^2 = 9,8696$.

The gravitational constant is not 6.67 but it is 6.9 (7/10 (Π^2)) and here the moon had an even bigger influence. It is a fortunate coincidence that we took water to be the measured calculation since water holds the combined value of 17,5 and that is half the value of either space (31) or time (Π^3). That makes a kilometre (1 000 m) one cube laid flat and since movement represents space-time occupation in a linear manner, it is the cube that went in a single line.

More of the same fortunate coincident is that we connected time to spin long before Newtonians came along. The Earth spins through space at 360° in one day and space represents 10, therefore there is 3600 minutes in the 7° of spherical angle moving to the outer rim representing again the seven. All this makes explaining matters a lot less difficult.

The value of the proton is not 2/3, but Π^2. The proton spins at a rate of Π in a dimension of 2. The neutron is not 2/3 + 1/3 + 1/3, but again it is $\Pi^2\Pi$ and the electron's 3, holds a dimensional implication, because time is in singularity and space is in singularity. Time is eternal and heat releases space, which is time from singularity. In an effort to make the understanding simpler, you have time at an eternal value and that makes space zero. $R^3=T^2$; $R^3/T^2=1$, Time and space interlinks because it is the same thing and heat, (matter in many spin rates) allows time to break free from eternity by allowing a distinguishing of the flow of events. Time in movement is the result of matter (which includes heat) to change their relating positions.

Think of a movie. The continuous flow of pictures indicating the change of the position of the photo's bringing about the concept of time. Play the picture too fast or too slow and it will be unreal because we know at what tempo matter changes its position in relation to all other matter surrounding it. That is time and that makes time irreversible, because the position matter holds in relation to each other (in considering it to be throughout the Universe), can never repeat once it has changed.

Newtonians, forget about time travel because just by mentioning such absurdity you prove what little you know about the cosmos. To go back to a certain time, you will have to redirect all matter in the Universe in a reverse, apply that reversing of all particles up to the point required, stop the movement of all particles and start time going forward. Before some Newtonian grabs for a calculator, remember, you that are doing the changing, is as much part of matter in the Universe, therefore your action in changing the direction will stop even before you start! It is silly to think people with healthy minds, acting like adults will indulge in senseless stupidity such as claiming to be able to reverse time.

What was within the Universe at the start will be in the Universe at the end. The Universe holds all; maintains everything and combines the lot. In Afrikaans we call the Universe the "Heelal". It is a combination of two words namely "geheel" and "alles". "Geheel" means everything and "alles means everything. Therefore the "heelal" directly translated from Afrikaans to English will mean the "Everything of everything". Nothing can be added and nothing can be lost. It is all-inclusive. With this fact so commonly known and accepted, how can the Universe grow?

How can the Universe expand? Well, it cannot, and that is yet another illusion the Newtonians create through misunderstanding. What is in it is in it and it cannot grow, as much as it cannot shrink. It cannot expand and it cannot demise. It is only a consistence of changing relevancies, where the relevancy flows away from one part of eternity or singularity (space) to another part of eternity or singularity (time).

Every aspect of the Universe holds relevancy by applying time to space and the time to space first claims space from singularity then controls space from singularity and influences space outside the direct contact with singularity. In every event the factor remains the same, as it is only the relevancy re-applying a dimensional influence on space-time.

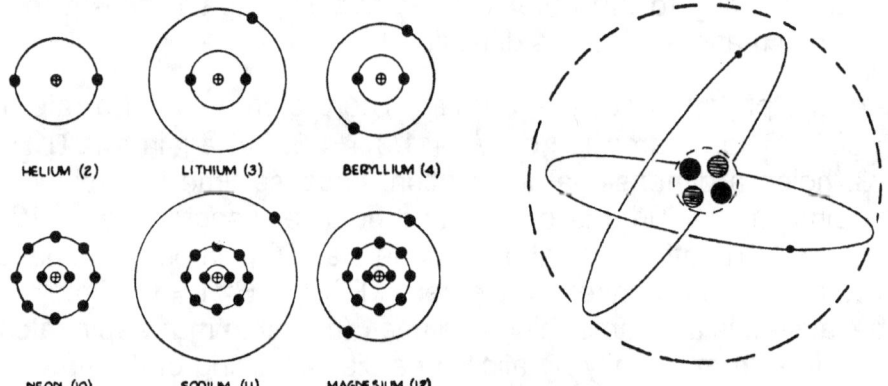

The   relation that an atom has with heat stems from the number of protons in the nucleus of the proton cocoon.

The key to the relevancy is heat and space. When matter heats, it expands therefore it takes more space. When matter is cooled it shrinks, therefore takes less space. That is the relevancy because matter in any form is heat. Heat produces the increase of space and reducing space produces an increase of heat. That is the relevancy. That is the secret of the Universe. That is the secret of gravity. That is the secret of momentum and every other aspect within the Universe.

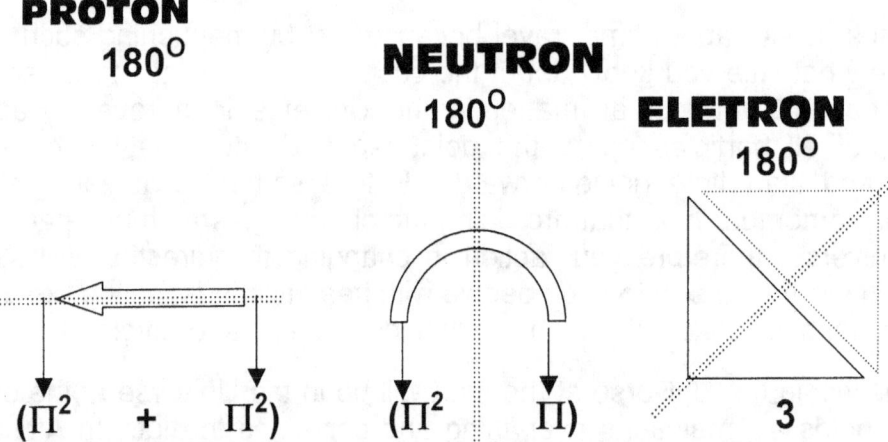

Time stood still in eternity, then after a command of the Creator, time started to move by overheating and eventually formed the relevancy of the proton $(\Pi^2 + \Pi^2)$ the neutron $(\Pi^2\Pi)$ and the electron (3). As a star returns time by depleting space to the dimensional increase of heat, space destruction is in progress and the star will abandon systematically some of the dimensions the atom holds. That is the relevancy. That will be whatever position there is in the Universe. In the depleting process of dimensional re- adapting, the star shall abandon aspects of space-time. The electron (3) may become obsolete, the neutron ($\Pi^2\Pi$) may become obsolete in neutron stars and even ($\Pi^2+\Pi^2$) the proton will become dysfunctional as space reduction completely disappears from the star's space-time occupation. However, those stars will be dark, and beyond our vision.

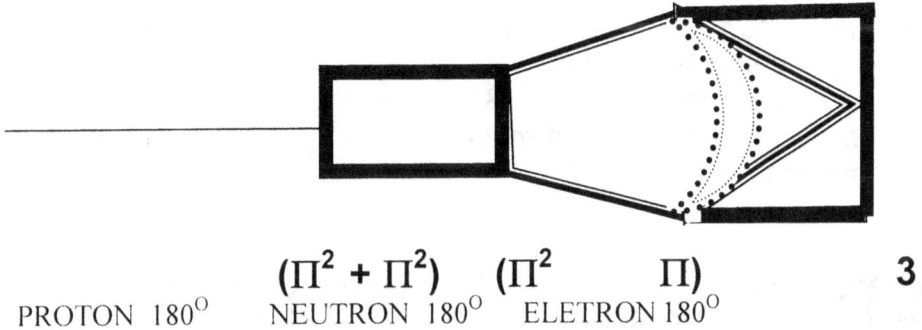

$$(\Pi^2 + \Pi^2) \quad (\Pi^2 \qquad \Pi) \qquad\qquad 3$$

PROTON 180°   NEUTRON 180°   ELETRON 180°

The relevancy holds value pointing the relation between the various dimensions as they are in the atom. The relevancy of ($\Pi^2 + \Pi^2$) ($\Pi^2\Pi$) (3) = 1836 will remain but the mass of the electron and the mass of the proton will change in every space that time applies. Cosmology thus far was incomprehensible because it was incorrect. When applying natural laws, it becomes so simple that a person as ordinary as I can understand and explain it.

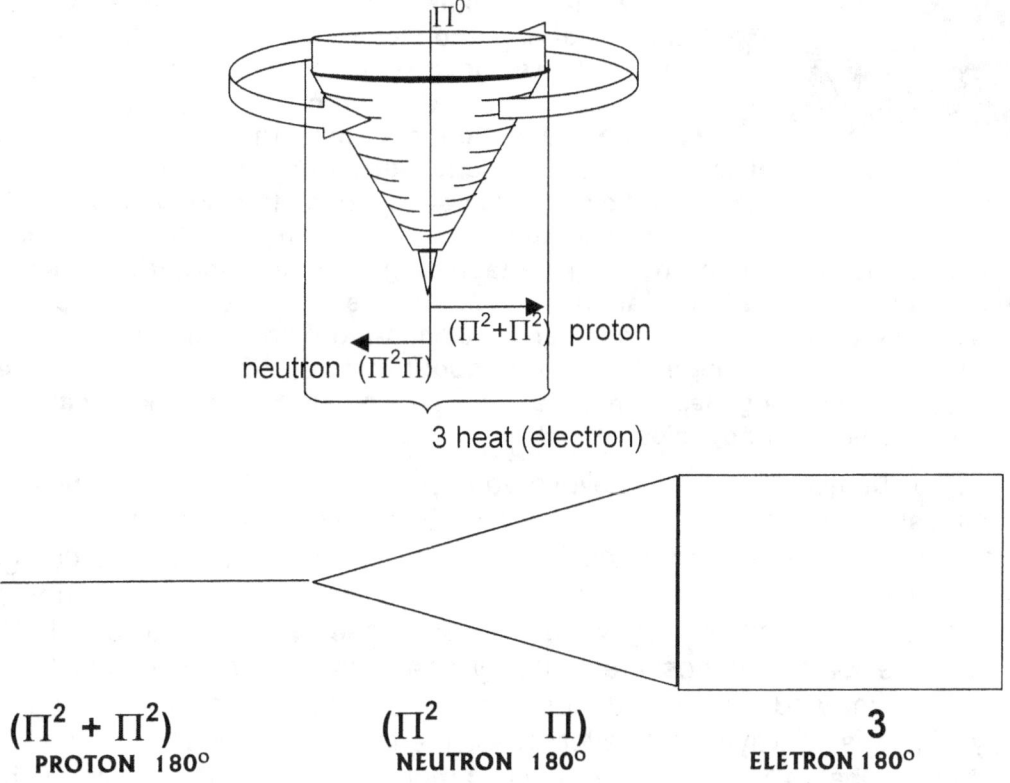

$$(\Pi^2 + \Pi^2) \qquad\qquad (\Pi^2 \qquad \Pi) \qquad\qquad 3$$

PROTON 180°   **NEUTRON 180°**   **ELETRON 180°**

The above indicates where singularity originates and how that establishes the factor in singularity Π. The Universe started from the factor in singularity Π. The entire Universe holds a spinning relevancy to all other factors in the Universe. If that were not the case, the Universe would not be there. The first person to consider the factor in singularity Π was Galileo. In the swing of the pendulum he saw singularity remain as one, that formed time, destroying space to maintain time. To prove my statement I shall very briefly indicate some barriers of motion science refer to as the Doppler effect, but Doppler used a slow moving train that at best could indicate two or three very minor moving limit.

If I can understand it, every other non-brainwashed human on Earth should understand it. The relevancy of $(\Pi^2+\Pi^2)$ $(\Pi^2\Pi)$ and 3, is a dimensional reduction of the flow of heat from space back to time. The flow of heat becomes necessary to prevent solid matter from overheating. By removing heat from the gas of space, through the neutron, to the solid of matter, space reduces as the intensity of heat flow requirements increases.

MY THEORY

When we look at the night sky we see images of stars. I am of the opinion that our vision of stars and our interest we show in stars is just what sets us apart from other species in the first place. We have the ability to recognise what we cannot experience and we are able to detect what is not in our immediate sphere. We have what makes us able to look beyond what touch and smell can confirm and acknowledge and to my opinion, that is what other species are unable to see. It might have something to do with the darkness of the night that is dark to us but is a clear light to most other species. We are able see what we never can touch though we can appreciate what we never can have. We interpret what we see without ever making contact to confirm and that gives us external knowledge and insight. Our vision about that, which we see tell us that there is more than the animal's concept of a plain survival on Earth where it is that you can eat or you can be eaten. Fathers show their children the constellations and although we no longer attach religion to our stargazing it never subdued the bliss we find in our astonishment about stars. That ability gives us religion as much as it gives us science and music, art and an ability to somehow reason by accumulative thought that brings about culture. We can build on the past by reason ad not by genes alone. We are what we generate and not what our accumulative genes of our species generate.

It is part of the human concept to believe your eyes. Seeing the sand dunes on Mars is equivalent to seeing the sand dunes on Earth by means of the television media and could just as well be of the Sahara. The Sahara is a place we can go and visit should any of us wish to do so, but the dunes on Mars are another problem. Visiting and confirming what we then see is not that simple to accomplish. The Martian dunes are not only space away, which means I can cross space in time and visit. The dunes of Mars is not even space away but is time away. There is no way I would ever cross time to see for myself what there is to see. That is what is wrong with science, amongst others. Science is of the opinion we see space. We do not see space. We see time, but it is not time we see, it is the distortion of time that we see. The "further away" we look the more time we see. However it is not time we see. It is the distortion of time that we see. The further something is away, the more it is in the distance, the longer it will take the light coming from the object to reach us. That means the longer it takes light to reach us the more time is distorted to put distance between what we see and what there is to see. It is not space that we see but the distortion or the compromising of time. It is the time delay between here where we are and there where the object is that we see and we do not see the object or the space the object has or even the space between us and the object. We see the time delay there is between the object and us. We see what was there in time gone by, however, we do not see what is there and we see space for what space represents to the Universe. We see space as time delay, time slowed down. That is what space is, space is time delay. That concept urged me to go and look for the beginning of space and the beginning of time and the origins of the concept space-time. Please allow me to explain the beginning of space by measure of time delay. At the time of the Big Bang everything was small...not so...it was as big as it is today. If the Universe was the size of a neutron, then we had no size at all. One cannot compare apples with oranges and see bananas. The space we see is the distortion time has to separate points of comparison.

In order to understand what I am trying to say I have to use a picture that is most probably not a true event. What we think we see is space. It cannot be space that we see. In the forefront we see a line that is a result from a comet travelling. Then there are pixels indicating lesser star structures and some clear dots indicating stronger light spots, which would personify larger stars present in that direction. The rest we see is the black of night. If the Big Bang theory is correct and to my thinking there is no doubt about that, then not to long ago there was a lot less space between the objects than is the case at the present. The space was less. That cannot be the case because if the space was less it would then take the light much quicker to arrive at the spot we are at present. The light coming from what should be the comet is relatively quick in reaching my location while there may be some of the faint dots that have light travelling a considerable time to get to me. I presume the comet is closer. Looking at the image of a roving planet it shows a structure filling space at intervals. The space it fills is a constant because the space does not change in becoming bigger or smaller. However, the space it is moving trough appears to grant the roving planet another position every time it is photographed. It is in the terms of time that the answer is. It takes a different period to position and obtain the light coming from the different position where the object is located.

If the prime object were the space as it is in the case of the space serving to fill the roving planet, then no changes would come about to the space. It would take as long to fill the space between the object and where I as viewer am in position with travelling time. It does not because the motion that the light has to endure is shorter or longer by time duration. It is the space that is constant, yet the time to travel varies. It takes time to cross the space whereas the space holding the object remains filled at an even volume every time. In the case of the planet space filling with material by gravity the same space filled without changes. In the case of the dark space, that space is putting time at a different duration to reach the location I am in. The "further" the object is in distance from where I am, the more time it would take the light coming from the object to reach me. The object will appear smaller as the distance increases but I know the space the object holds is filling the same volume as it does when being close to me. In the case of the space filled, that space appears to change but that space is filling a volume at a constant. It is when the space in which the object moves increases the space the object holds then diminishes. The space that the light has to pass through to bring me the picture of the object increases. That cannot be because the space is filled all the time by the same margin. It is the time the light takes to bring me the picture that increase and it is that light that shows me a diminishing space. It is not the space between the object and my location that increases, but the time that increases and by allowing the time to increase I allow the space of the object to appear to become lesser. The space the object holds has to remain the same and the space between the object and me cannot change by motion. It is filled by volume that motion cannot change. Only time can be affected by motion and since it is motion that is changing it can only be time the motion can change. The slower the motion the longer wills the time be that it takes the motion to negotiate the space. That black of the night that I see is not space that I see but is time that I see and the space I think I see is the retarding or slowing of time that I see. Outer space is not space but it is time that space retards and therefore space is not space but a retarding or a distorting of time. That means that which see thinking it is space that we see is all the time, time that we see and being time it has no outside because time is eternal. Space, being infinite interrupts time to give time in eternity duration value.

The relevancies we are about to address are about form. It takes us into a Universe when a line had the same value as a half circle and as a triangle does. It takes us beyond space to a Universe when time formed space. It puts the Universe beyond distance. It is what came about when space interrupted time to deliver us the black of night, which we incorrectly think of, as space. The Universe did not start small it started outrageously big. It is not expanding it is reducing. When the Universe started there was no outside to that which started because if there is an outside then what is on the outside of that which started. The Universe has always been an inside that went smaller. The limits grew smaller not bigger. The initial start had no limits. That which we think of as so small and tiny, so small it has no sides is so big it cannot have sides because it is too big to have an outside and all we see and all we cannot see fill the inside.

Where we are now in the Universe we are so much smaller than what was when the Universe was the size of a neutron or whatever it was. If the Universe as one block without limits had no outer limits and was the size of a neutron then it grew smaller because what was our size when the Universe contained all it had in a neutron. When the Universe was a neutron we were not even a thought. It is easy to lose perspective but perspective is all we dare not lose. That which took al the space a neutron could offer back then has no limits now and has no boundaries. It is too big to be cooped up by limitations and boundaries. We with limits and boundaries now have measurable quantity to calculate, but what was the Universe then has no calculations art present. Where there is no boundary to shift what shifts then and yet they say the Universe is shifting its boundaries because the Universe is expanding therefore it shifts! Where no growth is possible since it captured the growth at the beginning where too can it grow. The end of such a shift by what cannot shift to where no shift is possible will eventuality be what they named The Big Crunch even before locating the Big Crunch. It is like naming a baby even long before knowing how the procreating is taking place that will lead to impregnating of some member of the specie (which member it will be is still then still unclear at the time the name giving was undertaken) where it later on will lead to conceiving the baby ... that is the manner in which science dogma is enunciated but that is how clever those mathematicians are that knows everything there is to know on science. They can name a baby before even knowing what procreation is and that they do by calculating what they don't know anything about... like procreating the baby! It seems more likely that that which has no prominence finds prominence, which means the lot is shrinking. The Universe is surely shrinking to give us space to be.

When we altered the size of the moon in relation to the size Mars has what we did was change our relevance to that of Mars. We first brought Mars on a time line as close as it would be if it were hanging around in the space the moon has at present. Then we moved the time line back because it takes time to travel to the structure. Pushing Mars back does not increase the space, because eventually Mars fills the same space. It increases the time duration between Mars and us. It is not space we cover. If it were space then the time would be equal for light and for all to complete the journey in the same time. By changing the time the relevance change as to how long it would take to get there.

When we look at the images of the two solar objects i is so easy to put them out of perspective and in the same size, although we know they are not the same size. All one needs to do is just play with the dimensions and find the results. One changes the space they have to match and they are equal in size.

In cosmic reality the reality is quit substantially different. When we put our hand out we are able to touch...say the door we are immediately in contact with the door. It is the door we touch because it is the door we see we touch. Moving back one meter we find we are no longer able to stand upright and touch the door because we are one meter away from the door. We are one meter away because we can see we are one meter away from the door. We grew accustomed to this thought because Galileo's pendulum shows we are in time in space in the Earth timer in space. The time we will take to touch the door corresponds directly on Earth with the distance there is. Things change drastically when we leave the Earth or when we view object not confined to the Earth as we are. The truth is we are accustomed to think we are one meter away from the door we are unable to touch because we think we see the door is one meter away.

However that it is not the door we see. We cannot see the door because the door is not there for us to see. We see light banging on the door and as the light is rejected by the same door that the light comes flowing to us. We see the rejected light bringing an image of the door we cannot see. It is light we use and that we are used to of using to confirm what we see but such confirmation is what makes the most intellectual stumble. In quite the same manner we see the darkness of the night and observe such darkness as darkness. By darkness we interpret the meaning as that which we cannot see or that which we are unable to see. Reality tells me that the darkness is light that is too bright for us to see. Take an image of Mars with a close up view. Then reduce it and go on reducing it until it is so small it becomes invisible. The space filling darkness is not darkness filling space because the ratio of darkness increases as the ratio of light in comparison to the darkness reduces. The object does not go dark by moving back. It rather becomes more of the same when it blends with the darkness, which proves the darkness is not darkness but it is light.

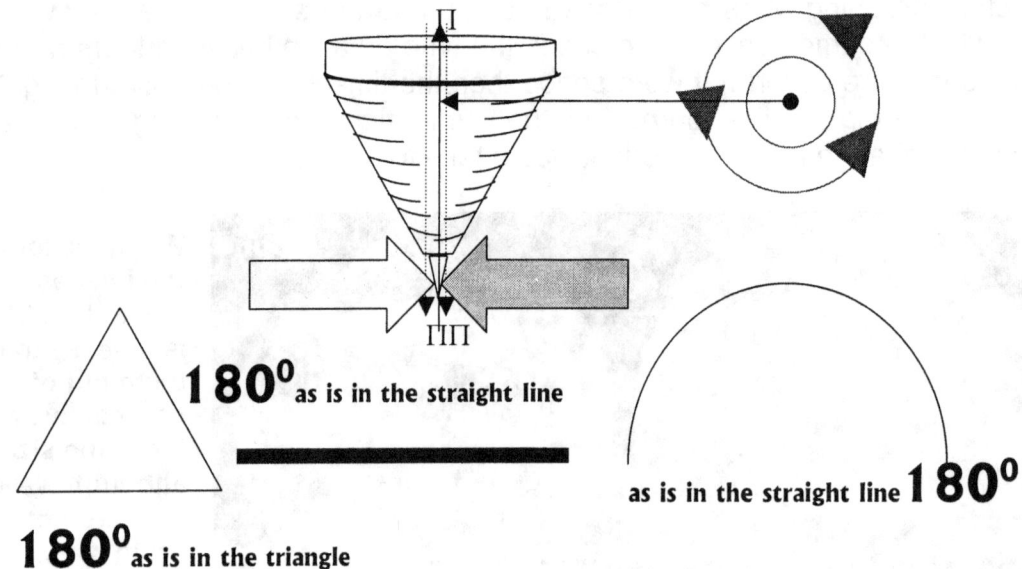

180^0 as is in the straight line

180^0 as is in the triangle

as is in the straight line 180^0

Einstein famously said there are matter, time and space and if I am not mistaken he said very little more. Is the space that Einstein would have reduced to a flat surface forming a flat Universe really space just because it is not time. Then on the other hand we must ask the question what is time then? What is our Super designer of space whirls really reducing when he reduces space to have the lot fit into not one but two Black Holes? What is matter and what is space? Looking at what the top tell us we have matter time and space a little confused. We find matter in time but also we find time in matter and that no one before realised! By reducing the space an object has the darkness becomes either more or less but the darkness promotes the object or reduces the object. The fact that large objects are close and small objects are at a distance we on Earth relate to more space and less space. The only factor that can produce more space and less space is time because time is irremovably connected to time. By reducing the share of the combination of space-time time must reduce or increase to allow space to do the opposite.

We have our focus square on the distance we find that part us from the object in our view. It is another culture thing from the time we hunted I suppose but we confuse distance with the time the distance really represent. By the end of the twentieth century every one has become so accustomed with photographs and lenses bring into focus and bringing objects close it became part of our breathing. We reduce the space in order to see the object better. Is that really what it is all about or is the issue significantly more complex.

This is the best example we may ever find of space-time. In the term we use space-time it is the time that is reducing the space. The time divides the space according to the time it takes to reach the space from where the viewer stands.

The time factor reduces the space factor as it divides the space factor into smaller parts of time holding space in eternity. That proves that all containing space is in fact time holding space. $k^0 = a^3 / T^2 k$

$T^2 = a^3 / k$ and most of all $k = a^3 / T^2$. The time T^2 it takes k to reach the space a^3. That is space-time and proves that space-time is not some mysterious cosmic scheme covering up of a Black Hole image the Universe devises to benefit only those with highly groomed mathematical skills. The more time develops and time pushes the object "further away" the smaller the object will seem in relation to the containing space or time. Outer space is not space but eternal time holding space.

The realising that only time can affect space by the measure of appearance is a huge step in the right direction. Space is a constant therefore time has to influence the appearance of space to become apparently more or apparently reduce to become less. Being big is a sure sign to the brain of an object being close. That would then appear as if there is little space between the observer and the object in observation. That is culture talking because space may appear larger or smaller but it can only be a medium of space that may allow space to appear. Space as such has the same measure and has the same prominence when measured. Time is the factor that allows space to reduce and even to reduce to the obscure.

Moving the object back into obscurity does not reduce the space the object has but puts the space the object has into a much larger definition in space in relation to the space I witness. The light streaming from the object will also fade into obscurity and disappear as the definition of the object declines in relation to the space it holds by comparison to all the other space in view. The light the object had did not decline or reduce but it diminished in relation to the gross of space holding light. In relation to the space out there the space diminished the light in relation to the darkness the light then offers. That way the light could only reduce by comparison if the light was less in relation to what the light is in the darkness we see. That means the darkness is flooding the light the object has and therefore the darkness we see is light. However, our relation to the light makes us in relation to little to be able to appreciate the light because as the object retracted from the position we had, we also diminished in space by the same measure. The space we hold therefore is too little to enable us to appreciate the darkness flooding us with light.

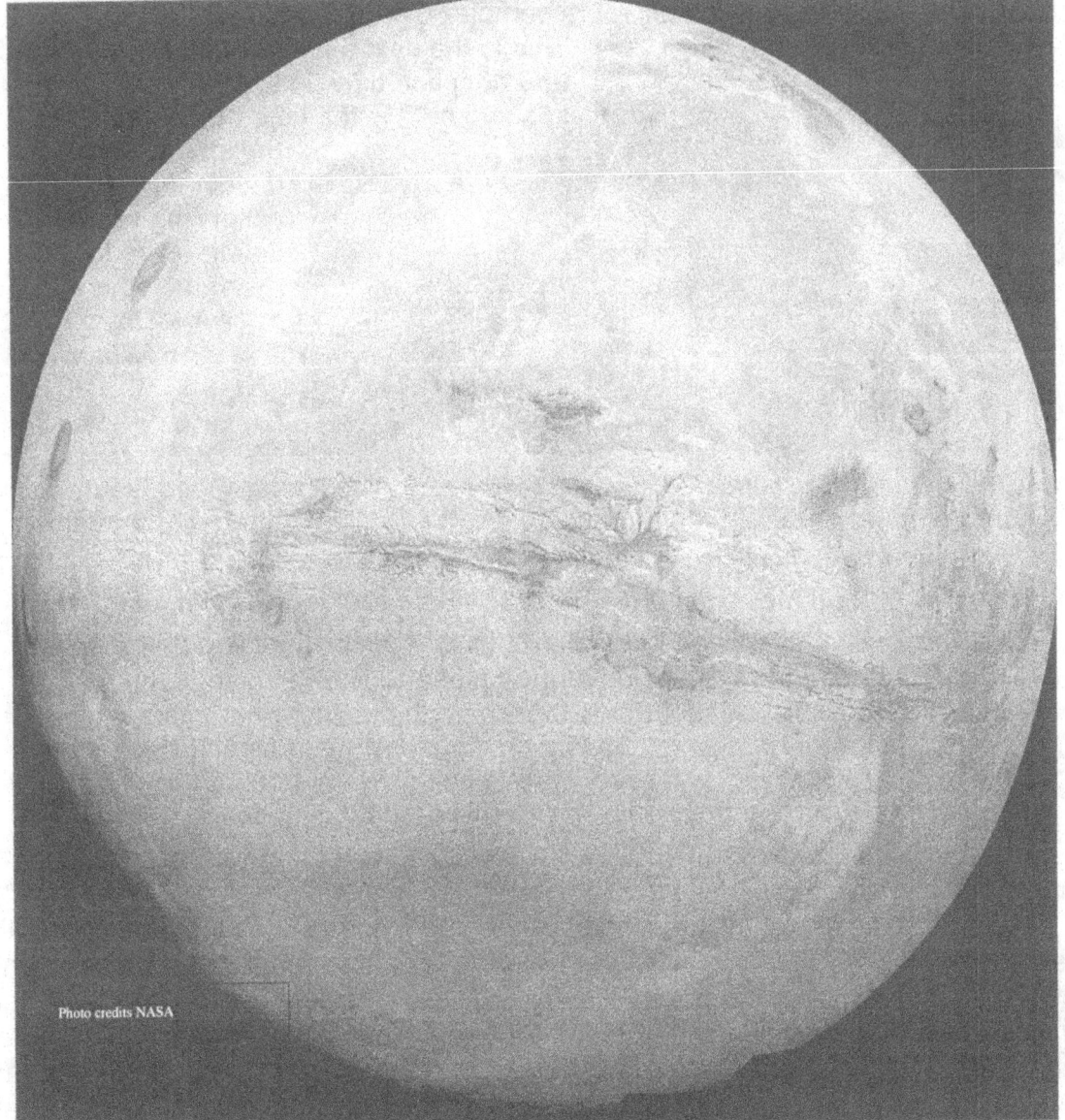

Photo credits NASA

Even the fact that the objects seem to be as near or far is defined by Kepler as a fact in $a^3 / T^2 = k.$ This answers the question a friend of mine, Johan Boonzaier, asked me one night around a dinner table and which that night sparked life into my first Afrikaans book. His question about the blackness outside that could never end, was the question which finally got me inspired and set me off writing these books...It was his question about what space was and why space in outer space can never end. That was Johan's question and my trying to explain what time was and what space was according to my opinion was what started me off writing my work. Johan, that which you asked me what it was out there in the blackness of the night, that is not space, it is time, and therefore it can never end. If it was space, then there had to be limits, borders and measurements defining it but because it is limitless it is time eternal. It is the line of time being over grown and stretched limitless but still an eternal line. Borders will always define space being infinite but time is forever eternal.

That is the easy part to figure. By moving the object back in relation to what we view is not diminishing the space the object has because the object will hold the same space it had before. The object is as big at present as it was at the Big Bang event because what was there was there with no adding. What is present in the Universe is in the Universe and no adding or removing of what is in the Universe is possible. If the Universe grew the object had to grow in parallel with the Universe because the Universe got somewhat bigger than the size a neutron has but so does the object have much more space that what the neutron has. The size the Universe had contained what was inside the Universe at the time the Universe went bang. In that there is little to no change possible.

Therefore moving the object that is in our vision back does not diminish the space the object has. What does increase is the time differentiation between the position I hold and the relation of that to the new allocated position the object holds. Should I wish to reach the object at the speed of light, I then will have to diminish my space to that of a photon to be able to reach the speed the photon has. However, in the event I do not wish to surrender any space I hold for reasons of the survival of life, I then would have to travel taking much more time to reach the object at the location the object holds. Then we get to the question of what did it take to shift the object into the allocated slot it now has in the distance it

Photo credit Anglo Australian Observatory

now has. It took time to shift because it did not take space to move. The time it took the object to move to a new location was added   between the moving object and me. It would take the same time but in a different ratio of moving to retain the time should I wish to move to the location where the object filled the same part of my view I have of space with the object in that space. That means since the Big Bang there was no space added but only time increased what parted the objects since then. Space is the retarding of time.

As the cosmos present its evidence, we can see from such evidence how destructive overheating is. Forget pressure, because Newtonians over simplify everything with pressure and exploding. That might happen to a drum they fill with gunpowder but that is not applicable in the cosmos. In the cosmos, unlike in containers, there is no retaining wall that sets limits to pressure inside the container versus pressure levels outside the container. The cosmos has no pressure or pushing or pulling. It has a flow of space-time by concentrating time and duplicating space as it is driving space-time towards the centre. In any picture about any star there is no containing wall that keeps whatever is inside, inside. There is no limit to what the wall if the structure can contemplate before bursting. In the centre of a star is a point holding singularity and since such a point has no space and is immovable, space has to compromise by flowing towards such a location. We regard what we see at night as space and how wrong can we be?

Have you as you sit reading this part at this minute sat back and gave a thought about the light enabling you to read? Such a thought brings to mind the most simplistic answer one can imagine. The light hits the page bounces from the page and contact the lens of my eye where the lens conveys the photons becoming electricity to a part of the brain that translate the electricity to an understandable message and that makes one read. It is as simple as that! Ever gave a deeper thought about light streaming across the night sky, coming from ends of the Universe we do not even realise it is there? How does the photons manage to convey one complete picture coming from as far apart and as wide an area as it does? With a few photons connecting the eye or lens no one ever noticed the wonder of light. The photons reflect a view that seems as if coming from all the billions upon billions of stars. But most is coming from darkness covering an area no man can measure. Yet how many photons can actually connect to the lens of the camera or to the eye? Still a few photons coming from a single direction directly ahead eventually tell the entire storey. It is very simple to take the process of seeing by means of photon conducting very lightly and I have never heard one of the Brainy Bunch really in sincerity uncover the process to its utter and full potential. It is impossible that light from such an array of assorted sources can simply come together at the eye lens and show a picture of objects spanning across a Universe as wide as our mind can receive where the objects they reflect is beyond human measurement and the quantity is inconceivable many.

If the darkness was the representation of "nothing", then that should be exactly what we must see, nothing but the stars. Taken from the top picture some stars and leaving the rest to nothing is what we see in the picture below. A blind person sees nothing but when we look at space, we see something that we think nothing of as we see as space. One cannot have the ability of sight and see nothing except by closing your eyelids and then you see nothing. But in that case you do not see "nothing" in contrast of "something" you see "nothing" without it contrasting to "something".

Nothing is all about not being and not "not seeing".

By the ability to see the darkness renders the darkness something other than nothing and that changes the acquired value of the darkness from nothing to something. There is an eternal difference between something in infinity and nothing.

The arguments introduced up to this part of the introduction prologue only touches the most basic aspects of my work and by no means can such an introduction secure an opinion. Yet, not once through all my long investigation in the past thirty or more years have I found any other person claiming such views that I have brought about even in this skimpy way as I do in the prologue.

Light is much more than the medium science takes it to be. Light connects the Universe in a way we cannot contemplate. Light being far apart originating from regions not in the same time or Universal space connects in a way that present us with a picture holding the Universe in an understandable content. From the point we stand and we watch the Universe the significance of what we see surpasses the sense of understanding of what we are experiencing. How can the few photons that our lenses catch coming from such an area as the night sky cover transmit the complete picture of what we see. Take a few seconds and study the picture of the night sky then rethink the picture applying the full content in the picture to what the size of you eyes is. Think how big the picture is that your eyes take in and translate that area to the size of your eyeball in an effort to determine a ratio. One will be forgiven if one thinks of the ratio as eternal to nothing. Yet a few pages back I showed that according to mathematics there couldn't be anything as nothing. Consider the path the light followed from the source connecting to light from all other sources where all particles of the other light may come from and bringing a full picture to the lens one use to look through. In your mind connect a line from every atom producing light and connect the lines to your eyeball and see how you can manage to fit all the lines, as small as the lines may be.

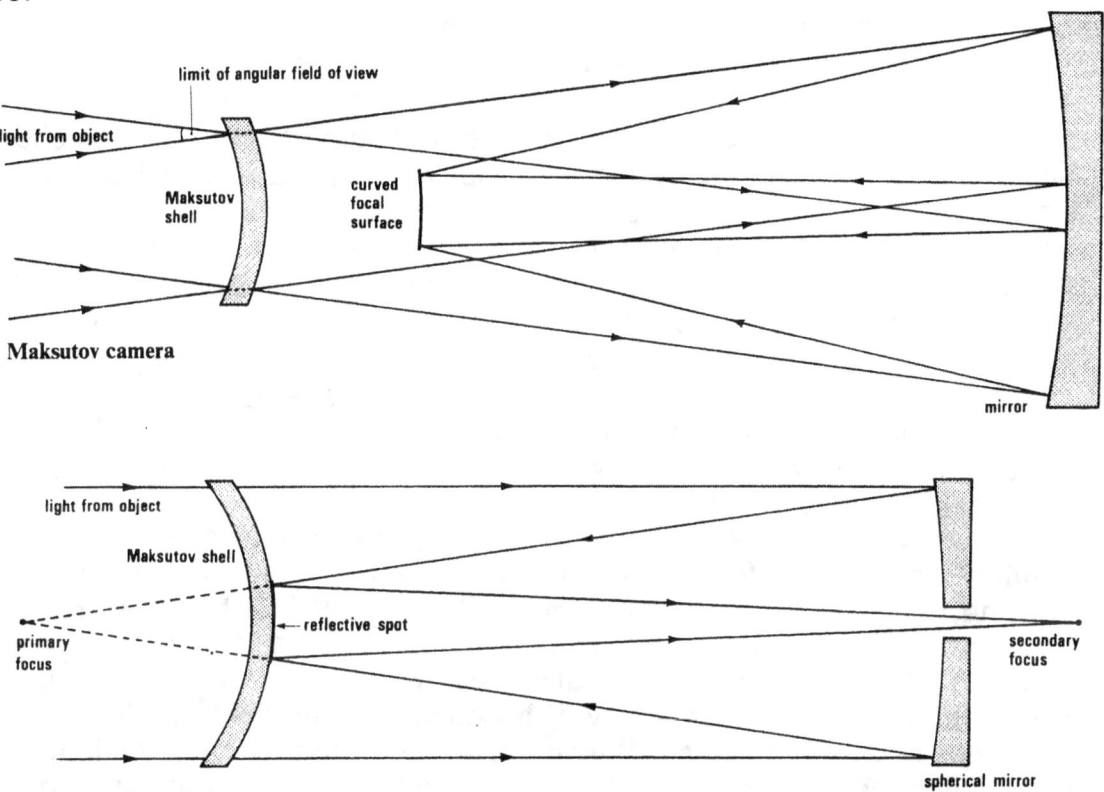

Maksutov camera

Maksutov—Cassegrain

If it is lenses that enable us to see what we can't see in outer space it also means we cannot see the light, which is outer space because we haven't got the lens to match the curb of outer space. Newtonians think of outer space as geodesic zero, with nothing in outer space but space. Geodesic zero means the light travels in a straight line from where it originates unhindered all across space to where the light connects the eye. Such an idea by itself is outrages because the stream of photons reduce in space to such a minute quantity that taken the area the photons travel and the space in vastness it covers, the chances of one photon coming across many hundreds of light years through billions upon trillions of cubic kilometres of space and selecting my eye to convey the electricity is less than infinite. Yet such conveying takes place every second of every minute. The position of the location of the second singularity, which is the precise duplication of the first singularity but in a diminished capacity, is obvious to miss when one is not applying a detective mentality, as one should in scrutinizing the cosmos. Culture will have us believe that when one sees a colour shining from an object the colour is associated with the object. Logic tells a different storey. A yellow dot is all the colours in the spectrum but yellow because it is disassociating with the yellow. That goes for red blue and all other colours we may visualise. I think the norm accepts this as scientific fact with very little argument or substantiating proof about that required.

If light came as individual streams of photon flurries, then our visage would translate that as such shown in the fragmented picture above. It would be a picture unconnected bringing across some photons in the manner where every object stands apart not being related in any way and that will be what we see, if it is anything that we see. That we know is not the case but that means geodesic zero is as much rubbish as anything Newtonians regard with simplicity and with careless thought. Geodesic zero means nothing and how can I see nothing as darkness because "nothing" is not darkness, nothing is "nothing" and the darkness I see is darkness showing the darkness as something. What then about colours that are technically not colours as is the case with black and white? White is simple. By spinning all the colours in the spectrum the colour white shines through. Black is quite another matter. A friend of mine whom is one of the best painters I have ever come across told me that one couldn't paint black but have to make black a dark blue to show shade on the canvass. That apparently is his success in achieving the realism. He also went on to explain how many variations of dark blue form the shadows in one simple tree. This remark set my mind in motion. One cannot see black because black has no colour to show, but black is the colour most prevalent in the universe. One can see only by colour and since black is not a colour we should not see black, but we do.

If it was true about a yellow object not being yellow and a red object rejecting red and therefore not being red, the same must be true about dark and light. The bright object rejects all the light just because it is light from the outside. If it is light from the outside it has to be very dark inside. That too would count for what we believe it to be dark stars. Such dark stars must be most brilliantly lit because they keep all the light to their inside and well protected. They are dark because they keep the light on the inside where we can't see it and therefore out of our viewing range. The stars have gone so cold the stars have to conserve all heat to

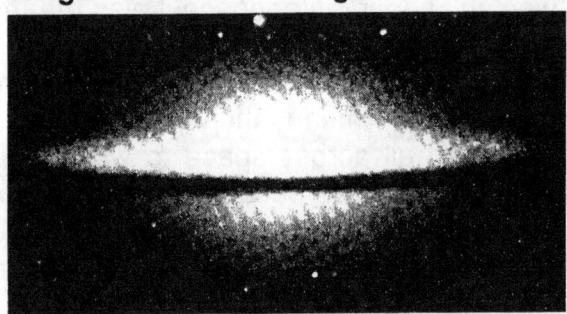

remain in gravity. With light being the highest concentrated form of heat, it stands to apparent reason that the light would be the energy of prime choice to contain. The same must then apply to outer space in that outer space is conserving all light and by keeping all light, outer space is brilliantly lit. We just are unable to witness the light because our position is much concentrated where as the light being dark is expanded to the full. We are able to see the galactica because the galactica represents highly concentrated light in one reduced area. The darkness contrasting the light we see as darkness because the light is expanded to the ultimate. The fact that we can see the darkness makes the darkness light, which we are unable to see. However with the space stretched to the maximum the lens we see the light by has as far as our position goes, not even slightly curved because we are so small. Now you go and tell any mathematician in charge of theories this much and see how far you can get convincing him about your view. They wish to manufacture and design space whirls and not see reason.

The fact that we see light means that the dark next to the light cannot be "nothing", If the darkness was the representation of "nothing", then that should be exactly what we must see, nothing but the stars. Taken from the top picture some stars and leaving the rest to nothing is what we see in the picture below. A blind person sees nothing but when we look at space, we see something that we think nothing of as we see as space. One cannot have the ability of sight and see nothing. It is light that we see and it is light that we use, which enable us to see. That proves the darkness that we see in outer space is light that we see without recognising it as such. If the darkness was the representation of "nothing", then that should be exactly what we must see, nothing but the stars. Taken from the top picture some stars and leaving the rest to nothing is what we see in the picture below. A blind person sees nothing but when we look at space, we see something that we think nothing of as we see as space. One cannot have the ability of sight and see nothing. It is light that we see and it is light that we use, which enable us to see. That proves the darkness that we see in outer space is light that we see without recognising it as such.

What puts us humans in a category one higher than animals (or so we like to think) is our ability to think about that what we can see. The less develop an animal is the more it has the attitude of eat or be eaten. The higher developed animals are the more the animal find reason to argue. One may teach a crocodile not to eat you if you start feeding the animal. That is a mindless reptile and yet it can think above eat or be eaten. What we see is not merely the truth and it requires reasoning to see the truth and substantiate between culture motivated observations and thought through decisions.

When the star is encountering adverse conditions, the flow will interrupt the even-handedness anytime will set a new standard. It happens all the time and every time the centre fails to set a standard that the flow of space-time in the star can meet. The relevancy of time sets in place a new standard and this comes about by the use of a principle we think

mostly of in

© Anglo-Australian Observatory

aviation. It is the principle we refer to as the Coanda principle. It is where motion creates a flow of space-time, which establish a centre and where that centre performs demands that the flow has to initiate. The containing of the space is as much set by the time of the flow as the retaining of the centre. It is a proven dimension implicating Kepler's vision of gravity.

It is the Coanda effect that produces gravity and it is the Coanda effect that is keeping the Universe together. Let us take the "Sound Barrier" from a point where we see phenomena apply laws that matter complies with. The sound barrier is a prime example of the relevancies, which I suggest, takes place. There are two points in singularity always referring to each other and one is expanding while the other is contracting. In nature there is the Roche limit placing a limit on the reduction of space and the inflow of heat to sustain proton cooling. At a point of $(\Pi/2)^2$ the reduction of space disallows any object to immediately compromise space claimed in time because the cosmic object cannot reduce space while an entry to its area demands such a time reducing in space claimed by the material.

The first question that one can ask is why would there be the value of $(\Pi/2)^2$ between orbiting structures positioning themselves in a time relation to space.

Every person knows about the entry restriction an orbiting spacecraft finds that forces the craft to comply with. The entry maximum is 21,991 and the minimum entry is 7. This is without doubt, the number of Π (21,99/7). The Earth holds its value to $4\Pi^2$ and when an object is not part of the surface of the Earth, even say a mountain; it becomes a holding value of 7. Later in this part I explain the sound barrier in more detail and the 7 will then become better understood. At this point one must see the Earth in the proton status of $(\Pi^2+\Pi^2)$ while acting as an atom. In this relation the atmosphere including all particles in the atmosphere will in relation be either Π^2 or Π. When water is in a vapour form, it will have a value of 3Π, having heat separating the water to the factor of 3. By dislodging the thunderbolt, the 3 receive a square value and displaces to the Earth in the linear light to time stance of 3^2. With heat (3) grouping by initial spin value, it will remove from space leaving the water to the value (Π to Π) and this will then give the water a relevancy of Π^2. The factor of Π^2 places the water no longer amongst heat as gas, but heat as a liquid (rain) or solid (hail). The relevancy of the water will change from 3Π to Π^2 placing the water's position from space (3) to liquid or solid (Π^2). Where does the Π that one find in the Roche limit $(\Pi/2)^2$ and the vapour (3Π) finds its relevancy to gravity? Every particle that enjoys space-time outside the Earth's structure ($\Pi^2 + \Pi^2$) will hold a neutron position of ($\Pi^2\Pi$). The Π^2 ends will be at the point where heat passes through the object directly to the Earth and this position of space-time relates to the neutron time link of Π^2. The space link of the neutron will then form the Π link. The value of the Π link we find to be $(\Pi/2)^2$, but the explaining to why it is $(\Pi/2)^2$ is rather more complicated.

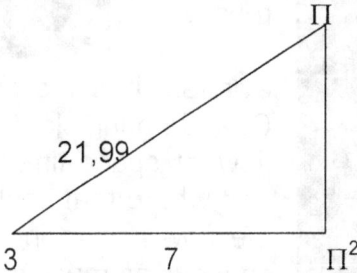

At the end of the space relevancy 3 where matter occupies space (21,9 / 7) is a border Π. That border is the exact point where space reforms to a square of time placing all matter (occupied heat) and heat (unoccupied matter) to a value of the square of time. That specific point is in relation to the square of the diminishing shield around the Earth. However it takes matter (R^3) from the 3 dimensional positions to the square (Π^2) in relevancy to time in singularity. With time holding space in singularity the 4 sides of Π truly relates to half of the total square value.

The "gravity" factor of Π^2 becomes one and only holds the square to the Π position as it holds space to singularity at a square (time dissolving space at a square) and the time value (Π) remains dimensionless in singularity at (1). It then is $(1)^2$ where the one becomes the space position ($\Pi/2$) representing time (1) at that point. This makes the position that time normally has Π^2 but directly links to the controlling singularity which we then give a value as Π^3 which then relates to the singularity position of space diminished from the three dimensional to times single dimension in the square. That makes the Roche limit hold the position of $(\Pi/2)^2$ when the neutron position of time (Π^2) links directly to singularity (1). This may only represent figures, something to accept through intellect but lying far outside the reality surrounding our everyday understanding.

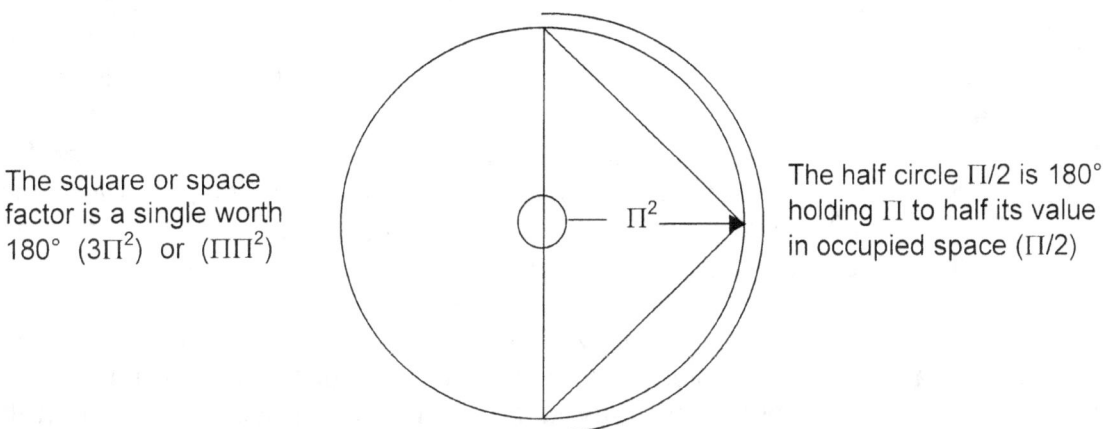

The square or space factor is a single worth 180° ($3\Pi^2$) or ($\Pi\Pi^2$)

Π^2

The half circle $\Pi/2$ is 180° holding Π to half its value in occupied space ($\Pi/2$)

We all think of many reasons why birds fly and when the pioneers of flight started experimenting the main presumption was centred on the flapping of wings. It is a natural tendency to presume that flight comes about from the wings "pumping air" in order to raise the mass and I am sure but not certain that such a thought had presented itself at sometime in the Head of Newton. Once again that is as far from any truth as is possible.

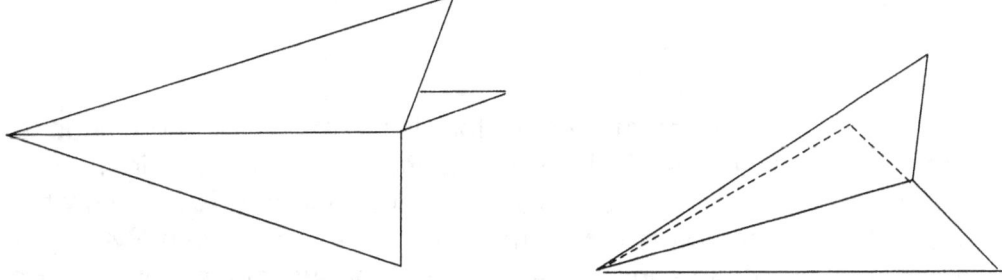

That is the shape needed to fly in the surrounding of the Earth that we presented with a name THE ATMOSPHERE. In outer space a flying object can hold any shape because any shape fits the three-dimensional. It could be round, oval, flat or any combination of all shapes and forms. The dimension is 6 with a linear value of 3 applying in one direction. This implicates the fact that wherever an object directs a direction in time revaluating positional change in space, it will forever be heading in three directions in the same time. The only control to direction is in the concentration of heat at a point opposing the direction of travel.

By applying the release of heat, one is applying "a directional change". One is producing space that never existed before, because the release of heat will produce space. To the object in travel another dimension brings about influence to this three directional space application. This application of the heat brings about a fourth dimension that applies directly to time. I shall come back to this point duly.

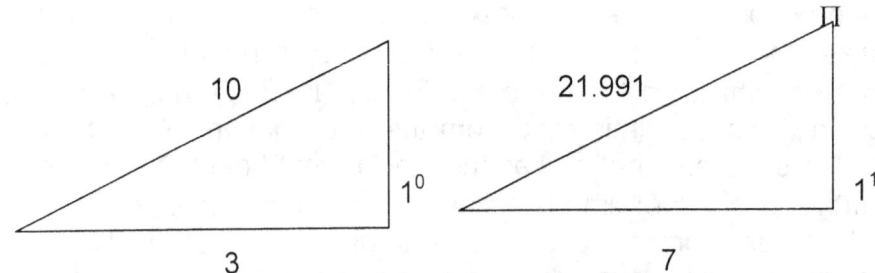

Without the application of specific heat, the object remains in the three directional moving of six possible directions. The value of space unoccupied therefore remains $\Pi \Pi^2$, as it was before the "Big Bang" event, whichever "Big Bang" you wish to refer to, because there were many. But space unoccupied holds time to the value of 10 to 1, and as also indicated, holds space to Π. Therefore unoccupied heat holds the relation to space in applying 3 directions of influence $(10)^3 = T^2$ $(R^3 = T^2)$. Always part of this equation is the dual function of space in $R^2 = T$ while at that very instant one have R=T. Therefore in space in time you have $10(10^2) = T^2$. Applying the fourth dimension does not bring about another part in the six dimensions, but actually cancel the influence of the one dimension by favouring a specific dimension. Bringing about the fourth dimension will lead to halving one dimension because of favouring the direct opposing side. Then the equation of influence becomes $(10/2) (10^2)$ and this means the implication of any heat, be it heat, be it light, will apply under the factor of $5(10^2) \Pi$.

$3\Pi^2 / 5 (10^2)$
$2(5)(10^2) = 100$

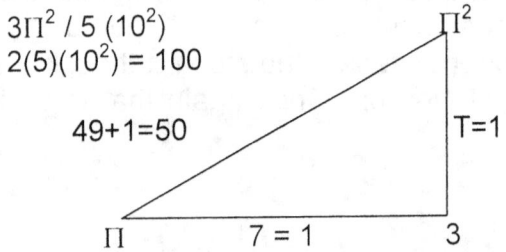

$49+1=50$

50 = 5 (10^2) where the complete   is Pythagoras be comes a factor $\sqrt{100} = 10$ the value of space.

Let's run through the sketch once more after the deliberation of the previous few pages. The fact of this comes as 49 plus one becomes 50 and that is in the three dimensions of space $R^2/7$ where 7 holds the relation to one and R/7 again where 7 relates to one. At this point it is most important to remember that Pythagoras works on the application of the sum of the square of the two sides. When seven has a direction in the fourth dimension applied to it, the opposing dimension will be one and this applies in time relevancy, therefore the interchanging in time between infinity will place matter at $7^2 \times 1$ relating to circular and $7^2/1$ with $7^2 \times 1 = 49$ plus one (singularity) always being a factor of one. Space in time however, never can be a cube, it will always be a square with one side pointing the direction of time from time to the past (1) to time to the present (1) to time to the future (1). This means with space in singularity in relation to time in singularity there will always be $R^3 = 1^3 / T^2 = 1^2$.

I am the first to admit that the calculations introduce with the atomic factor relation on the Universe has little mathematical value other than indicating some relevancy and some predictability. But you cannot use it in precise calculations because the heat on the day of the day will apply different measuring standards to what science perceive as constants. A kilometre will be slightly longer or shorter an try to tell a pilot travelling at two thousand five hundred kilometres an hour a few meter nearer or further has no importance! For instance launching an object into the outer space will require an atomic relevancy of $4\Pi^2(7(3(\Pi^2)\ (\Pi^2/2)\ (\Pi^2)\ /$ $3600 = 11,2$ km per second. Should you wish to make it an applicable factor, one will have to multiply the second object's proton relation with 10 making the total formula stand at $\$T= 4\Pi^2(7(3\Pi^2)\ (\Pi^2/2)(10\Pi^2)/3600 = 112$, which will indicate the cosmic factor value of independence, the same value that applies to Newton's $F = G\ (M_1+M_2)/r^2$. By using $\$T= 4\Pi^2(7(3\Pi^2)\ (\Pi^2/2)(10\Pi^2)/3600$ you cannot put a rocket into orbit that is true. However you can establish an indication to the position in the cosmic space and time holding matter. This is not an absolute accurate science, but it helps with indication of relativities in the cosmos. When dissecting such formulas and the factor that arrive from it, one can judge the conditions applying.

In this case:

$4\Pi^2$ one proton holds absolute cosmic superiority
7   the spherical value influencing the formula
$3\Pi^2$ the two objects will relate $(\Pi^2 + \Pi^2)$ where the one holds a proton position and (Π^2) where the other holds a neutron to time position.
$\Pi^2/2$ the minor object will be totally dependant on the "gravity" of the major component.
$10\Pi^2$ the final position of the minor structure (Π^2) as an independent cosmic (10) object in space in time.
3600   braking down the kilometres / how we work in to kilometres / second. This I use to evaluate speed, although I must add we adapted the value of time and space to apply to degrees (360° the circle of the Earth) and kilometres, being the same as $1m^3$ that rolls into a straight line, where the measurements apply both ways and in that manner it brings confusion to something we regard as proven.

By using the factors one can establish the planets and their moons or satellites in the position they hold, as well as how they came to be there. By applying it as an indicator helps to establish knowledge that was previously not there to help. But let me assure the Newtonian Order of High Priesthood, for all their precise calculations, they are far worse off when looking back at the nonsense they came up with. Man has NO MEANS of precise calculations as far as the cosmos goes for ONE MUST NOT VIEW THE HEAVENS IN AN EARTHLY MANNER.

The "mass" of the proton is easily identifiable, and so is the "mass" of the electron. Separating the two known factors, is the neutron that proves to be rather un-defined. This indicates that the balance is within the neutron, determining the electron value. I say this, because it is not the amount of electrons that determines space, but the amount of protons in the specific space.

It is the value of matter, be it surrounded by heat solid, liquid or gas, that differentiate between time in space and space-time. Without applying heat release the space in time would be $10(10^2)$, which is the circle we think we see space to hold. This is where astronomers make their biggest judgment error, they look at space thinking they are viewing the cube of space, but they are in fact viewing the half of the base times the square of the cube $(10/2)$ (10^2). What they think they see is 1 000 instead they are seeing 500, but as I will later explain, they are not viewing 5 (10^2) they are not viewing light setting the time factor $3\Pi^2$ but they are in view of light in space (Π) being in time $(\Pi^2=1)$ the connection to space is $\Pi/500$. That means whatever the radius between the objects the light reaching it will have a comparable velocity of the distance of travel divided by 500. This will be discussed later in more detail as well. From this there is a measured point where the dimension of space in heat forming gas ends and space in "liquid" forming the neutron begins.   The dimensional time application in space diminishing starts at that point. We think of it as the Earth's atmosphere. This no person, no force, no money can recreate. I have read articles about some Newtonian Wizard that plans on building some ship where this ship will "create artificial gravity by centrifugal spin."

Again and again Newtonians show just how little they truly know science. The depletion of space through spin increases, the Earth brings on acting as the nucleus of an overgrown atom. The point Π indicates where 3 lose one side of its dimension becoming Π. In that space all objects are at the mercy, not of a FORCE called GRAVITY but in a six-sided three-dimensional container of matter not relating directly to one specific point of singularity. All cosmic objects are cosmic atoms and all cosmic structures hold claim, control and influence on all surrounding space in its immediate surroundings as well as the cluster effort of the group – unit stretching as far and wide as the observer wish to take the relevancy. However in the Universe as it stands there is no collective standard all including and all counting general gravitational constant that has an equal value and measurement every where and wherever Newtonians may go Newtonians will encounter this constant. At that point a dimensional change take place where the Universe sacrifices one dimension from 3 Π. In that space objects which wishes to fly, need a replacement for the dimension sacrificed. The aircraft needs wings to fly. Hot air balloons and the Zeppelin Hydrogen flying cylinder use the natural tendency that hydrogen provides by enlisting concentrating heat to maintain at the value of Π and not reducing to Π^2. Aircraft of any kind, that wants to maintain a vertical flight would have to apply three directional wings for stabilizing because of the loss of one dimension. There may be the possibility that with enough application of heat at various points, the introduction of the fourth dimension (heat application allowing space creation) may even bring about that stabilizing may occur in such a manner. Although the prospect seems light, however with no research to widen the view on the possibilities available, it is hard to tell.

What is overall important though is that the point the Roche limit indicates, being $(\Pi/2)^2$, the dimensional change takes place. Any object holding space-time in own value, will either reduce space occupation, or enlist time duration enhancing, or most probably a little of both as it will cover itself with heat. This is the result form the atoms on the outer edge of the atom, repositioning space-time occupation where the normal application will be in the dimension of the Titius Bode law developed to a certain degree. This change will bring about that the atom will revalue its position to incorporate the Roche limit in protecting its own space it holds in the time of that specific space. The value of the space-time of the atom of the aircraft therefore will reduce its electron position, as it locks itself in a position under the cover of the heat shield. The point of the Roche limit is as much part of the Macro Cosmos as it is of the Micro Cosmos.

At this point of $(\Pi/2)^2$ the reduction of space through positive space-time displacement will start. Any other cosmic solid that will not comply with the flow of heat to the object will remain outside that area. This means in more suitable language, that the second object will block the flow of heat to the first object (A). To object A everything outside its proton sphere of $(\Pi^2+\Pi^2)$ must either be Π^2 or Π. If the object is in a neutron time position it will hold the value of Π^2 and if it is in a space position it will hold a value of Π.

Should the object B maintain a density where it will become a singular value to the proton value of object (A) in as much as maintaining its own proton value of $(\Pi^2+\Pi^2)$, it will cause overheating in the proton heat flow of object A. The demand in heat supply of object A will either remove the heat object B needs for cooling, thus increasing object B's chances of overheating, or it will start overheating itself. By overheating the space object A occupies, will increase because the overheating will demand more space occupied. By increasing object A's space-time occupation, the demand on heat will increase as the point of space reduction shifts further away. The increase to the value of the $4\Pi^2$ factor, will increase the $(\Pi/2)^2$ linear factor.

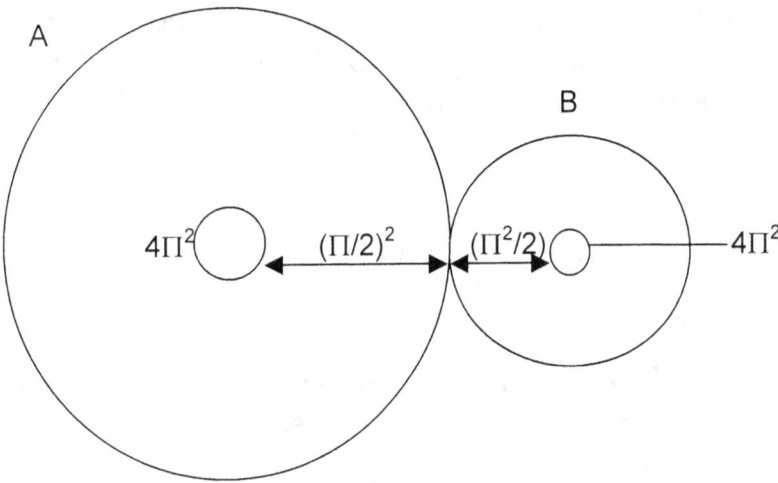

As the demand for heat rises, the competition for heat sustaining will also rise in the occupied space of both structures' time zones. That is the Roche limit and that is the purpose of the Roche limit. At a point where the space reduction and demand on heat supply starts in all earnest $(\Pi/2)^2$ seen from the view of the object in question $(4\Pi^2)$ it will either reduce the other object to a suitable value of (Π^2) or Π, but it will not allow any object to compete with it's demand $(\Pi^2+\Pi^2)$ on the supply of heat. It will reduce the competing object's value of $(\Pi^2+\Pi^2)$ as it relates to its own value down to a mere manageable (Π^2) or (Π).

The reducing method object A will provide, is removing as much heat flow from object B in order to get object B to overheat. Thus object B will increase its space-time value from a solid ($\Pi^2+\Pi^2$) to that of a jelly (Π^2) or even better still to a liquid (Π) by removing all gas (3) from object B in its demand for heat. If object B can comply with its own demand for heat, it will remain in space of it's own, providing it's own time within the cosmos ($\Pi^2+\Pi^2$).

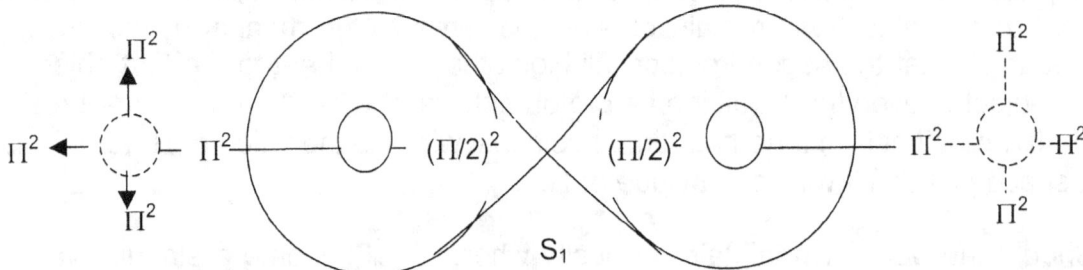

Therefore one can observe that the Lagrangian point of S_1 is in fact the electron position because the two objects hold the status of a compound, sharing space-time and heat flow. In other words, the two objects become two super sized atoms forming a compound in space.

By the same measure a Lagrangian system becomes an enormous five-electron atom with all the cosmic objects working in conjunction as a unit with one mean goal and that is self-preservation through group preservation. This we shall come to later.

If the superior object does not take the minor object as a threat blocking its supply of heat flow, it will consider the minor object as a neutron that can help it grow. Even if the two objects maintain a relation of ($\Pi^2+\Pi^2$) it will hold the other object to a value of ($\Pi^2\Pi$) and outside the combined effort they will jointly relate to the rest of the Universe as 3. Therefore seen from the two separate views there is the following atomic order:

$$(\Pi^2+\Pi^2) \quad (\Pi^2\Pi) \quad S_1 \quad (\Pi^2\Pi^2) \quad (\Pi^2+\Pi^2)$$

Proton   Neutron $\downarrow$ Neutron   Proton

Electron

But placing the two in a battle for survival the relation becomes super straightened where the one will not give any way to the other object's demand for gravity then the relation becomes that of

$$(\Pi^2+\Pi^2) \quad (\Pi^2/2)^2 \quad (\Pi/2)^2 \quad (\Pi^2+\Pi^2)$$

. Proton

Space-time establishing

At present that value supersedes the space value of 10Π completely, so it holds a relative position far above the geodesic value (10Π). That is not all, because the one object acts as a neutron to the other object, both claim a proton flow of heat from space, holding a time of $(48,7 + 48,7) = 97,4$. That places the active space-time development with the combined structure to an atomic era that was relevant just after the "Big Bang". Heat became a gas at the atom value of $7/10(\Pi^6) / (6 \times 10)$ and darkness parted from light at the atomic value of $3^2 (\Pi^2+\Pi^2) = 98$. As you can see, the two objects push each other back in space to a time that applied three eras ago.

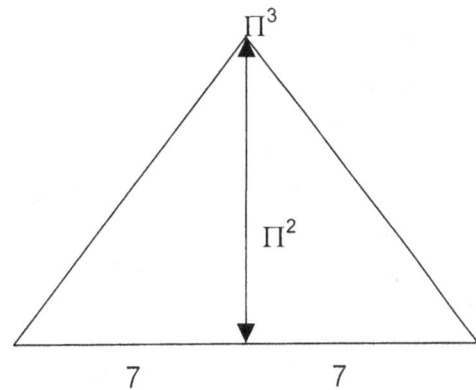

7 are matter   and   7 are matter.

They join space-time, therefore the matter factor is the same. This is where one can visually see the one object, filling the space of the other object's atmosphere.

$$7 \times 7 = 7^2 = 49$$

That is matter Π^2 (time) times matter (49) = 483,61.

This will bring about space-time development that applied when the Universe was fresh and new, hot and lively and full of spinning power. As the objects cannot destroy one another, there is the Hubble Constant affect that came into place. Each structure had to build a Π value, to maintain and secure its relevant position until the end of the final eternity arrives. Said with using other words, each object will secure its place in future as a Black Hole or Proton Star. By producing a value of Π separating the Π^2 from three and thus ensuring itself an own atmosphere it will have to revert to space-time development that even preceded the "Big Bang". I will produce once again the value of Π from its position of Π^2 (time) and matter. Matter is seven and time is Π^2. Both hold claim to the same space, therefore the development of matter and time will produce Π and through that the Titius Bode Principle.

As this is all under the law of Pythagoras the law will evidently place a square root to that value of 483,61 and therefore $\sqrt{483,61} = 21,991$. This leaves the space value of the Roche-limit, as it develops into the Titius Bode law giving them a shared value of 7 (matter) and 21,91 (space) the value of $21,991 / 7 = \Pi$.

Then the relation becomes

$(\Pi^2+\Pi^2)$ $(\Pi^2\Pi)$ $(\Pi\Pi^2)$ $(\Pi^2\Pi^3)$ holding space (3) still outside. They therefore will share space and that sharing will continue till times end. We know by now that matter is 7, and space is 3. holding time to a relevancy in singularity of 1. Sharing the space means that 21,9 will become (10)   space to the one side

1   to the instant position of time (k^0)

,99   lost to space depletion $\Pi^2/10$

7   the relation to matter.

Through that the Titius Bode law comes into affect of 10/7 or 7/10, depending on whether space or matter holds a superior position to time. From that stance, all objects will relate to one another by the value of $\Pi^2\Pi$ and seen in a whole sale total 7/10 or 10/7.

This does not stop at that point, because this will affect the energy called life, and the extensions life develops because it has the ability of manipulating space-time. When an object relates to the cosmos, it relates to it from the point where the object is a proton $(\Pi^2+\Pi^2)$. With that the objects holds everything else to the factor that it uses when depleting space through increasing time. That value is Π^2. Because the object holds the value of matter and matter parts time (R^2/T) from space (R/T) separating the two factors in singularity $(T \times T)$. This positions matter in space (R^3) in time (T^2).

Matter holds the value of 7, always 7, in the position of space (7+3 = 10) and the combined space-time value will relate to time in singularity during that specific instant of time duration ending one eternity with the intervention of infinity (1). Placing all this array of words into a mathematical solution or formula R^3/T^2=1. Space ends the eternity of time to the value of 1. (Time is the movement (spin rate) of heat in space). When a structure holds heat to time in space the relevancy becomes $4(\Pi^2+\Pi^2)$ therefore when the object retains time to space it will have a value of $4\Pi^2$. This puts all the sides IN TIME SHARING WITH THE OBJECT AT $4\Pi^2$ to the Π^2 of positive space-time displacing. All objects will relate to the $4\Pi^2$ in the same way. As I showed, a few paragraphs ago, if space-time extends beyond the critical of $(\Pi/2)^2$, it will attach as the Titius Bode Principle of 7/10 or 10/7. When an object is in the atmosphere (the atmosphere being the neutron value of the Earth) it holds a position of $4\Pi^2+\Pi^2$. That means the object is completely secured by the positive space-time displacing of the Earth. When not secured it becomes $4\Pi^2+\Pi$ and I have shown that Π is the concentrated value of matter $(7)^2$ in time (Π^2). The atmosphere, we know through actual orbiting entry, holds an entry limit of 21,991° as a maximum and 7° as a minimum. That means to become part of the neutron status of the Earth, the object has to be space (21,991 or less) and prove to be matter (7) before the Earth will accept it. If holding a position of less than 7°, the Earth will discard it and if it is more than 21,991 the Earth will find the relevancy to be higher than the space it holds in a neutron time.

That places the object in a relation of $4\Pi^2 - \Pi^2$ (because it is not part of the Earth) in a position acceptable matter holds (7) within the confinement of Π (21,991/7). That means the object is part of space (22,991) acting as matter (it holds an acceptable own proton structure) 7 relating to the Earth in the position the Earth allows of $3\Pi^2$. With the space position of the matter in the parameters of 21,991 it relates to the Titius Bode law as a factor of one. The object has the space value of 10 plus the space value of ten, in that instant of time (1) complying to the Earth's space (10) reduction (Π^2) formulating $\Pi^2/10 = ,99$. That makes the object complying with the full agreement as laid down by the Titius Bode law . The object is, no matter where it is, travelling at a rate of 7 $(3\Pi^2)$ in the space of the Earth (21,991). This will be agreeable to the parameters of the Titius Bode law as long as it remains within the space depleting "gravity" limits of less than Π^2. In accordance to the Lagrangian atom layout, anything less than 5Π is manageable and is in effect less than Π^2. When it exceeds 5Π it will start opposing the dimensional equilibrium space holds of 10Π, therefore it will (according to space) exceed the linear point of R/T, which is $10\Pi/2$ (space going in a straight line). By exceeding the straight-line value of $10\Pi/2 = 5\Pi$, it will then categorize itself in the position time holds to space, and that value is $10\Pi^3$. By exceeding 5Π it will start to defy space and this will automatically bring about an individual space-time relevancy of $\Pi^2\,\Pi$, establishing its own proton position in the confinement of the Earth's space-time of $7(3\Pi^2)$. Thus while it holds a relevancy of Π and Π has the relevancy placing it at a value of Π^0 because the neutron space link to time is $21,991/7$ in the atmosphere, the Earth will tolerate individual movement of up to $7(3\Pi^2)\,\Pi^0$ to $7(3\Pi^2)\,5\Pi^0$. Beyond that point, problems start arising as the object is not complying to the Titius Bode law any longer.

Moving beyond $7(3\Pi^2)\,5\Pi^0$ the Roche factor comes into effect and this is all in the space-time depleting zone established by the neutron atmosphere of the Earth. That means the Earth still holds its own $21,991/7$ neutron base, but the object breaking the Titius Bode law bring in its own value of $\Pi/2$. This value of $\Pi/2$ is in the neutron parameters of the Earth where the Earth removes the linear factor value of the neutron from 1 (Π^0) to Π. That means there is a dual for supremacy of a neutron position is as much as ($\Pi \times \Pi/2$) becomes $\Pi^2/2$. That places the sound barrier at $7(3\Pi^2)\,(\Pi^2/2)$. At the value of $7(3\Pi^2)\,3(\Pi^2/2)$ more implications will come about, because the object will establish an own space within the space-time of the Earth and this will lead to the object whether joining the time of the Earth $4\Pi^2$ or securing itself an individual space and time. Obviously the heat supply will be insufficient to bring about the value the object needs to place it in outer space ($4\Pi^2(7(3\Pi^2)\,/\,(360° \times 10))$) so the aircraft will forcibly join the core of the Earth, crashing on the ground.

From the sound barrier and the effect the sound barrier has on matter, one can measure galactica and how that "growth of space" seems to become a reality as the Hubble Constant indicate.

When the object holds a position above Mach 1 and Mach 2, the temperature to the outside of the aircraft rises dramatically. This is the same as that which we may find to be the value of the part that we can see from the outside in the core part of the galactica. The structure of the craft becomes a solid or a proton. The craft holds the value of ($\Pi^2 + \Pi^2$) in relation to the Earth because it maintains an own space value in the related time of the Earth. The craft applies a demand for additional heat, because to all cosmic purposes, it shows a higher resistance in sharing space and time. Place this in a human context and the aircraft has a higher mass. Science holds the movement of the structure in the category of momentum, but momentum is only linear "gravity" and mass is circular "gravity". In order to claim the space-time occupation that the craft demands, will mean an excess of protons in order to bring about the space-time occupation required for that strong relevancy to heat. The cosmic law does not allow for artificial heat production and life has no normal place in the cosmos. To the cosmos it is all a relevancy of supply versus demand.

In order to stipulate where I refer to space –time my referring to space-time have to be by the use of a signal. Therefore I decided to use the $ as a sign by which I shall refer to space-time.

That is where the Mach principle finds a solid foundation. What separate time and space in joining singularity is the matter and matter is heat. However Mach placed all emphasis on mass and mass is only the resistance of matter in parting with own space to join another object's space. Part of the resistance of mass or parting with individual space is to apply the Roche limit that will remove the neutron value of the object from $\Pi^2\Pi$ to a value of $(\Pi/2)^2$. This will improve the density of the proton considerably because it removes the Titius Bode principle of matter in space as it eradicates 7 to 10. I have shown how 7 to 10 are the development of 21,991/7 and this of course is Π. By removing Π from the equation there no longer exists space in time but a value of half time to time $(\Pi/2)^2$. The influence that this brings along is a huge increase in time value of occupation of space-time that the object initiates. Where the object normally holds a relevancy of space to time in as much as $(\Pi^2+\Pi^2)$ $(\Pi^2\Pi)$ (3) = 1836, the atom adopts a new relevancy of $(\Pi^2+\Pi^2)$ $(\Pi/2)^2$ = 97,4. The normal mass of the proton will be (within the confinement of the Earth) 1,672648 x 10^{-27} and the electron will be 9,109534 x 10^{-31}. With the alteration of the neutron and electron relation the "electron" value of the total object becomes 18,85 times larger. That means the relation between the proton and the neutron has a normal relevancy of 1836, but is only 97,4. That means above and beyond all the electrons' space-time demand, the structure as a whole, has an immense demand of 18,85 times larger. All this happens within the atmosphere of the Earth.

The structure holds a normal value of $T = 7 (3\Pi^2)/10^2$ if one wishes to replace the Earth confinement of matter to the outside. Therefore it holds a relevancy of $T=2,07$ that makes it a cosmic relevancy that helium holds. Helium is an element that surrounds itself with heat in a gas value. When the structure reaches an own velocity of $T = 7 (3\Pi^2) (\Pi^2/2) /10^2$ it then will have a value of 10,25 which holds a comparable mass (space-time integrating resistance) equal to that of Boron, at 10,811. At mach 3 the space-time demands will become $T=7(3\Pi^2) 3 (\Pi^2/2) /10^2= 30,778$. This places the relevancy where that particle can demand own space in own time due to the density it acquires from space. It has all the properties that a star in space requires, but it will still be confined to the Earth. In that way the object then must acquire a relevancy of the full Roche limit, without the protection of the Titius Bode law. The Newton cosmic relevancy will reduce to nothing $T=(\Pi^2) + (\Pi^2) + (\Pi^2) = 9$ and subsequently it will find itself in a huge struggle to defend its space-time or demise its position. With a relevancy of $3\Pi^2$ it holds the same value as liquid heat, which is light. The $3\Pi^2$ comes from the proton $(\Pi^2+\Pi^2)$ adding another neutron space value of Π^2 and this will then be (proton $(\Pi^2+\Pi^2)$ + neutron (Π^2) = $\Pi^2+\Pi^2+\Pi^2$ = $3\Pi^2$, the same as the photon at 29,6. Beyond 29,6 the structure will have to claim individual space-time value. That is the same value as that of cosmic structures still within the inner-core of the galactica. The structures in the galactica holds a value of 7/10 $(\Pi^2\Pi)+(\Pi\Pi^2)$ = 28,8 and to space it holds a relevancy of $3\Pi^2$.

That actually means that the objects that are still within the galactica core, shining because of the outside value of the particles being $3\Pi^2$, are in fact burning silicon, or if you insist, it is glass. In the very centre, where professor Hawkins presume Black Holes to be, is the very opposite, holding structures that have carbon inner core and therefore holds a neutron value of $\Pi^2\Pi$, with no proton core development at this stage. They are still in eternity beyond the "Big Bang" and therefore beyond light. This value brings us right back to the aircraft that applies an own space-time relevancy of Mach 3, or $7(3\Pi^2)2$ $(\Pi^2/2)$. This relevancy places the object holding a star status core of $4(\Pi^2+\Pi^2)= \Pi^2\Pi / 5$ that is the value of a star holding an element of carbon. That is still two eras away and they are not yet even in the present times in our field of vision. To the Earth holding an iron core $4(\Pi^2+\Pi^2)$ in a galactica *7/10) an object in a space time position of $\Pi^2\Pi/5$ is completely out of space, out of time, out of era. It will apply the Roche limit with such ferocity, as its own space-time occupation will allow. According to cosmic law that object has a relevancy of less than that of the Roche limit $(\Pi/2)^2$. The space-time relevancy that the Titius-Bode law allows is (7/10=0,7) and (10/7 = 1,428).

With the fact of the speeds I use in the sound barrier explanation, all the figures are only relevant below 500 m above ground level, and ground level being sea level. Any point above 5×10^2 becomes a changing factor to the relevancy of the Titius-Bode Application. In a previous part I pointed out about the inclination of the atmosphere and that is true, but that view apply the way we see physics through the eyes of Newton. That mentioning of the changes of relevancy is only an introduction, in the same way I introduced the relevancy of the Roche-law and the relevancy of the Titius-Bode law indicating the way Doppler interprets the Titius Bode law.

The relevancy has a lot to do with that of Π also playing its part. That part is rather a bit less straight forward than the facts I mentioned up to this point and in that sense it becomes a little bit more complicated because time in the square starts to mingle with space in the cube. We are dealing with the space part of space-time that has no time value outside the Earth's atmosphere. The limit to the sound barrier being 5×10^2 is because at that level the relevancy of linear space (5) holds space (10) in time (10^2) to a relevancy of 1. Just like the Π^0 that I explained. Above that the next level of factor interference by relevancy changes comes about where the 5 (half that of space placing space (10) to a linear value of half). This is the same application that time has in $\Pi^2/2$. However, time cannot exceed Π^2, but as space only holds a relevancy because of the heat (matter) that it holds, space can then receive a cube value. All of this applies within the atmospheric space of $360° \times 10$ (space) being a relevant 1.

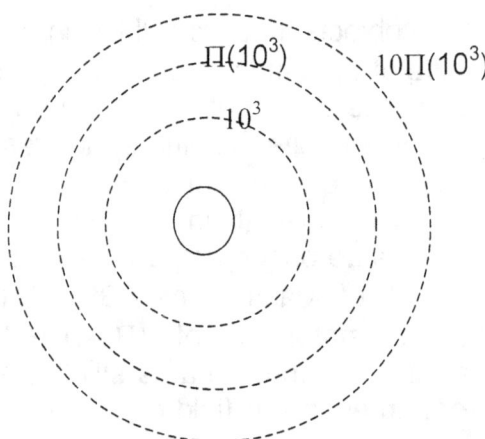

Beyond the atmosphere the Earth's atmospheric space-time or concentrated space-time loses its influence and it becomes a factor enforcing the Roche limit. For example, when the aircraft wishes not to crash but instead flies nose up, pointing in the direction of space, the factor of 360 x 10 will start applying. Then the equation changes somewhat where the effort will have to overcome that disadvantage as well. The space-time occupational need of the craft will be $7(3\Pi^2)\ 5\ (\Pi^2/2)\ /\ (360 \times 10) = 0{,}855$ km/sec. That falls way short of the escape velocity requirement of 11,17 km/sec and the Earth circular displacement will destroy the craft's linear displacement in seconds. Coming back once more to the relevancy applied by the atmosphere holding the escape barrier of (360° x 10) in a relevancy of 1.

The atmosphere is space (R^3) in space-time and for that reason astro-physics and physics can never apply similar equations or formulas. At a relevancy of 1 to space ($R^3 = 10^3$) a barrier will apply reducing the actual velocity to a value where space restricts the craft by 6. In reality it still is the same as what it is at $5(10^2)$ but at $5(10^2)$ there is no actual space, it is time to space (10^2) to half or linear space (5) and at 10^3 it becomes a cube of air that holds the craft in the air. The next barrier of factor changing relevancy will be at an altitude of $\Pi(10^3)$ holding space (10^3) to a time value of Π. At an altitude of $10\Pi(10^3)$ that means space-time in space (10^3) the relevancy decreases to the extend that the aircraft will need to maintain a relevancy of 10 times that of ground speed to stay in the air. What this means is that at that altitude Mach 1 becomes equal to $T = 7(3\Pi^2)\ (\Pi^2/2)\ (10\Pi) =$ 32231km per hour. If you wish to reach Mach 1 at an altitude of 31 000 meters above sea level, you have to reach a velocity of 32 231 km/h. For that reason and that reason alone, N.A.S.A. cannot bring an aircraft to the altitude of 31 km above sea level and then let it quickly slip through the space barrier of (360° x 10) without the Earth noticing it. When the American Generals boast about x15 reaching Mach 5 or Mach 7, they only boast about how little they know.

Should the aircraft wish to break free from the atmosphere while applying a linear displacing position, the craft will need a velocity of $7(3\Pi^2)\ (\Pi^2/2)\ (10\Pi)$ x 3 (Mach 3) leaving it with no less than a speed of just meter 1 00 000 km/h required and even at that velocity I am not sure it will pass the Π influence of 21,991/7 barrier that shield the Earth remaining totally intact and unscathed. I wish the alien hunters that claim aliens can come and go, as they please will use their brainpower in the direction of common sense and not common fantasizing. No way in heaven or hell can there be a craft that will slip in and out of the atmosphere at leisure to the demand of its alien pilot. The aircraft has to have a structure built at least from Californium 98. That shows how ridiculous the alien detectors are and they do not even know their own stupidity.

Bringing in reality again I would like to focus on Einstein's view of the curvature of space-time. Einstein made the human misjudgement we all do, by placing the focus he holds on the space his is in, in the time he is in. The aircraft travels at Mach 2,7, in relation to the space and time in compliance to sea level on the Earth. That means Einstein is standing on the Earth on a beach and has the binoculars focused on the craft flying at an altitude of 31 kilometres straight above him. The craft is maintaining a velocity of 2 700 km/h. On the ground the aircraft would be relating to a certain area of Earth every second. At the altitude of 31 kilometres that relevancy no longer apply.

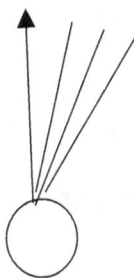

On the Earth the aircraft holds a position that allows 250 meters of soil to pass him by every second. However at 30 kilometres above if he wanted to have that same distance of soil to pass by every second his relative space-time displacement has to increase by a measure of $(7(3\Pi^2)\ 2,7\ (\Pi^2/2)\ X\ (10\Pi)$ in relation to the $7(3\Pi^2)\ 2,7\ (\Pi^2/2)$ that is valid on Earth. That will then still hold a relevancy of 750 meters of soil passing under the wing of the craft. The effort by the craft therefore has to increase, not only by 10Π but another (10^3) to comply with the space orientation factor that the craft needs to sustain the equivalent space-time displacing. One factor of the space-time will remain constant, while the other two will have to vary. Should the energy remain at one, the velocity will reduce and the time to object relation of the craft will reduce. This means that in the triangle of space-time the triangle's 180° will maintain form, but the half circle's 180° will bend out of shape and this will bring about the straight line time has to hold in order to hold the value of 180° will also bond. When the triangle maintains 180°, none of the other two factors will apply their standards of value any longer. The circle will become oval and the straight line will become a semi-circle. From Einstein's perspective on the beach he will see time bend, and he will see space bend, because he focuses on the constant of the speed of light. He places the speed of light (which is a three dimensional triangle of (3) space (1) relevancy Π^2 (time) in the value of the factor remaining at 180°. That means time (Π^3) goes skew, and so does space, to prevent the constant of one to go skew. He allows the variables to vary in a dissimilar fashion to what the cosmos grant as the variable thus will vary. The cosmos holds Π^2 at a constant, always, no matter what because Π^2 come from the proton and the proton's space is the last space that time will comprise. The evidence is in the Black Hole where the Proton Star bends the 3 of space completely around, (the 3 is 3 sides of a possible six that will relate to any other object in the Universe) therefore he turns the 3 factor inside out. This evidence we see as matter spirals towards a centre point.

At the same time it completely demolishes the relevancy factor of 1, because of the repositioning of space to matter. This will lead to one of a possible two relevancies going astray. The total relevancy of the combined proton value places Π^2 at a number higher than we can ever calculate. When Π^2 changes its value to comply with the totality of the combined proton cluster, and space spirals changing three around, the time contact of 1 will maintain its position on singularity. Therefore matter in space will reshape, and gravity will reshape to maintain a constant to singularity. In the speed of light the 3 factor will change and the gravity factor will change in order to maintain the contact with singularity.

Einstein allowed singularity of space holding (heat) to bend and the singularity of time contact (1) to bend in order to have matter $(3\Pi^2)$ maintain its norm. We all know from what the Roche-limit proves, and the Titius Bode law proves, and from what we can see happening to our craft, that matter in space $(3\Pi^2)$ changes shape, form and norm, keeping the singularity of time-space-unity to the relevancy of one. There is no curvature of space-time, only of matter occupying space-time and when focusing the constant factor on the variable, the true constant will seem to bind.

Again I say: There is no curvature of space-time, only the revaluation of matter, be it unoccupied space-time as in that we regard space heat to be or in the occupied space-time as that we conclude matter to be and both forms part of the same unit. By applying our focus incorrectly to matter, science concludes that space is growing. That is as inaccurate as Einstein's curvature of space-time. Matter can relocate and revaluate its relevancy, but space and time remains the constant of eternity in singularity. This means that saying the space of the Hubble Constant is growing, is just as inaccurate as saying the Sun is rising each morning.

The influence of the Titius-Bode law may have one other extremely significant influence on the speed of light. To every possible way I look at it I am convinced that my view is correct. The part of the Titius Bode law which science confuse with the Doppler effect has to play a part as the Doppler effect can only apply under the conditions laid down by the Earth's space-time displacement and that renders the opportunity of precise half circles following one upon the other. In the Titius Bode Principle however, the matter holding a relevancy to the heat in that space determines the value of space-time.

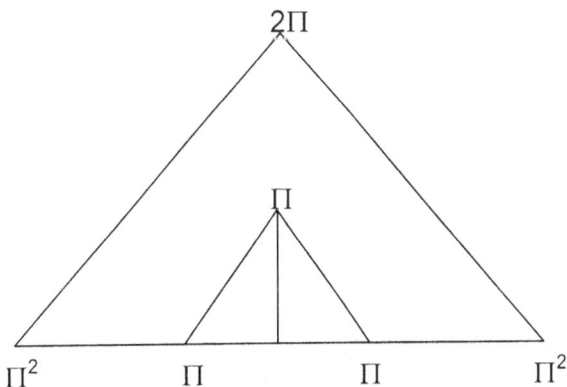

The combined value of matter holds the space component at 3 pointing out the point of origin where the particle was holding space-time before growth started. This will give a value of 3 and the point of relevancy through time by position allocation will be 4 ($\Pi^2+\Pi^2$) this will then be the seven Π 's in the law. Therefore the 3 always determine a position of origin and the 4Π a position of location.

That will bring about that all particles orbiting any structure will see the position of shift through time advancing or the Hubble growth in a manner of time to space. As I have indicated the progress to matter's relation to space will be the 7 that matter forms (7Π) in words the seven pi and space (10Π) in word the ten pi. Matter will always be in a galactica submitting the influence it holds to the heat space has to provide, maintaining the cooling of matter. The density factor favours heat in space and heat in space will conduct existence (mass) to space depletion (gravity). In the case where matter lies outside Galactica the opposite is true and therefore the opposite will become the Titius Bode controlling factor in as much as space (10) becomes relevant to matter (7).

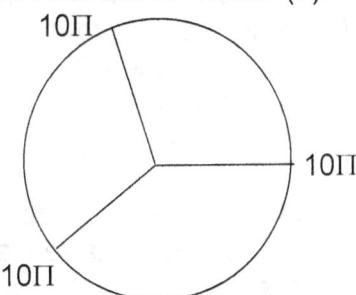

The metric system provides a vehicle that makes space and time measuring far less complicated, because we base the space measurements to that of water. One cubic meter of water is one ton with one meter in the cube moving one meter per Earth velocity which all connects nicely to the earth governing singularity. Space holds a relevancy of 10Π. That will be the value of the full field of the spherical composition of space.

To that reason time has the relevancy of 10Π^3 and space holds the relevance of $10^2(10)$ ($\Pi^2\Pi$) where the second 10 is the matter's composition of 7Π+3(Π) = 10Π, but because Π only connects the fourth dimension to the time factor it falls outside the fourth dimension therefore space holding matter carries the total of $10(10^1)^2$ in the full circle.

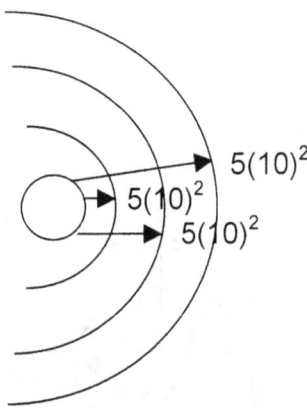

The linear factor of light travelling from the Sun will be the distance between the Sun and the Earth. The light density that reaches the Earth is the density that remained after the rest of the wave (3^3) went to space in the form of radiation (3^2) and heat invisible to the naked eye, the heat that holds the universal gas that covers all objects, the same heat that one day must all return through matter to time. When looking at a sphere the three we see, has a lot less dimensional value then we can see from a cube. Even looking at a cube from only one side, will still have the support of half of the other sides that provide enough stability to the cubic orate so that we can keep on seeing the one side.

To introduce my theorem, in short I wish to bring a very short overview, before we start with the complicated cosmic laws and definitions I am forced to use to defend the point I make. I take the most common phenomenon on Earth and build a theory based on that principle. The basis of my theory is that everything is heat, be it solid, liquid, or gas. Culture tells us that hydrogen is a gas, or that gold is a solid. Our biggest drawback we can have is precisely such wrong cultural conceptions with no base of proof. Hydrogen is much less a gas than gold is a solid. We connect a cultural conception to what we may presume as facts, but that does not make it a fact, it only makes our conclusions an elusion.

Everything except life uses an energy driven by heat removed or heat expanded. It is all a balance where the flow of heat started at a point and pushed in a direction. This brings me to my first definition. What is energy? I shall prove that **energy is the interaction between heat creating space and by demolishing space, heat concentrates. All energy in the Universe holds relativity to this, no matter what.**

The Universe started from something I named time in singularity. Time in singularity is the only constant because there is always a direct line of contact to time factor in singularity Π and time in singularity is the end product of everything that is cold, frozen beyond space. We view the ultimate freezing point to be outer space at a temperature of $-273\,°C$. Heat will always move, from the hottest area to the coldest area. If the outer space were the coldest point there is, the Earth would be frozen solid, because heat will never arrive at the Earth from outer space.

We would have a continuous flow of any heat the Earth may hold to outer space. Such is not the case. We may view that heat flows from the Earth at night to outer space and that is correct. However there is no flow of heat that will bring about temperatures to fall to limits even close to outer space. Nothing on Earth can reach –200° without the interaction of human life. That means the Earth will always be a place colder than the Universe, just because we have more heat on Earth than in the Universe. If it was not the case, the Earth should be at least as cold as the Universe because all heat will flow from the Earth to the outer space.

IF THE OUTER SPACE REGION WERE COLDER THAN THE SOLLID STRUCTURES, THE COLD WOULD DRAW HEAT TO THE HOTTER REGIONS. THE NATURAL TENDENCY MUST APPLY WHEREVER THERE IS HEAT VERSUS COLD.

The opposite is happening

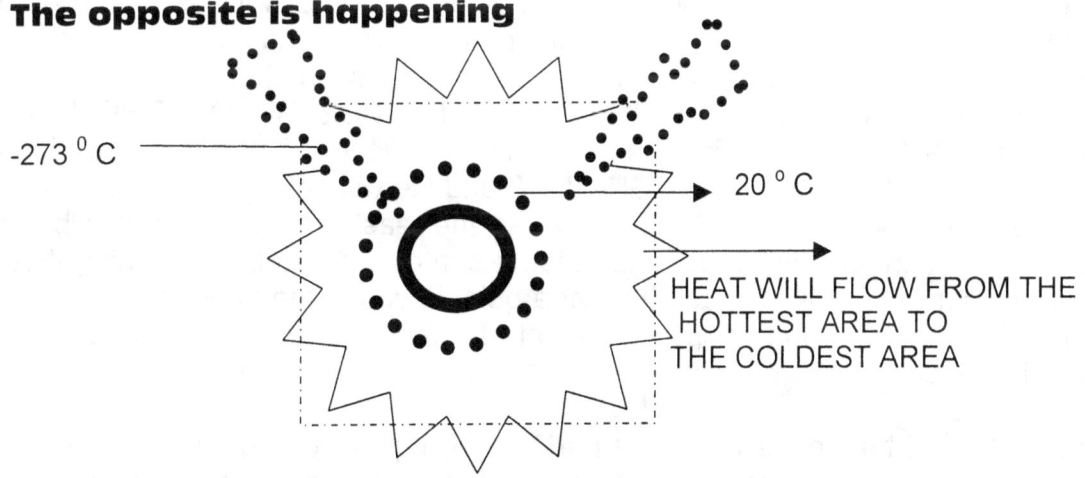

NSTEAD OF THAT, THE NATURAL LAW PROVES TO HOLD HEAT TO THE STRUCTURES. THAT MEANS GRAVITY HAS SOMETHING TO DO WITH HEAT CONSENTRATION. THAT ALSO INDICATES THAT GRAVITY IS THE RESULT OF SOMETHING MUCH COLDER THAT IS NOT IN THE VISION LEVEL OF THE HUMAN EYE. THE ONLY SUBSTANCE HUMANS CANNOT VISIONALISE IS THE PROTON.

Why would the Earth appear hotter, but is colder than the Universe? It is the fact of gravity, a term I reject because of the connection science applies to gravity being a force. If it was a force then something is pulling or pushing and since we do not believe in spirits creating magic forces, and no one can detect anyone applying any force, we have to dismiss the connection of a force or forces. For the moment in order not to confuse any reader in the introduction, I shall remain using the term gravity but in doing so, I dissociate myself totally from the implication that science connects to gravity.

One may say that this is the effect of the Sun shining on the Earth, and one may be partly correct, except when considering the wider picture. Measure the space the Sun pours light into and divides that area the Earth holds in the vastness of space. I truly think no ordinary pocket size calculator will provide a realistic reading. Yet the Earth is many times hotter than the rest of space, if it was merely the Sun shining on the Earth, the Earth was no factor to consider in relation to the space out there. That alone cannot account for the difference there is in temperature.

This brings us to gravity, (not a force). My definition about gravity is that: **Gravity is the reduction of space to concentrate heat,** therefore the more gravity there is, the more heat there will be. The Earth, or any cosmic structure claims heat from outer space by concentrating heat. You may shout that fusion is the blaming contributor to the heat in the sun, but I shall denounce that shortly. The more gravity a structure holds, the hotter that structure is, Pluto is the coldest place in the known solar system, but it also is the smallest place, and it also has the least total gravity. The one scenario compliments the other scenario, you may be of the opinion that PLUTO IS THE SMALLEST BECAUSE IT HAS THE LEAST GRAVITY, AND I MAY BE OF THE OPINION THAT IT IS THE SMALLEST BECAUSE IT HAS THE LEAST GRAVITY, and we will not agree at all, although we are using the same words. If Pluto has the least gravity, it will also be the coldest planet, if gravity concentrate heat, it can only do so by something that is much colder. This will increase heat at a point because the effect behind gravity is producing a reduction of space, a point of reference that is much colder than any place else.

I started off at the beginning of this part, by saying everything started at a point, that was so cold, space froze to singularity. Let us test this statement: When an element heats up, the space it occupies, increases. When an object becomes colder, the space it holds reduces. The less space per atom there is the colder the object must be. Anything can freeze rock solid, as everything can boil to gas. This means all elements in nature, is neither gas, nor liquid nor solid. It is the space that is between the elements that allows the elements the form they hold at that moment. By reducing space, space has to concentrate because to concentrate is to reduce the solution. By reducing space, we find more heat. If you find more space by increasing heat, and you find more heat, by reducing space, then heat and space is the same thing.

THE FURTHER ONE RISE ABOVE SEA LEVEL THE COLDER IT BECOMES, AND THE DEEPER ONE GOES INTO THE EARTH (NOT THE SEA) THE HOTTER IT BECOMES.

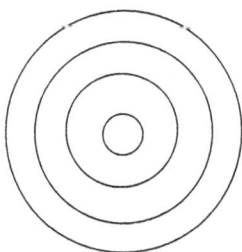

You may think that at this point I reached a point of becoming ridiculous. By concentrating matter to a solid, the matter holds less space in relation to the heat surrounding the matter and since matter is pure concentrated heat, the heat actually increases because the space around the matter is less dense heat than the density of heat the matter holds. Therefore, matter is the ultimate concentration of heat, because it is heat-frozen rock hard.

The colder anything is the less space it requires and matter requires much less space than does space require space, If mater is heat frozen rock hard, then although having a higher concentration of heat, it also have a point within, that is much colder, and the flow of heat towards that point of the ultimate cold, concentrate the heat. To cool any overheating object, you increase the flow of air. The air is space, and that space increase brings about a reduction of heat by increasing space. That means the space -air ratio changes and the product is heat reduction. However one has to look past the obvious to the factors not red dally in sight, It also means by increasing the flow of heat, the heat factor reduces and that will lead to a decrease in heat availability, with a lesser heat factor.

The lesser heat factor will provide a faster relevancy to the flow of heat. That means the faster heat flows, the colder heat becomes, because the more space there is to allow the flow of heat. Looking at a rotating object the relevancy of space reduces by the factor of four to the halving of the radius. If one draw a presumption that the same volume of heat flows through less space, the flow of heat must become faster by the square of the decrease of radius, and that reduction leads right down to the point of singularity. As that point still connects to singularity, a point without space because it froze space out of existence, the flow of heat away from such a point becomes "over-spaced", because the point of singularity is without space. All things "over-spaced" are also overheating. Therefore, as everything away from such a point is hotter, it will also hold more space to heat because the more space means the overheating factor is higher. That is the cosmos, the place hotter than singularity and therefore more space to hold heat. This applies in the precise way inside an atom. The atom holds more heat to less space and therefore it has to be colder on the inside than what it is on the outside.

While outer space is growing the growth can only come about from singularity, which was out of control and expanded without direct influence of contracting. The contracting at that pint was secured in the atom and all heat was stored in a secluded space by using the Coanda principle. In this the cosmos gathered what was beyond control and expanding into the oblivious.

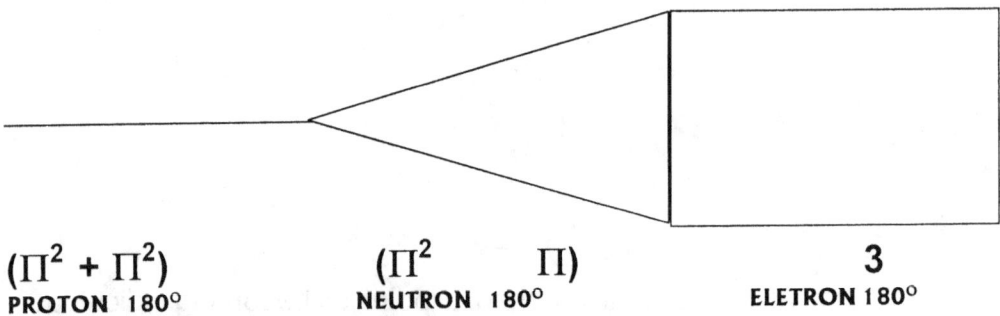

$$(\Pi^2 + \Pi^2)$$
PROTON 180°

$$(\Pi^2 \quad \Pi)$$
NEUTRON 180°

3
ELETRON 180°

We view the atom always from the outside, which means from the electron. The fact that matter to the inside is dense, only means that the matter to the inside of the electron is frozen solid. Frozen solid by itself has some terminology, some concept we have to define in order to establish the true meaning. The frozen state of matter within the atom, also apply a frozen realization bringing across implications that are to our mind completely alien. To us something frozen solid means that the density around the atom hold the least space we can imagine.

This is correct, but it is also relevant because what may be frozen to our view is still liquid to nature. Even freezing an object to −273°, only means that the heat surrounding the atoms is at its limit of forming a gas. It is at its limit, but still it is a gas. If it was more than a gas, the electrons will touch, because the electron is the state of heat where heat becomes so cold it turns to liquid.

If there is only liquid heat separating electrons, the electrons from different atoms will connect and establish a fluid, linking the atoms in space-time. Linking atoms is better known in science as fusion. The atoms melt together, they fuse, they bond. Hiroshima and Nagasaki as well as the Bikini Islands bear testimony of such an event. The cooling of the atoms have no more gas-heat flow in maintaining the atoms at a relevancy below the cosmic relevancy of 112, and the whole atom turns liquid. The liquid forms energy as the liquid heat arrives at a higher spin rate than the frozen state will demand.

Energy is the release of heat to gas, forming space, and the space creates the nuclear winds, accompanied by radiation and light. Radiation is only intense spinning gas and light is a heat droplet. Therefore the heat becomes condensed heat in the form of heat "Vapour" and heat "drops". This is the closest we can come to see what is on the inside of the atom. The rest our mind must tell us, and not our eyes. When heat travels, it becomes more condense, holding less space, and is therefore colder. Behind the electron, the movement of heat is faster than the electron, because it spins with an ever reducing radius, therefore as the radius reduces to the atom core, the spinning of the heat holds four times less space for every time the radius halves in length.

Having an ever-increasing spin rate, the heat moves faster, condensing even further as it progresses to the proton core. That makes one realize that inside the atom the heat contracts, holding less space, and by holding less space it must be in a more frozen, or colder state. This is about the same as feeling the core of a radiator of a cooler unit, and thinking if there is that much heat on the outside, how much heat will be on the inside. Well, the similarities may not be that correct, but the principle is very much corresponding. Matter is frozen heat because it is spinning through less space.

I shall make a statement, that will surprise every reader, I think. Space has no value, because space is the product of time in singularity, a part of singularity, a by-product of time in singularity that has no cause to be except to hold heat. When boiling water one finds bubbles in the water. No one blew bubbles in the water, it obviously could not enter from the top, because it will eventually rise to the top and it did not enter from the sides or the bottom of the pot, the heat that forms the pot is watertight as much as air tight. Yet there are newly formed space that came from nowhere and will disappear into "thin air" as soon as it leaves the pot through the top of he pot where the water ends in space. No one can detect the space but in water vapour and between the water vapour is no space, there is only heat. If it was not heat, the water will be less hot, and be a liquid. This proves that space has no value, except for the heat that forms the space. This is another point made, which proves that outer space is actually less dense heat and matter is more dense heat.

At this point I arrive at the biggest bone of contention I have with Newtonian science. It is not mass that produce gravity, it is the density of matter that produce gravity. When one cubic meter of water forms vapour, it will be a cloud with as much mass as the equal cubic meter of liquid water but it will float in mid air. It will hold in mid air because it holds more space, therefore less dense heat, than would the water have because the water has a liquid form. The mass of the vapour is the same as the mass of the water, yet gravity applies less to the vapour because of the abundance of space allowing less density. About this I bring proof that Pluto has substantially more gravity than the sun, calculated in the manner the Newtonians do.

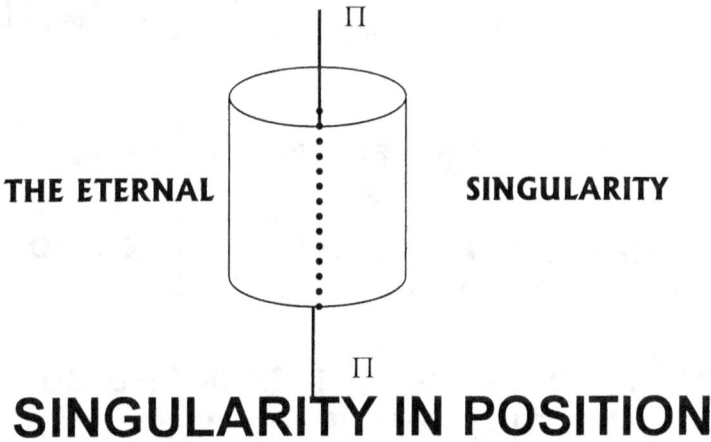

THE ETERNAL   SINGULARITY

SINGULARITY IN POSITION

As a principle when any point expands the expansion is evenly in all direction simultaneously. That configuration only Π allow by dictating the form of the sphere. The fact that space holds 3 and time holds for in relation to a centre is the proof of the formula we use to calculate a sphere. Therefore only the sphere could form when creation initiated by motion. The formula is then $a^3 = (4\Pi\ r^3) / 3$. Although it was running on the line the line through the interlocking of the numbers that applied Pythagoras was forming sphere that kept space apart from spheres that kept space within. But as the Roche limit came into use and not the dividing of five, we can be sure that only form applied and at that point space was not yet a part of the Universe, just as singularity at present is not a part of the Universe while it is well establishing the Universe.

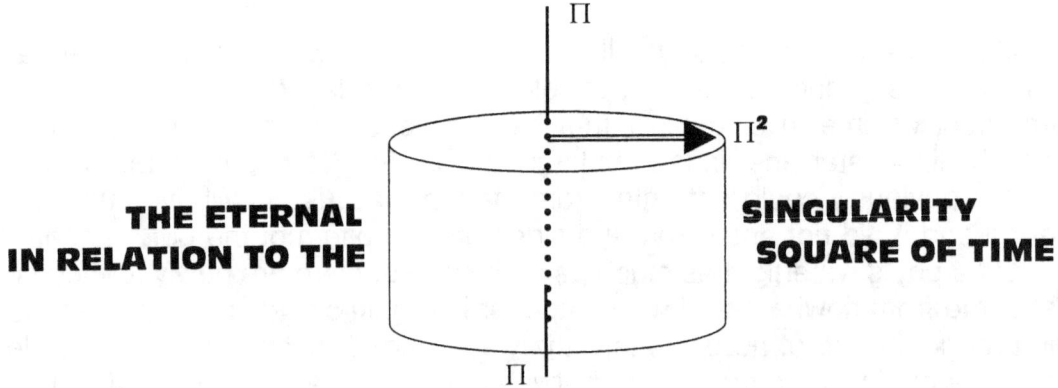

THE ETERNAL IN RELATION TO THE SINGULARITY SQUARE OF TIME

SINGULARITY CONNECTING TO TIME IN THE SQUARE

Of course, the Newtonians make one great blunder, which I also prove, but more about that later. It is not the space that one should calculate, but the density heat has within the space in question. All space holds time therefore it is space-time. Space does not apply and can disappear, if not for heat within that space, allowing space to be. This then dispute all claim that Einstein made about space-time curving under gravity applying. Space does not exist without heat, and gravity concentrate heat, therefore space has no bending to it, and it all depends on heat. Gravity concentrates heat by reducing space. That brings us to gravity and time. What is time? If the Newtonians wish to bend time, they should at least find out what time is before they can start bending it in all forms and shapes.

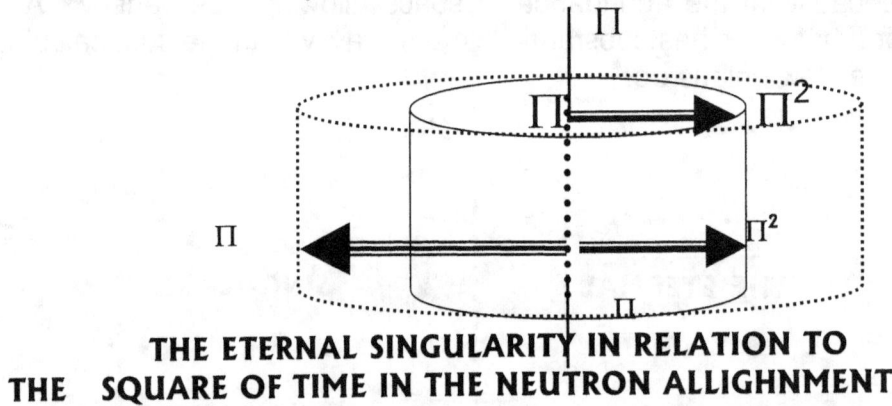

THE ETERNAL SINGULARITY IN RELATION TO THE SQUARE OF TIME IN THE NEUTRON ALLIGHNMENT

SINGULARITY CONNECTING TO TIME IN THE CUBE (R³) OR Π^2 Π. THIS IMPLICATE THAT TIME IS THE PROTON AND SPACE IS THE NEUTRON, AND IN THAT VERY CONCEPT IS THE "MYSTORIES OF STARS AND STUFF"

To find what time is, let us see what the standard is we apply to create a sense of time. The Earth holds a position relating to the sun, and that position is different every second time ticks by. In summer, winter, autumn or spring, the position is always different. One may say that I am wrong when I say time can never repeat, but the truth is that because the Earth never repeats any position in which it was. While changing the orbit the Earth follows around the Sun it is also following the orbit the sun has around the Milky Way and as much as the Sun changes the location it has every cycle, in the same manner does the Milky Way change the orbit location of the solar system by the growth of material expanding. There is no chance that the Earth will ever be in the exact spot twice. Newtonians, as usual never cast their eye past the obvious. The Earth is not the only planet in the solar system, there are nine other planets relating to different positions as they combine a relation with the Sun as much as hold individual and connecting positions according to the sun.

In this context, the Sun is only one measly star, of countless many , all circling about a centre point around the Milky Way. In that respect, no point that any star holds, can repeat that position in accordance to the combining relative positions all the objects hold in relation to time past, time present and time future. The Milky Way is one measly galactic amongst countless others all repositioning their position to the structures within them as much as with all the other galactica the Universe may hold. In that respect I would love to see how Einstein will bend that lot! Light has nothing to do with time, because every photon is just another relative position filled with liquid heat, changing its position according to all other universal occupants, how large or small they may be, because the combined effort of repositioning create the variable one find in the flow of time. Running up and down a chamber between two space whirls will not reposition every object the Universe may hold. That shows how ridiculous science has become, with their view on time travel, holding two space positions in one instant and the rest of funnies the Newtonians declare as science.

Time and space are one and the same in singularity, with heat in many forms separating the singularity of space and time. Neither space, nor time can bend, because they are concepts existing beyond the point where heat starts and heat ends. Space does not exist and heat in various forms apply time, therefore time in singularity holds a value of eternity and space, where heat in various forms relocate their positions, holds infinitely as every repositioning heat produce apply an infinity, by breaking the monotony of eternity.

This is then my third definition: Time is the spin rate (movement) of heat in space.

This brings us to the fact that I mentioned about heat having many forms. This then is where I dispute Einstein's other theory, the one about the critical density of the Universe. Einstein said that a point has to arrive where matter will contract, because of some force that will start applying. One should view the time when this theory came into place, more than the theory itself. A person by the name of E.P. Hubble brought "indisputable evidence" about the Universe drifting apart. I shall at a later stage dispute that statement, however, at this point, I wish to bring to your attention, and the crisis this finding brought about in the circles of science.

In view of this critical density, one has to think when was the Universe at such a point of critical density. If Newton's claim does apply, then there is no Hubble Constant, as there is no Big Bang. However, everything about the cosmos point to a widening shifts in matter, where the space between matters is on the increase. This can be detected to a position where the Universe was infinitely small. What better time will there then be, than that point for the critical density factor to kick in and produce an implosion the Universe will never forget. If the force can contain and reduce the space separating matter, and the matter were locating in a space of non-existence, the mass of the Universe then concentrated in a space of infinity allowing the force the ultimate opportunity of endless strength. The force at that point must be so great, that no force can bring about any separation. All matter the Universe at present contains, it contained back then in infinity to space. There is no adding of matter. There is only adding of space. The Universe did not collapse, it "exploded". Why then would it explode at a point it had all the reason to collapse by the mass of gravity causing an implosion, well, that is if Newton was correct.

The answer is so obvious, it is a joke. Newtonian measuring and calculation methods, by the way, do not apply to the cosmos, and science should keep mathematics out of it. The answer to the Hubble explosion that started at the Big Bang, depends on the increase in space. A few pages ago I showed that space has no value, it is a form of singularity, therefore adding space is the counterpart of matter reducing space with heat concentration.

The more space gravity will reduce, the more space heat will apply as heat in space reduces to heat in time. By adding something that effectively does not exist in any case, and can appear or disappear at the beg and call of heat, it will and can bring no change to the cosmos as a whole. It has no effect other than to indicate the moving flow of time.

However, why the desperation then and why did the institution of academics go into a spin about E.P. Hubble's discovery? All of modern science finds its base on the fact that Newton proclaimed where he never produced evidence in support but still declared that a force named gravity will reduce the space between objects in relation to the products of the mass of each body. The space separating particles will reduce in accordance with the gravity the mass will apply. The emphasis falls on the reduction of space, and yet, undeniably, Hubble proved that the space does not reduce to the contrary it increases with the event of time. This cracks open the completely complex issue about the misconception they call the science of physics. Again, I repeat, Newton is as clear as daylight when true forces apply, but only life can apply any force!

Only as far as Newton is cosmic outlook goes, no one can understand Newton. Anyone that understands Newton will not accept Newton. To accept Newton's application of cosmic law, you brutalize and bully students into a state of accepting Newton and then have the student believe he understands Newton. It was at first a fashion that later became an industry. If you do understand Newton, you will realize that you do not understand Newton.

Through the centuries, across the world, in all countries, they institutionalised Newton. All of science will fall flat, because the opposite of what science use to base their many calculations on comes to nothing. The Universe expands it does not contract. Therefore, who better to get than the man that already produced the relativity theory to come up with some cockeye notion about Newton's law that will still apply one day? With the help of another misconception introduced by an Austrian, E. Mach, Einstein recalculated the theorem about the critical density of matter. This was all to one big effort to save Newton's claim on which rests all of science.

To distract all attention about the folly, they started a major attempt in deceit, (wittingly or unwittingly I will not know) whereby they calculated the density and mass of matter contained in the Universe to reapply in order to reconfirm the truth about Newton's claim. At this point, there is only ten percent of the required mass floating around to save Newton's claim. Again, another attempt was launched to save the industry where they will locate, so called dark matter, in a hope that enough dark matter will save science. This is another fiasco where the poor and insignificant tax payers hard earned money is thrown into a bottomless pit that will bring no conclusive answer.

The conclusive answer is in rejecting Newton's claim about the cosmos. We see the Earth and we see life on Earth as part of the Earth. In the obviously short sightedness science have, they wish to project everything found on the Earth as a direct translation to outer space. Outer space holds no life.

Newton's other laws do apply, as it is, precisely to point the genius made, and a genius he was, about that there is no doubt. To get a perspective on Newton, see the apple that fall as life, not as the cosmos. See an aircraft, a boat, a building, and a dam as an extension of life not as a product of the cosmos. That is the only way to save science and Newton, because that is the only thing wrong about Newton.

Newtonians' physics belong to life on Earth and other laws applying other dynamics apply to the cosmos.

As much as life does not belong to the cosmos at large out there, the calculations and mathematics life can use has very little significance but to bolster the ego of those mathematicians that consider their mathematic skills far more that that which the Earth may appreciate. Those mathematicians should put their enormous skills to better use and build a dam or a skyscraper.

Gravity is just the retracting of cold and the expanding of heat. Heat finds new locations in order to distribute the overload while cold is the point that found relief and contracts because of cold. Gravity is the interaction behind games heat and cold play.

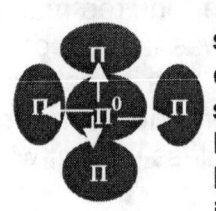

By moving from 1^0 to 1^1 and from $1^0\Pi^0$ to $1^1\Pi$ requires space. Yet such moving did not leave the realm or the domain of singularity. The motion was still within singularity because moving involved forming a relevancy between heat and cold between infinity and eternity, between space and time and most of all producing what will in the far future develop into a Universe that can even be a host for life albeit on a very small spot for a very short while in relation to the vastness space has and the duration cosmic time has. However, the distribution of the heat overload required a simulation of motion, which was interpreted as liquid and serves as such since then.

Relevancies came about when the dot moved away from the spot but had no space to move into. All that was possible was to charge singularity by relevance to comply in being activated into complying. Charging came in the way of distributing the heat overload to four set points forming time. That formed the basis of one solid controlling four liquids relevancies. Space-time is motion and movement are all the same things only separated by dimensions and dimensions are formed space, where the dimensions become space being in motion and the space is motion by contraction or by expansion but because time is almost eternal at k^0 our perception of the universe we are in is a stable and steady eternal structure. Gravity is motion and motion creates space to the third by the third in the third that interacts with one but establishes ten.

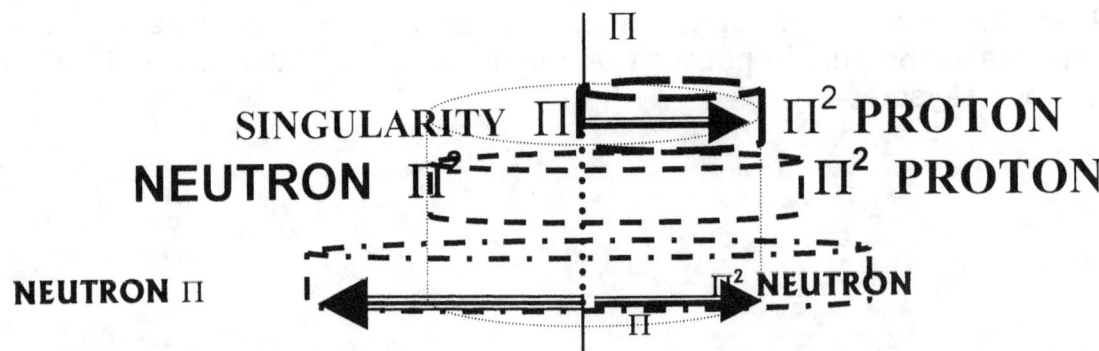

The cosmos holds no constant and that is the only constant. Every aspect of the cosmos is relevancies where matter in different forms, form different relations to other matter also in various forms. As I indicated about time, where time is an ongoing repositioning of relevant matter locating relevancies in the position they hold to the time they apply. The cosmos cannot grow, as much as the cosmos cannot shrink. It is a never-ending flow of changing relevancies, where singularity meets singularity as much as space meets time. The point in singularity I named time, is the point where time started and where time ends as much as where time will finally fulfil its reducing of space. Heat made space renegade and time slowly contains space by reclaiming heat. That is the cosmos.

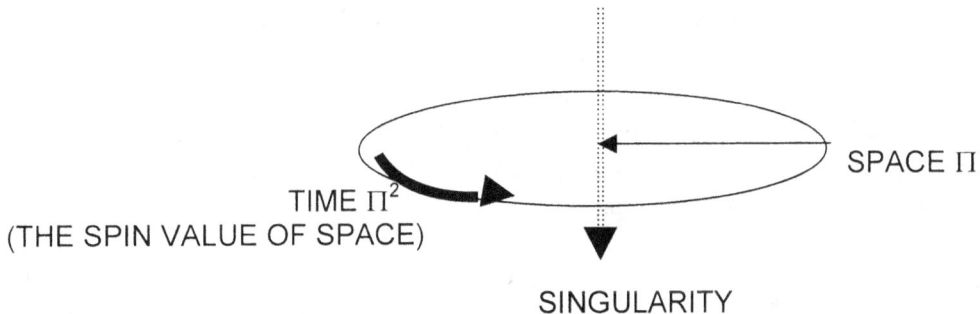

TIME Π^2
(THE SPIN VALUE OF SPACE)

SPACE Π

SINGULARITY

The reclaiming process has various stages, all forming separate dimensions. The following description is wholly inaccurate, but in view of maintaining simplicity, it is also the best I can do. Mass is a door, a hole or a gate through which heat flow from space to time whereby it produces gravity by introducing the events forming time. This is a relevancy. There is in this relevancy time, which I gave a value of Π^3. The value is rather complex in explaining while I am in effort to attempt in this letter to keep information as simple as I can manage.

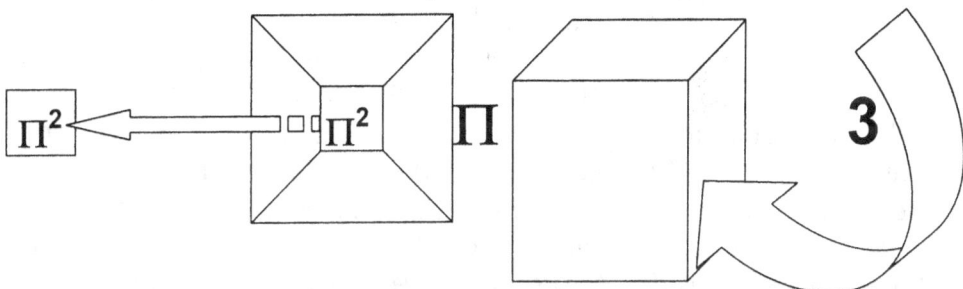

Connecting to time in singularity there is a rotating door that is two protons, one linking to singularity where time and space in singularity meets, while at that very moment, another proton, also Π^2 connects time and space where time and space holds heat. This I called the double proton ($\Pi^2+\Pi^2$) and the heat is in the densest, most frozen form heat can be outside singularity. As I said, this is a revolving door where the one at one point dips eternity into infinity, while the other takes heat to dip when it is the next proton's turn to dip into singularity. The value of Π^2 is also rather complex to explain in the introduction.

Behind this is the neutron double gate where gravity removes one dimension. The proton removes one dimension to take heat to the dimensionless ness of singularity; while at the same time it removes one dimension of space by applying the square of time (T^2). The neutron removes one dimension of space, by applying gravity in a two-prong third dimension of $\Pi^2\ \Pi$ where Π^2 is the circular part of gravity and Π is the linear part of gravity. Following this dimension is the fourth dimension where the six sides an object can have, will only hold three sides in relation to any actual relevancy, since we can only be in one part of the Universe. The other three sides will apply to the other part of the Universe that we do not occupy at that precise moment in time. The conclusion to this "port" figuration is $T\ (\Pi^2+\Pi^2)\ (\Pi^2\Pi)\ 3 = 1836$, the difference between the mass of the proton and the mass of he electron. That is the relevancy to all space-time.

Time started … only God knows where. I can take you to a space-time relevancy of $\Pi^3 x \Pi^3 x \Pi^3$ and I do in The Thesis, but without detailed explaining it will only be numbers. The explaining and calculations are tedious as they are complex. I can also take you back to a relevancy of $(\Pi^3/2)$ $(\Pi^3+\Pi^3)$ but without the proper explaining it is senseless. I can take you through many more relevancies but they will be as senseless as the next.

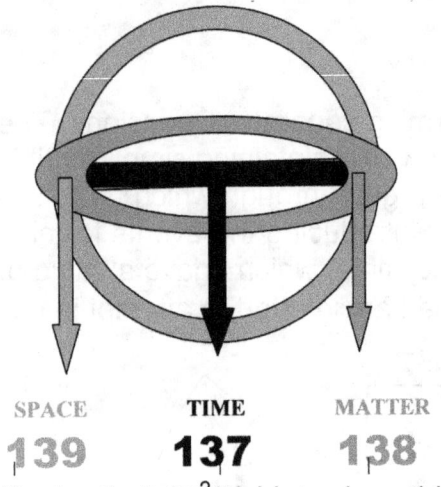

SPACE   TIME   MATTER
139   137   138

I wish to start at 7 $(\Pi^2+\Pi^2)$ = 138 also 7/10 $(\Pi^2/2)$ $(\Pi^2+\Pi^2)$ = 136 and 10 / 7 (Π^2) $(\Pi^2+\Pi^2)$ This is where the Universe received a sphere in formation, and although space was still part of singularity, the Titius Bode law on matter (7/10) as well as the Titius Bode law on space (10/7) formed a conjunction with the Roche limit in occupied space-time $(\Pi^2/2)$. At this point matter formed 7$(\Pi^2 + \Pi^2)$ where 7 is the value of matter and during this the prelude to matter in space (7/10) as well as space holding matter (10/7) became part of the occupied position of the Roche-limit $(\Pi^2/2)$ I introduce this aspect, because the Titius Bode law applying to matter (7/10) and to space (10/7) as well as the Roche-limit holding position outside occupied space-time $(\Pi/2)^2$ is of all importance to the understanding of how the Bible recalls the Creating events. On that part, I shall have to apply some technical aspects, in order to combat the onslaught I know that the Newtonian atheists will launch.

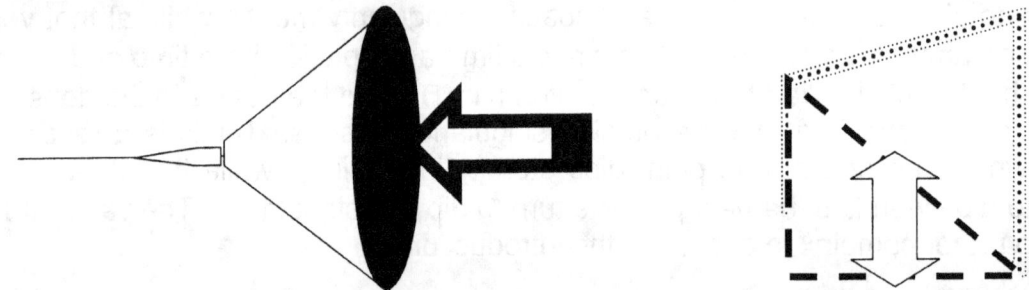

WE HAVE NO POSITION IN THE UNIVERSE EXCEPT FOR ONE MEASELY DISCARDED STAR-CORE. THAT IS THE ONY WE SPACE WE CAN OCCUPY, THEREFORE WE SHOULD NOT SEE WHAT WE WISH TO SEE FROM OUR PERSPECTIVE, WE SHOULD SEE WHAT THE COSMOS TELLS US TO SEE ABOUT IT SELF AND THE RELAVENCIES IT COMPILES IN TIME MATTER AND SPACE.

The cosmic Calendar as I call it, indicates space-time relevancies where time overheated, as it poured heat into space. At first, it was solid, dense, matter that did not receive cooling, therefore it became the jelly liquid state of the neutron, and later the liquid form of flowing heat. Finally, at the point where the Big Bang arrived, heat got its final form, being gas. Through this event the final element formed, the one we and all other elements, and element combinations occupy. Every star in the cosmos is a proton, with its atmosphere (be it gas or liquid or singularity) acting as a neutron to the value of $\Pi^2 \Pi$. The Π holds the dome of the sphere at a relevancy of 21,99 /7 and to the outside is space, already in a square of 10.

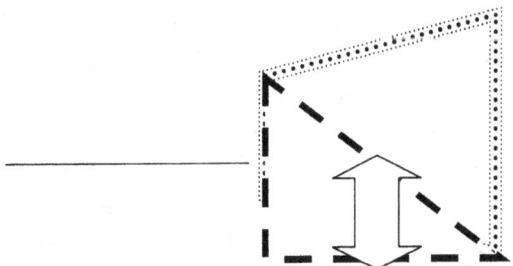

WE MAINTAIN A SPACE-TIME POSITION SOMEWHERE BETWEEN THE FOURT AND THE THIRD DIMENSION, NOT BELONGING TO THE FOURTH AND NOT BELONGING TO THE THIRD DIMENTION EITHER.

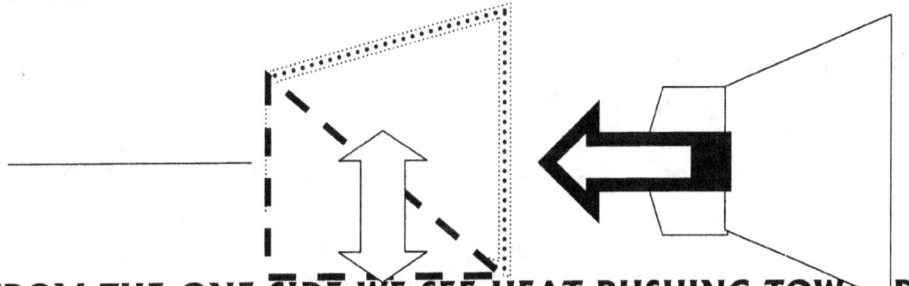

FROM THE ONE SIDE WE SEE HEAT RUSHING TOWARD US AND MOVING HEAT WILL SEEM COLD

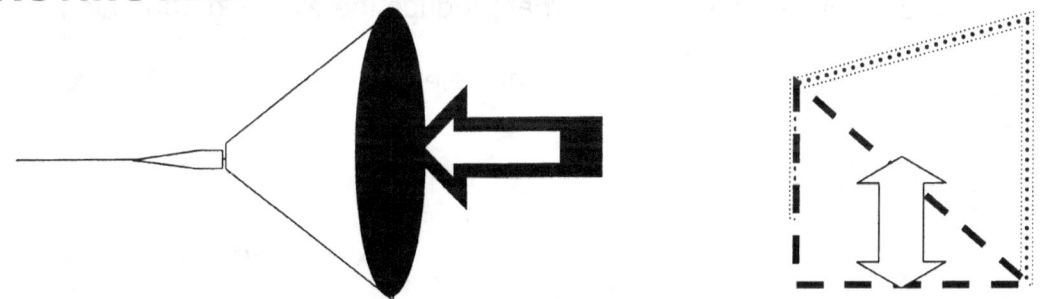

FROM THE OTHER SIDE WE SEE HEAT RUSHING AWAY TO A POINT, MUCH SMALLER AND THE CONJESTION ALLOW US TO BELIEVE MOVING HEAT SEEMS HOT.

There is one more relevancy I think is worth explaining in this introduction, and that is the relevancy of the speed of light $3\Pi^2$. The 3 holds the photon's relation to space, being part of space, where the proton (Π^2) draws the photon 3 as it connects the photon to time in singularity. The photons do not merely fly around because they have nothing better to do. They move in a straight line, to the nearest proton, that connects them to time and because of the liquid density of the heat, the photons can only break free from the wave, by entering the neutron through the electron (3) to the atom at Π^2. The rest of the wave will pass, on its way to the next proton to which the photon connects. In this way, "gravity cannot bend light" but the density of the proton numbers to the space occupied by matter, will influence a position of the wave of photons, disturbing the wave where the proton density is so great it can influence a complete sector of the photon wave. This then will bring about a disturbance in the density of the wave of photons at one given sector, and as the photons are liquid, they will then re-apply equal density throughout the wave. These are all relevancies that space-time holds to space-time, much in the same way as logs are relevancies numbers hold to the basis of 10.

In order to explain events according to the Bible as the Authentic Author reported it, we first have to consider the relevancy time has on space. When two points are 2000 light years apart, even though each one relates to time at that very second, time to one another is 2 000 years in the future. The time travelling at the speed of light will take 2 000 years to reach the other point. Therefore light leaving point A, will relate to a Universe that had the development and growth applying to a Universe that has a space-time development of 2000 years later. In real terms, the explanation is much more complex, but that I leave to the Thesis.

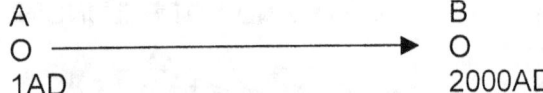

```
A                          B
O ──────────────────────▶  O
1AD                        2000AD
```

By the time the light that departed from point A left point A, space-time had a specific relevancy to development. It leaves A now, but it is 2 000 years to the future of B. It will reach B only in two thousand years. The same goes for B. Light is only a messenger, and if the messenger delivers a letter written 2 000 years ago the news apply to the recipient as if it is breaking that very minute. 2×10^9 years at present. The message however, relate to events 2 000 years to the past of its current position where it is 2 000 years in the future to the events it holds as news happening now. Time is a constant that is true, but in the constant time remains at a relevancy to the heat that produce the events determining time.

Let us place time at a factor of one to the relevancy of heat in whatever form forming time in space.

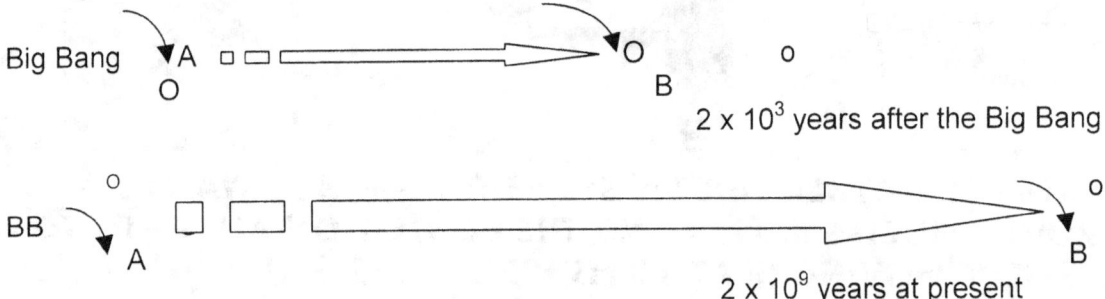

Big Bang A ▢ ▢ ══════▶ O   o
                        B
2×10^3 years after the Big Bang

BB ▢ ▢ ══════════════▶ o
                        B
2×10^9 years at present

If A and B rotated at an even pace as time to the constant of singularity suggest, then, because A is rotating around a much shorter radius, the value of time in A will be pretty eventful since A has a relativity, which comparing much more action to much less space. On the other hand from the position B finds itself in, B will be to A one million years to the future and B will therefore be invertible. Therefore we can see NOT SPACE but TIME TO THE PAST as light, but we cannot see to the future, because light does not travel back in time. Therefore no messenger will bring news from 2 000 years to the future, light can only move with time remaining with time. Hubble said the further objects are away, the more they appear to go faster, and the most furthest point, seem to rush and run away from the rest.

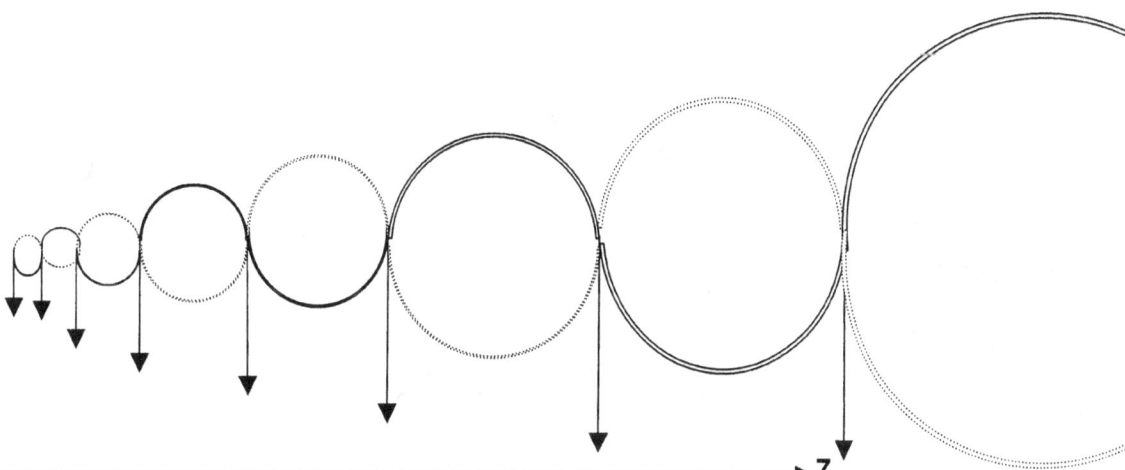

AFTER EVERY PERIOD, (WHAT EVER CONNECTION ONE WHISHES TO APPLY TO THE PERIODE; THAT HAS NO RELEVANCE) THE DURATION OF TIME MUST BE DIFFERENT TO THE PREVIOSE AND THE FOLLOWING TIME. IT IS EITHER THAT, OR THERE IS NO BIG BANG AND NO HUBBLE SHIFT. THE INVERSE SQUARE LAW WOULD NOT PERMIT A CONSTANT TIME DURATION, WHILE THE RADIUS OF SINGULARITY π EXTENDS.

It is all due to relevancy, in time having space as a product of time. Space is not growing at all. It is time becoming quicker in duration of time. If time remained at a constant and the Universe grew to Hubble's claim, it would have taken the Earth but a few seconds to rotate around the sun, seen from our view, that we find the relevancy of the distance and therefore the year rotation circle to be much smaller than we presently have. However, we maintain a constant of one to singularity, the same constant of one that time then had to singularity. Therefore, space had to be smaller, and time according to our view moved much faster. It had a speed relevant to our standards we apply at present, that things exploded, because we cannot even follow the events with our eyes. The truth is the space and the time both are in singularity, therefore the radius back then were the same as at present. However, the relevancy matter had to matter, were much more compact, and instead, they maintained the same positioning alterations in a much slower duration of time. If A and B move at the same speed, with the closer proximity of matter to matter at point A, point A will experience in slow motion, when relating events taking place at point B. From point A looking at B, light will not move because from point A, the light travelling at point B will exceed the speed of light one million times in relation to the speed of light at point A. Therefore, from point A the future at point B is invisible, because from point A, time travels one million times faster than the speed of light at point B. This is the consequence of the Hubble shift, and you cannot have the Hubble shift, without the effect time in duration has on space. Science either has to abandon the Hubble shift, which it cannot, because the evidence is there, it is clear and it cannot be ignored. Therefore, if you apply the shift Hubble saw, you have to adapt that shift to the implication altering with space changes. It is not the space that changes because space cannot grow to more or less; it is the concentration of heat in the space that changes the duration of time. It is because of this that "more gravity will effect time", as Einstein correctly pointed out, because gravity concentrate heat, and heat produce time in motion.

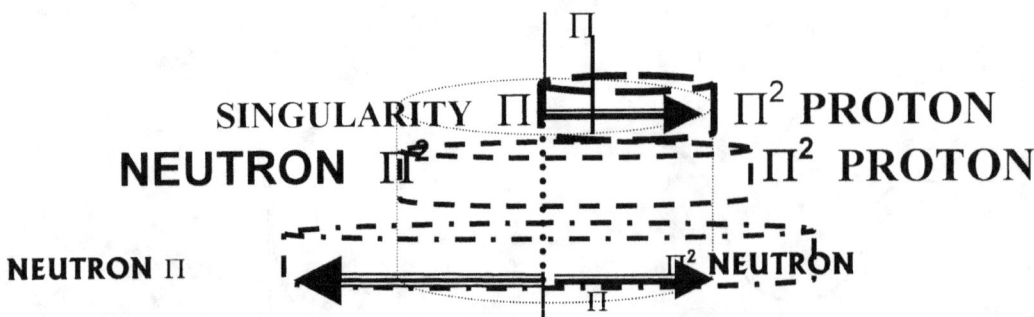

ALL COSMIC OBJECTS ARE A CLUSTER OF PROTONS W CALL SOLLID MATTER, HOLDING A GAS SPHERE WE CALL THE ATMOSPHERE THAT IS A DIMENSION AWAY FROM THE HEAT DIMENSION WE CALL OUTER SPACE.

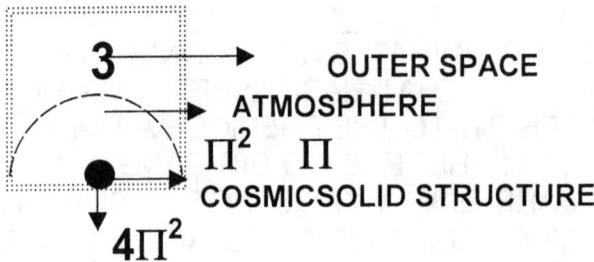

Once again, please I emphasize if your brain is still intact enough to accept this is not for your concern. It is only for the brain washed Newtonians that lost all acceptance through mind control.

All that the following calculations prove is that space holding matter to time where space (10) interacted with matter (7) in the already established Roche limit $(\Pi/2)^2$ to form the point where the Newton time value (Π^2) transmits heat to the double proton $(\Pi^2+\Pi^2)$. Heat in a jelly form separated from heat in a more solid form, forming liquid heat (later to become the gas we see as space) and matter, the third dimensional form of matter. You may feel free to see it in the same manner as the way a farm cream separator spins milk to separate the cream and at the same time, unlike the separator does, form bits of butter and buttermilk. In essence, there is very little difference.

The explaining may be rather excessive to non-Newtonians, but please remember that unlike the Newtonians, you are still blessed with an unblemished brain functioning well enough to accept.

Time and space holds the same value because time is space in singularity and space holds time outside singularity. Time and space is one thing. Time holds the value of singularity outside space therefore it is Π^3.

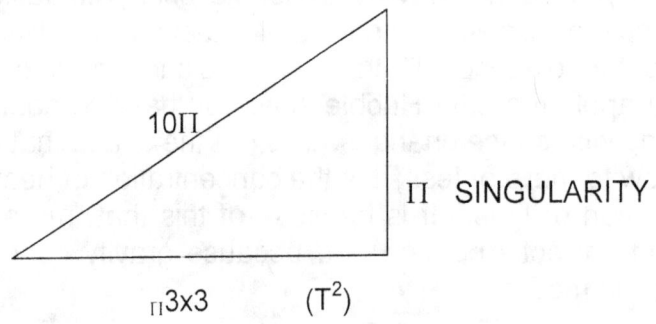

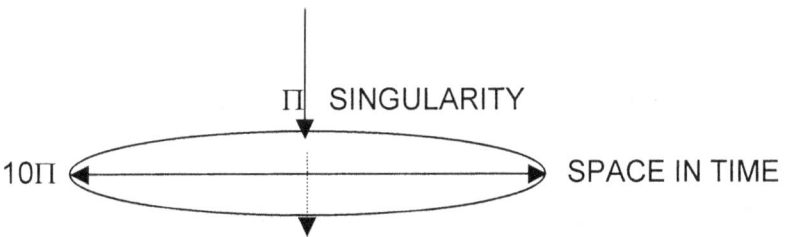

Time has always a value of square (Π) (R^2/T). The linear value time holds to space (R/T) has the value of Π. Therefore space has the value of 9 plus one more value presenting singularity $\Pi=1$ making it 10Π. Since Π^3 holds one identical position in space than space does, it also has to become $10\Pi^3$. This sounds irrational because from our position of time in space and not time as space, singularity as such by itself is irrational. To form the second proton of the double proton $10\Pi^5$ divides into $10\Pi^3$ as heat transforms from space to time.

From the position of time space will always hold Π^3 to Π. This produces time to space (one of the protons) and time in space (the other one of the protons). After developing the double proton, space holds a value of 10Π or $10\Pi^3$, being only in "space to be" 10Π was still in formation. With Π^3 coming as a result of $\Pi^3 \rightarrow (\Pi^2 + \Pi)$ matter formed within the limits of the space-time developed. This lead to a total dimensional possibility of 7. Matter is in a dimension one less than time, therefore in that space the three of Π^3 has the dimensional attachment. Because of this, and the fact that time ($10\Pi^3$) was producing space (10Π), the overall number had to conclude 10Π or $10\Pi^3$, and which ever makes no difference.

$$10\Pi^3 \div 10\Pi = \Pi^2$$

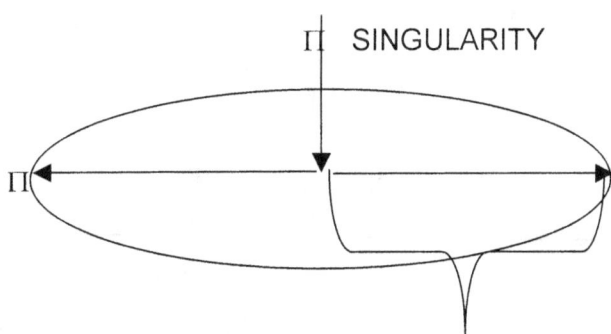

In one dimension space became 10 and in that dimension matter became seven. In order to separate matter (7) and space (10) through time (the spinning of matter) (7) in space (10) and space (10) spinning the matter (7) the following result came about through the application of the Roche principle $(\Pi/2)^2$

Then space mingled with time $10/7$ $(\Pi/2)^2$ $(\Pi^2 + \Pi^2)$ and space separated from time $7/10(\Pi/2)^2(\Pi^2+\Pi^2)$. In this, the result was that matter formed to space allowing the neutron a time value. Time divided into space $(10/7) \div (7/10) = 2,04$
$(10/7) = 1,4285 \div (7/10)\ 0,7 = 2,04$
Space divided into time $(7/10) \div (10/7) = 0,49$
$(7/10)\ 0,7 \div (10/7)\ 1,4285 = 0,49$

Then space multiplied by time $(7/10)\ \text{x}\ (10/7) = 2,04$

(7/10) 0,7 x (10/7) 1,4285 = 2,04

That brings about that the Roche Principle
worked both ways (double)   1. (7/10) 2,04 x $(\Pi/2)^2$ = 5,033
and with multiplication   2.(7/10) x (10/7) = 2,04 $(\Pi/2)^2$ = 5,033

Resulting in the combined value of   5,033 + 5,033 = 10,066

On the other side the other combined value came to   0,49 + 0,49 = 0,98

And the result from this product was

       0,98 x 10,66
=   9,8696
= Π^2

And with this the neutron's time end connected, established the link to either
protons when the proton connected to space.

From this the neutron developed the space value of Π. The process is as follows:

The square of time (Π^2) multiplied by the square of matter (7) became in fact one
part of the double proton. This is better explained by saying that when one proton
(Π^2) delivers heat from space, dimensionally reconstructed. In real terms the
space proton holds the value of $(7)^2$, which is 49. From Pythagoras, with the
interaction of dimensional equalization the factor becomes the opposite of the law
of Pythagoras.

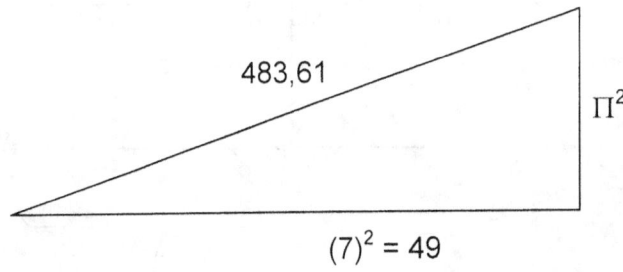

483,61

Π^2

$(7)^2$ = 49

$(7)^2$ = 49 x Π^2 = 483,61. But at this stage matter was in the relevancy of $(\Pi/2)^2$
therefore the factor that $(\Pi/2)^2$ represents, holds the value of 483,61. To get to
the resulting dimensional value of 483,61, the square thereof becomes a factor.

    $\sqrt{483,61}$ = 21,991

As 21,991 is one half of ($\Pi/2$). Π Therefore, must be that value matter holds,
which is 7. When matter divides into space (7 ÷ 21,999) the result from that is Π.
Through this the neutron (which is matter) holds $\Pi^2\Pi$ as a factor.

Therefore, for the first time the reason why Π is 21,999 ÷ 7 = 3,146 becomes clear
through the application of true astro physics, as derived from pure cosmology,
and not a lukewarm deduction of Newton's Earth physics projected to the cosmos

where it does not apply in any event. Through this one must never instantiate dimensions from the equation, otherwise the cosmic calculations becomes Newtonian physics. Again I stress, there is nothing wrong with Newton's physics, as long as the application of it links to life's accomplishments or some derogatory thereof.

Although Π^2 is in a square time takes the square to singularity, being a straight line of 180°. Matter in time holds the half circle value of 180°, where matter is the one link $(7)^2 = \Pi^2$ and time is the other link (Π^2).

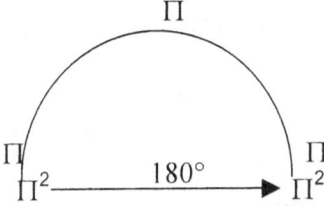

Through that, the value of the triangle in matter, space and time also holds a 180°.

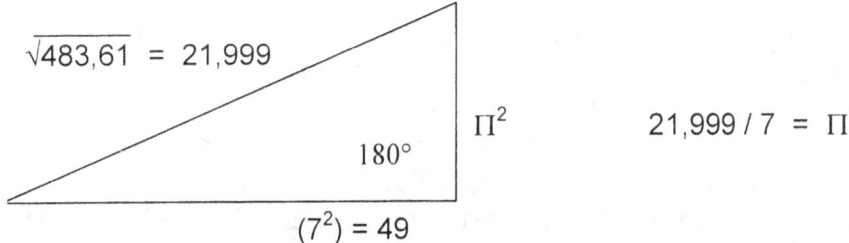

$\sqrt{483{,}61} = 21{,}999$

$21{,}999 / 7 = \Pi$

$(7^2) = 49$

Remember that this is time astro-physics, not merely the projecting of Newtonian concocting directly from Earth to space. You have to use your mind; not just your eyes. Everything, every object in the Universe is an island in a dimension of its own, created to the specific relation it has in time in singularity and space in singularity $\$T = (\Pi^2+\Pi^2)\ (\Pi^2\Pi)\ 3$ and $\$= R^3/T^2 = 1$ where $\$ = R^2/T \times R/T = 1$.

No person can deny the fact that the Earth is a sphere, excluding outer space, where our need to apply entry into the sphere of inclusion $(\Pi^2\Pi)$, and the law to abide by is the four cosmic pillars. You have to abide by the Roche limit where rules allow you entry, or destruction. No Newtonian can deny that. When you cannot deny the fact that the Earth is excluding space as it is including time the rest is beyond denying also. One has to seek the evidence where the evidence is, where one can locate such evidence and above all, read the evidence correctly. The evidence proves the existence of a binary before the Earth came to be. The four inner planets are left over parts, a reminder of a star that uses to be part of the cosmos.

The spinning of the atom around the point of singularity holds the position of singularity and singularity cannot be without spinning in the Universe. Any object in rotation holds one (imaginary then, if you wish) specific position, a point so precise that does not spin, turn or move.

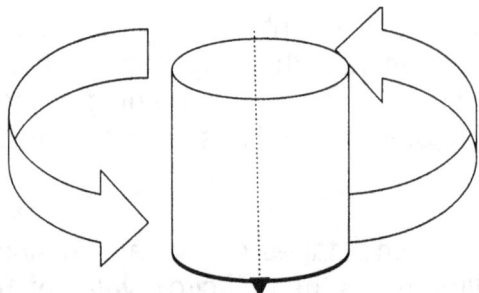

That line that cannot move is singularity. Any point away from that point is space wherever space may be, in whatever form it may hold and whatever density forms that space within that space. The moving of matter gives time (the motion of the space less spinning) its value away from eternity. As much as the rotating line is in the imagination, the line is in eternity and the line is in reality. That understanding of existence of such a line comes from human genius, the way the human mind works in understanding concepts. The line you shall never see feel or touch, BUT ONLY KNOW that it is there because your intelligence permits you the insight to form the concept about the line being there.

THIS FOLLOWING BIT OF INFORMATION IS ALL THE RELIGION I SHALL EVER SHARE WITH THE WORLD AT LARGE

The line holding time to eternity is as much present as the space as the space that is not there. The line is as much understanding as it is accepting through being human. That line that is never there and is forever there is the same as the presence of GOD, which is never there but forever there **FROM ETERNITY TO ETURNITY THROUGH OUT ALL ETERNITIES (EVERY ROTATING ATOM, PARTICLE IN SPACE). I CHALLENGE ANY ATHEIST TO PROVE THAT, THAT ETERNITY IS NOT THERE.**

The Bible says God is in eternity, from eternity to eternity, through out all eternity, and yet again, the Bible proves the to be the best book ever to be written on science that there ever was.

WHAT THE BIBLE SAYS IS THAT IF THAT LINE IS THERE, THAT LINE PROVES THAT GOD IS THERE, IN A LINE ETERNALLY PRESENT AND AT THE SAME TIME NEVER PRESENT TO OUR KNOWLEDGE. HUMANS HAVE THE CHOICE OF BEING ANIMAL AND NEVER UNDERSTAND THAT CONCEPT OR BEING HUMAN AND UNDERSTAND THAT CONCEPT. YOU CAN BE HUMAN IN BELIEVING THROUGH INTELLIGENCE, OR YOU CAN BE ANIMAL, BEING AN ATHEIST AS ALL ANIMALS ARE, WITHOUT THE BRAIN POWER TO UNDERSTAND.

I CHALLENGE ALL: PROVE ME WRONG!

That is all my religion I can ever share, because from that point my religion becomes far to complex to ever share with any one outside my faith. Science must realize by disproving religion they are only proving their own lack of understanding (therefore being animal) of true science.

All atoms are spinning; therefore, the line of singularity must be present. If all around that line is spinning, that line too must be spinning. The rotating value of any object spinning is Π, because Π puts space in dimensions of the square. Anything in a sphere or in a rotation holds the value of Π. That line therefore is Π and any point away from that line is also Π.

The point of Π comes from within every atom holding Π and that value of Π extends, while remaining one, in different rotating velocities. Each spin in velocity stands apart from any other point in spin velocity. Therefore the group of atoms forming that cluster of protons will also spin, at different rates in accordance with the value of Π as much as in density as in mass.

It is here, at this point, where proof comes in about every orbiting matter holding an influence on the cosmos as an atom. Every star holds singularity as much as the protons within it holds singularity and reflecting the combining effort of singularity and therefore it is spinning. The group effort of the combined density provide density provide the star its atom status, by giving it a spin value, A cloud of hydrogen do not spin, therefore it is not a combined atom, but is merely a group of atoms spinning in individual time alone. That means every spinning object is a sphere (a Black hole is spinning even by not spinning it runs so close to singularity, it cannot spin any longer to our concept. All we can see is the in going matter that picks up momentum as it will eventually exceed the speed of light just before entering the non existing Black hole atmosphere.)

To calculate Π follows the following method:

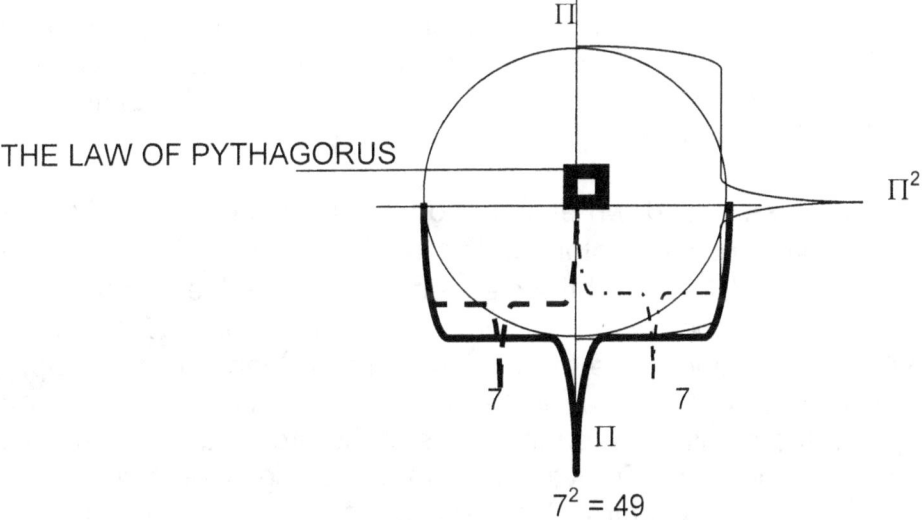

THE LAW OF PYTHAGORUS

$7^2 = 49$

The singularity in the eternal gives time the value of Π^2 while the matter that is spinning gives the time occupied by heat the value of 7, and that means the square of 7 is 49. To get the value of Π to space in the occupied space, it will be the combination of time in singularity and time in occupied heat that will position the value of Π to space occupied.

$\Pi^2 \times 49 = 483.61$.
Since time in singularity is in a square with time occupied, (the double proton) the law of Pythagoras plays its part in dimension equilibrium

$\sqrt{483.61} = 21.991$
The triangle forming the fourth or electron dimension therefore holds 7 in relation to 21.991 and that equals the value of Π.

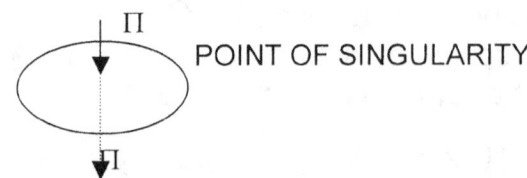

POINT OF SINGULARITY

THE POINT OF SINGULARITY IS WHERE THE RADIUS HAS NO VALUE, BUT THE CIRCLE REMAINS BECAUSE THE CIRCLE IS THERE.

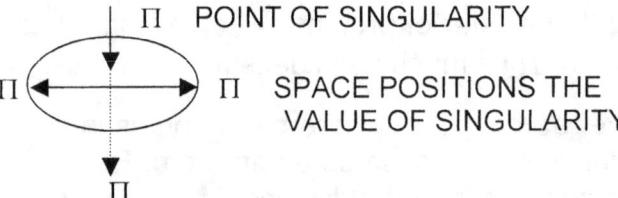

Π   POINT OF SINGULARITY

Π   SPACE POSITIONS THE
    VALUE OF SINGULARITY

THE SPACE THAT SINGULARITY TAKES ARE THEREFORE Π TO BOTH ENDS.

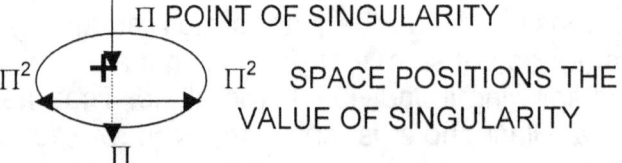

Π POINT OF SINGULARITY

Π^2 SPACE POSITIONS THE
      VALUE OF SINGULARITY

Looking at the Universe, the Universe is within every atom. Every atom holds the Universe to the point of singularity. From the point of singularity, every proton demands space. With the demanding of space, the atom controls space. That would be the proton. The proton is in total dominating control of space that forms the neutron the neutron holds space from the point of the proton Π^2 to a point of the electron 3. The proton forms time (T^2) and the neutron forms space (R^3). There is no large unified all-inclusive Universe out there that will be either filled with matter, or not filled with matter.

The centre of the Universe runs down each proton in the proton singularity. That is the centre of the Universe. The proton holds a double value of ($\Pi^2 + \Pi^2$), the neutron holds the value of ($\Pi^2 \Pi$) and the electron holds the value of three. This combination forms the relevancy of the Universe. $T = (\Pi^2 + \Pi^2)(\Pi^2 \Pi)\ 3 = 1836$. The test in proving this relation is the calculation there of. Many sources supply different readings. The accuracy of the results is beyond question, because there is so slight varying in the result between the mass of the proton and the mass of the neutron, that it is insignificant. The variation of the readings has to come from another factor that influence the control study results. The only flexing of the result therefore must come from a flexing of the heat supply during that precise time of calculating. That is the only variation that may influence the test results. I do not wish to elaborate on that point at this stage, other than to say it is a relevancy, and as with all relevancies the result may be different according to variations in the relevancy influenced by heat occupied and unoccupied changing the dynamics at that precise time.

I have also stated that the Universe has a precise centre running down the centre of every proton, proton cluster and individual atom. That centre point holds space occupied to a dimensionless point in singularity.

THE TIME THAT THE SPACE WILL REQUIRE IS ($\Pi^2 + \Pi^2$) TO BOTH SIDES. THAT ROLE THE PROTON TAKES. SINCE NOTHING IN THE UNIVERSE CAN

BE IN TWO POSITIONS AT THE SAME INSTANT, TIME COMES FROM THE PROTON ALTERNATING POSITIONS TO SPACE AND TO SINGULARITY.

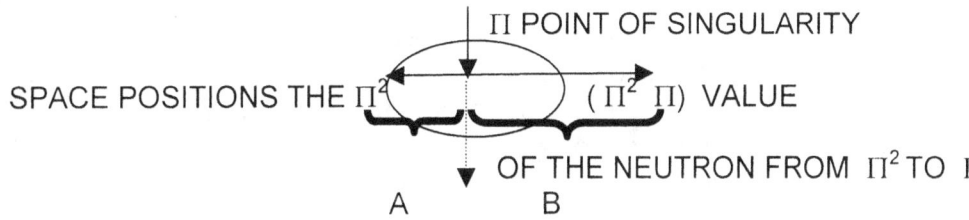

Π POINT OF SINGULARITY

SPACE POSITIONS THE Π^2 (Π^2 Π) VALUE

OF THE NEUTRON FROM Π^2 TO I

A   B

(A) SPACE THE PROTON DEMAND
(B) SPACE THE PROTON CONTROLE.

Holding Time at a very precise measure is the labour of one proton Π^2. While the other proton remains in contact with singularity at a value for that eternity of Π^3. While the one proton holds the time in eternity Π^3, the other holds equal value outside singularity at the time demand (proton Π^2) in the space control (neutron Π^2 Π) to the value of heat unoccupied in space (3)

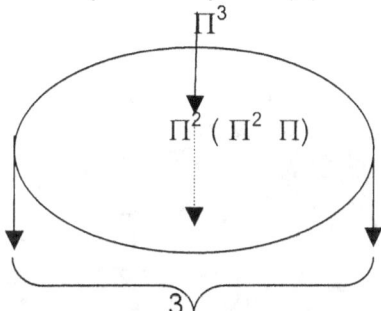

Π^3

Π^2 (Π^2 Π)

3

In this there are still a natural balance forming due to the interaction that the proton Π^2 and the neutron Π^2 Π hold to singularity. This is the point where the Roche limit holds the space dividing in order to prevent any destruction of singularity to any of the structures.

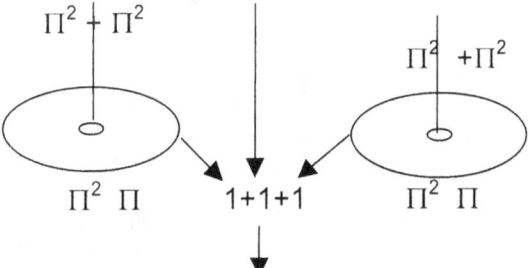

$\Pi^2 + \Pi^2$ $\Pi^2 + \Pi^2$

Π^2 Π 1+1+1 Π^2 Π

SINGULARITY EFFECTING HEAT TO THE VALUE OF 3

The unoccupied space-time is not distant from singularity because it too holds a position to the cardinal singularity, therefore the heat allow the spin of two atoms to maintain heat control while spinning in opposing directions. The cluster of atoms remain the same in domination and control to space-time, be it in a cluster of atoms the size of an atom or a cluster of atoms the size of stars.

Where two cosmic atoms control space in a joint occupation of space, the space domination will have to set in borders to defend each one in singularity. It is the singularity within each object that gives that object validity to exist in the cosmos. It gives the object independence as it makes that object the Universe. From that point holding singularity comes to space that controls of the atom. The atom is no longer in overheating but is controlling the heat by accumulating heat in order to grow. It is the atom that grows and not the Universe that is growing. The atom no longer is in the heat expansion, which brings devastation but that control which the atom influence put the limit of the heat under its control. The atom is now the epitome of gravity by expansion control as well as contracting control. The Universe became the atom and the atom's diversity became the Universe. Every star in every galactica and every galactica becomes an atom through the discipline that the atom brings to space-time.

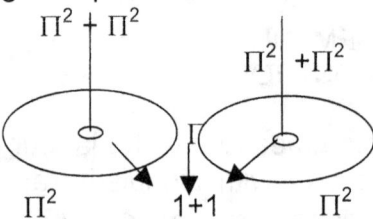

SINGULARITY EFFECTING HEAT TO THE VALUE OF Π

The atom forms the Universe as it shapes the Universe because the atom is the epitome of the Coanda principle. The atom spins and the spin provides an accumulative that drives stars, which in their turn give an accumulative spin that, drives galactica and in all that there is a centre governing singularity controlling every spin just as the simple top showed us.

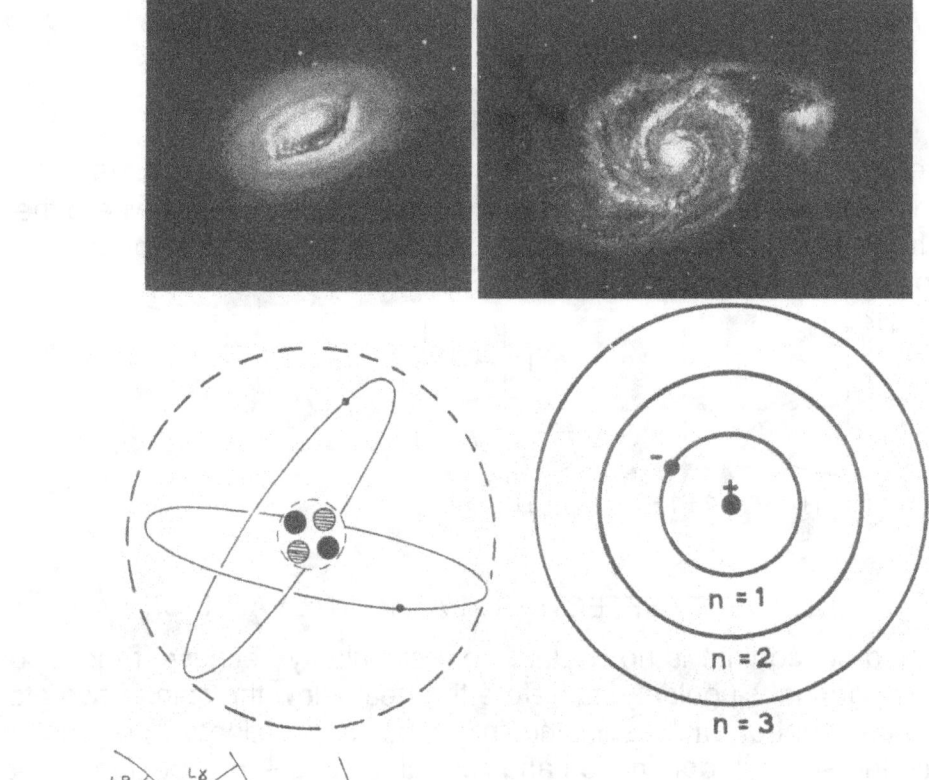

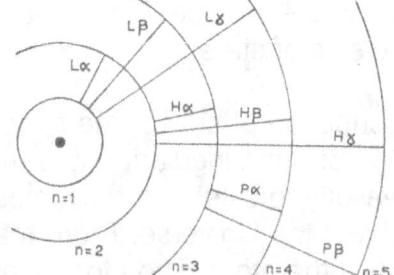

The control the atom has in regulating heat is plainly godly simple as it is Godly genius. The control the atom has on heat by expanding everything outside while accumulating heat, as a time retardant inside is that which creates the Universe just as much as it is that which produce the Universe. By producing space in expanding when heating the atom not only control what belongs to the atom but also that what is in the control of the atom and what is not in the control of the atom since the entirety of the Universe is the world of the atom.

By the time the Big bang realised into a 3 D Universe gravity was C^2. That gave the speed of light a relevancy of 1 and space a relevancy of C^3. With the neutron being space and the proton captured in singularity Π going to Π^0 the entire Universe broke into 3D by the margin of C and in C the atom was captured in space in timer and in gravity. That explains why the Universe was the size of a neutron. The atom is in accordance with the Coanda principle moving at c with the electron excluding all else from the space $a^3 = C^3$ in relation to the centre C which the motion of the speed of light confirms at c keeping the gravity that applied at C^2. Space is created from one position to another position and the duration it takes to complete the distance is time.

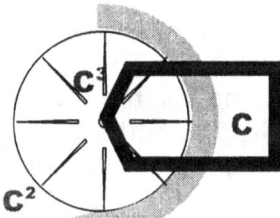

That the Coanda effect proves as water flows past a round object and the contact the flowing water makes diverts from the normal route the Earth gravity will enforce. Gravity is the very

The C^2 value Einstein came up with has no bearing on the speed of light, but is equal to the gravity displacement the neutron has when entering the atom. It shows the stopping of the flow of space –time and the atoms overheating in the process when the two atoms touch and stop one another's space-time flow bringing about cooling. The C^2 is the gravity at the point of the atom presented because of the neutron centralising the gravity C^2 at the rate of the electron C^2. What Einstein saw was gravity at the point where gravity equals the speed of light, but it has nothing to do with the actual flow of light going square. It is the flow of space-time. Going nuclear takes the atom back to what happened at the start of the Big Bang.

same but it is the recalling of the space by creating motion in the space. By duplicating space through motion in relation to singularity the flowing water diverts from the normal route. By recalling the space it is also reducing the space because it is counter acting the time expansion provides. That then is clarifying the reason why gravity will always on the limit be stronger than light. At a point it slows the time component down to such extend the space reduces faster in that time than what light can produce motion. This is why the Coanda effect applies. $a^3 = T^2 k$ then $k^3 = k^2 k$ and this is showing that the space k^3 is equal = to the motion $k^2 k$ of the space k^3 seen form one specific point. As the example of the atom in motion through space showed motion changes relevancies. In that it changes the relevancy of space-time affected by the altering of the motion and not applying relevancy of singularity. The Coanda example shows that motion establishes singularity, dominating space-time, controlling motion and the direction thereof. That means the motion establishes the space-time and the position as much as the direction of the flow of space-time. Motion creates space-time as much as space-time is supplying motion to form space. The only way to enable that to become a reality is that motion creates space as much as space follows the direction of motion. That is what Kepler said when Kepler said the space is equal to the motion thereof $a^3 = T^2 k$. There is no solid a^3 Universe but all interrupted by positional changes that recreate the space in the and according to the new direction singularity will create as singularity allows space to flow by motion by fragmenting space into time sectors.

This became possible as the Coanda effect brought the speed of light as gravity (**k**), which brought the motion of the Universe in contraction to the speed of light T^2 that made the Universe be the speed of light a^3, which is exactly and specifically what the **GUT** theory proves.

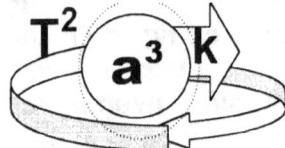

Space is the forming of motion **k** because space is the liberating of time from singularity where the hottest part of space will find a way to move away from the rest of the cosmos and that motion forms the time component **k**.

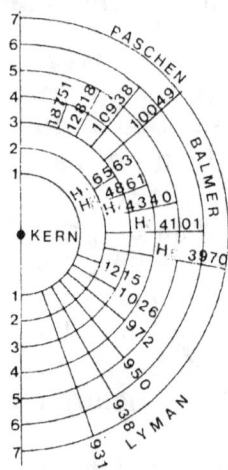

The Lyman series show precisely how heat excites the electron orbit and the heat increase brings about space increase. Heat is equivalent to expanding into space. We find the same degree of excitement in atoms when they are charged with more, heat, however the charging becomes a widening of the orbit circle of the electron. The duplication that increase amounts to more space within the set time and the duplication extends the space the material fills. In the case of the top the top wants to lift to a higher band of spin in a similar manner as the electron does when the electron is exited with more heat. The principle being the Coanda effect remains the same as the spinning motion excels the gravity and the gravity becomes stronger and more aggressive towards the governing gravity of the Earth.

The relevancy between the two cosmic atoms reads as follows:

$T =(\Pi^2 + \Pi^2) (\Pi^2 \Pi)$ 1 $(\Pi \Pi^2) (\Pi^2 + \Pi^2)$ THE 3 NO LONGER HOLDS POSITION

$=(\Pi^2 + \Pi^2) (\Pi^2)$ $\Pi/2x$ 1x $\Pi/2 (\Pi^2) (\Pi^2 + \Pi^2)$ INSTEAD THE Π SUBSTETUTE

THE BORDER
$= (\Pi^2 + \Pi^2)$ $(\Pi/2)^2 x$ 1x $(\Pi/2)^2 (\Pi^2 + \Pi^2)$ ALL BARRIERS COME INTO EFFECT CLOSING RANK ON THE BORDER THAT MAINTAIN SINGULARITY.

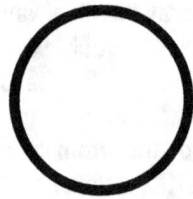

When the one cosmic atom cannot defend the singularity of self-conservation, the other object will destroy the singularity by establishing overheating with in the singularity of the minor atom. I think NASA refers to this as "blowing bubbles or blowing heat " but do not quote me on that. I just found it amusing at the time that the best brains in the world would come up with nonsense like that.

To the one object, anything distant from its proton cluster is space, space in whatever form. By destroying the singularity, the space becomes heat either under its influence or under its control. On the one side, space holds the value of Π and on the other side it also holds a border of Π. The time however changes to defend the singularity therefore the neutron time square then holds position where space normally holds position.

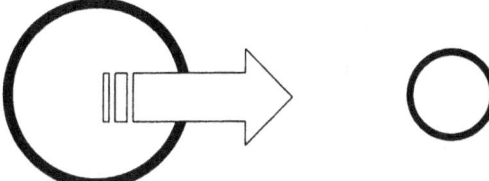

SEEN FROM BINARY MAJOR EVERYTHING STANDING AWAY FROM ITS PROTON CLUSTER IS 3, THEREFORE EVERYTHING IS 3 Π^2. The 3 are space and the Π^2 holds value to the space. This means that when the space does not relate to the proton cluster holding the Π^2, the Π^2 also becomes space at a value of 3^2.

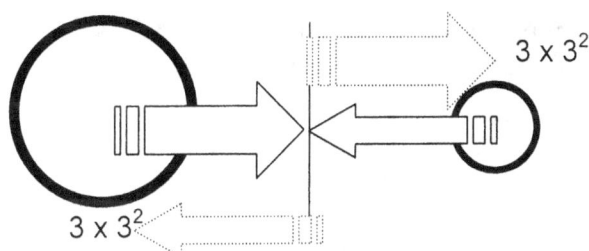

The value of $3\Pi^2$ holds the seed of light in relation to singularity, while the cluster of photos form space in three dimension at 3^3. Where Π links to Π^2 the value of displacement (moving through space) is well above the speed of the photon and the photon connecting to the proton is at a relative displacement of 29.6. The wave holds the relative displacing value at 27. This then bring about that the wave will always shine red in approaching any object and will always loose luminosity. With this the photon always holds two positions to space-time one being in direct contact with the proton at $3\Pi^2$ and in direct linking to the wave of photons at 33^2. By the way, the consequence resulting from this is that the electron will produce two signals under close inspection

LINE HAVE DUAL OR MUTUAL SINGULARITY

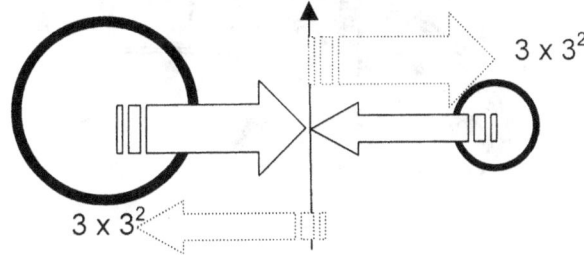

The mutual line of singularity $(\Pi/2)^2$ holds a position referring to each objects individual line of singularity in the value of $(\Pi^2 + \Pi^2)$.

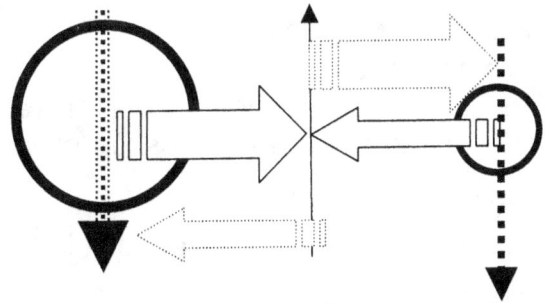

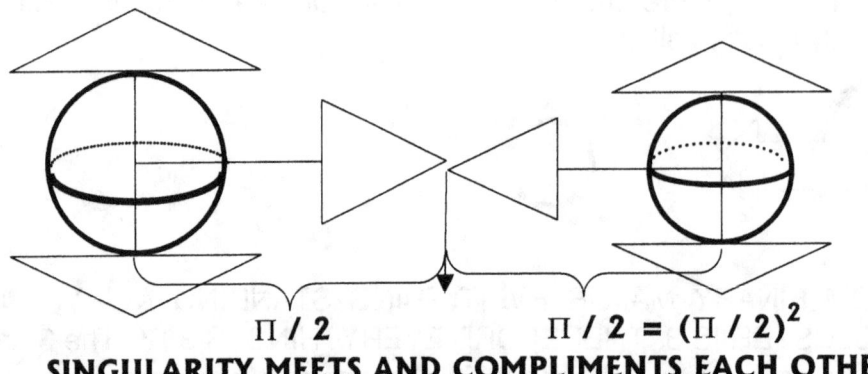

$$\Pi / 2 \qquad \Pi / 2 = (\Pi / 2)^2$$

SINGULARITY MEETS AND COMPLIMENTS EACH OTHER.

Coming back to the calculation of Π, the Titius bode law flows from the figure of Π

THE LINE OF DUAL OR MUTUAL SINGULARITY WHILE EACH ATOM PROTECTS ITS INDIVIDUAL LINE OF SINGULARITY.

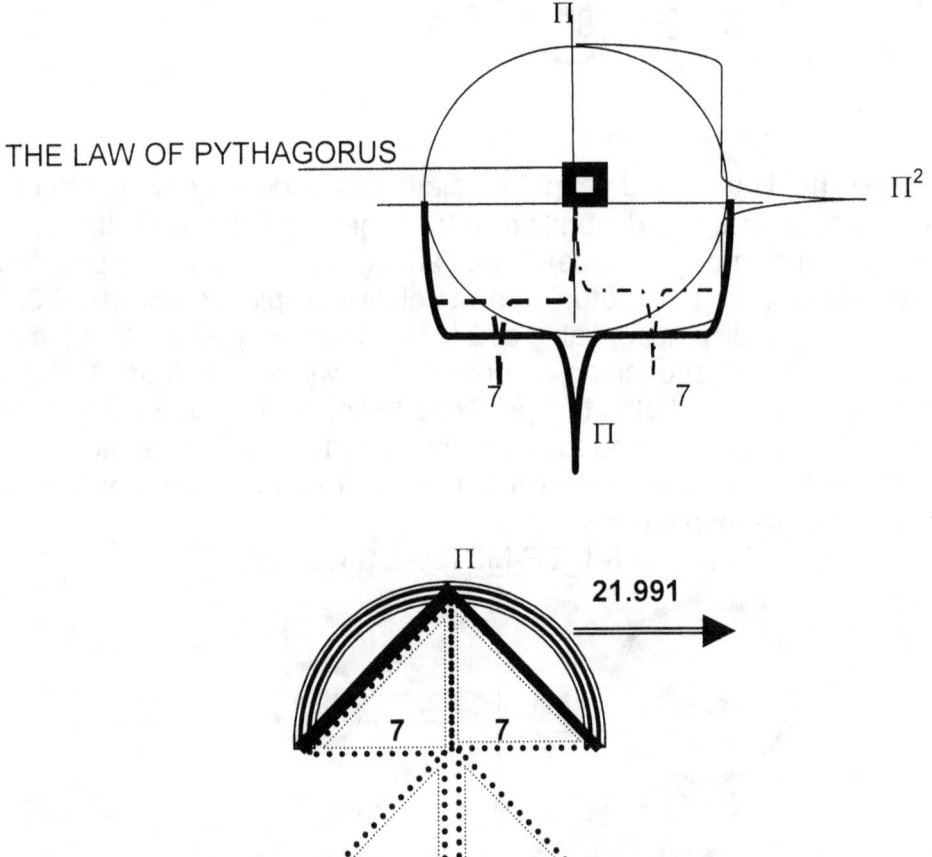

The combined value of 21.991 holding the circle is a combination of $10 + 10 + 1 + \Pi^2 / 10$. A more specific explanation about this aspect is elsewhere. At this point, I wish to (once again) confirm the value of space being 10Π, and how that proves the effect the Titius Bode Law holds on effects science merely brushes off as gravity pulling or pushing or even shoving.

At one point in the vertical, the Earth does not rotate at all and at that point is the value of Π in singularity.

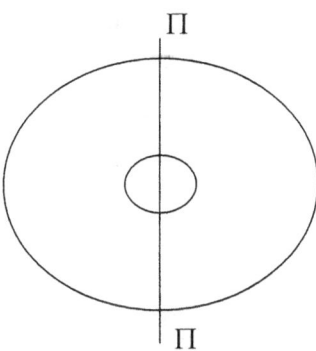

In the calculation of the specific value of, Π I showed that Π takes on the value of seven (7) in the form of matter.

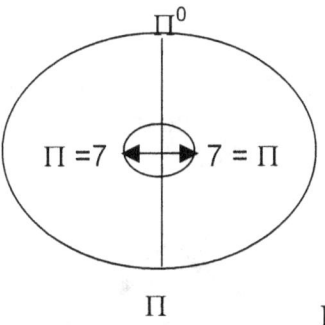

I also showed that 21.991 relating to space replace the value of space in the number of 10

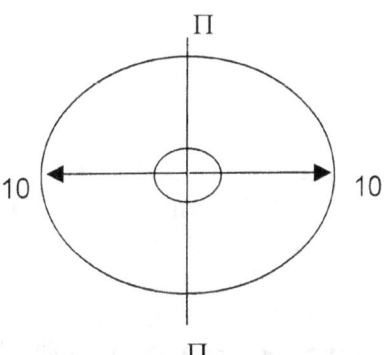

This means that the value of space in accordance with the Titius Bode Law effecting the Earth, the value of space-time is $\Pi = 7 / 10$, at the Arctic and at the equator the value of space-time is $10 / 7 = \Pi$.

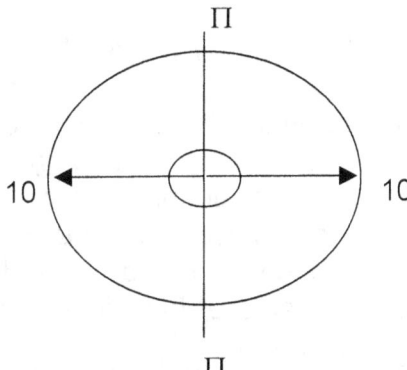

At this point, I wish to raise an illusion that the Earth is "bulging at the equator because of the slinging effect of "gravity" by means of a centrifugal force flinging matter away because of the spin. However that is not the main reason why I show this because at one point later in the book I shall show how the Titius Bode Law brings about the effects of La Nina and also El Nino, the weather phenomena science so desperately tries to connect to "GLOBAL WARMING".

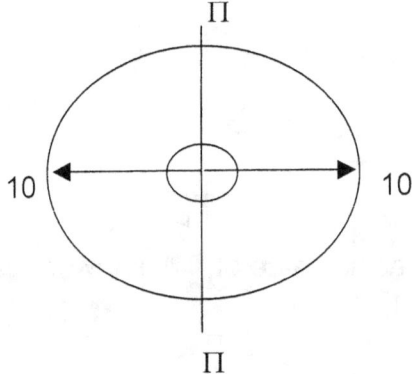

At the Arctic the space-time is at a minimum of 7/10 because of the position, it holds so close to singularity, and at the equator the point space holds is at a maximum, because of the same reason.

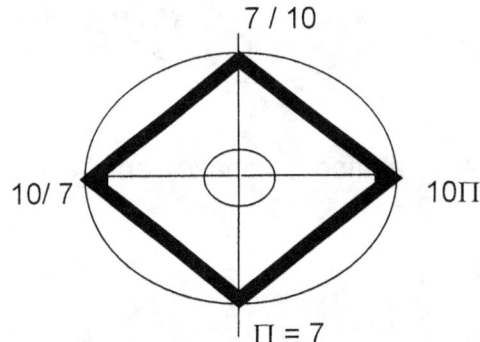

Therefore the Earth will be at a larger diameter horizontally (10/7 = 1.42) than at the vertical diameter (7+7)/10 = 1.4. With the atmosphere (space-time) at the Arctic regions at a value of 7 to 10 and at the equator at a value of 10 to 7 we can see where heat accumulates through the natural comic dispensation and location of heat in space-time. The equator holds a much higher relation to heat unoccupied than does the Arctic region and this will allow aircraft to hold a much higher speed in relation to actual time taken to fly at the Arctic than at the equator. Flying over the Arctic will be much quicker than flying over the equator, although actual speeds may be the same. It is all due to the relevancy of space-time.

The Titius bode Law holding 7/10 and 10 / 7 extends its influence much wider, but the influence of extending as far as the Earth is concerned is s drop in a bucket to the effect it holds on us through what the Sun provide in the manner the Sun's influences the Earth where space-time heating goes. Changes in the Sun influence and the space-time in the atmosphere of the earth which in turn heats the Earth core and the heat accumulation that the Earth uses as a driving fuel which the Earth reserves to influence the motion is cardinal to changes occurring. As much as the Earth claims its value of the Titius Bode Law, the Sun takes its share to a much higher proportion and similarly, the moon also takes its claim on space-time by a smaller margin. This brings about (partly) why the Earth / Sun has an oval orbiting pattern and the moon forms high and low tides. To explain the Titius Bode principle in detail one must once again return to singularity and understanding that is as simple as it is totally uncomprehending.

There is singularity at Π^3 in singularity at Π^0, which forms matter we named the proton Π^2 and that is another name for spin through exciting motion. Connecting eternity to the proton Π^2 is another proton connecting time to infinity as the one proton alternate positions and with the alternating comes infinity breaking the eternity into fragments. The aspect just mentioned is only the time aspect because this relates to the neutron that holds space to time. Getting into the thick of this will hold no purpose, but in THE THESIS, I elaborate how the neutron has no mass because it is space, confirming the matter of space to time.

The one proton takes time to eternity, by joining space in singularity, reducing space that holds heat in unoccupied matter. This dipping into singularity reduces space by one dimension to zero dimensions.

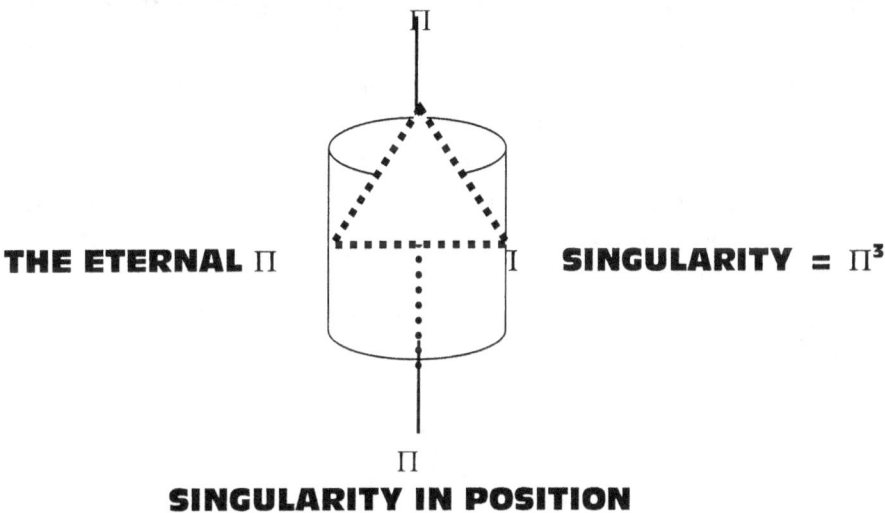

THE ETERNAL Π **SINGULARITY** $= \Pi^3$

SINGULARITY IN POSITION

This procession leads the dimensional reducing of space to the value of time in the square. The proton dipping into singularity effectively destroy space, and that leads all the way up to the Hubble shift, the Big Bang and all there is because the atom is the Universe, remember!

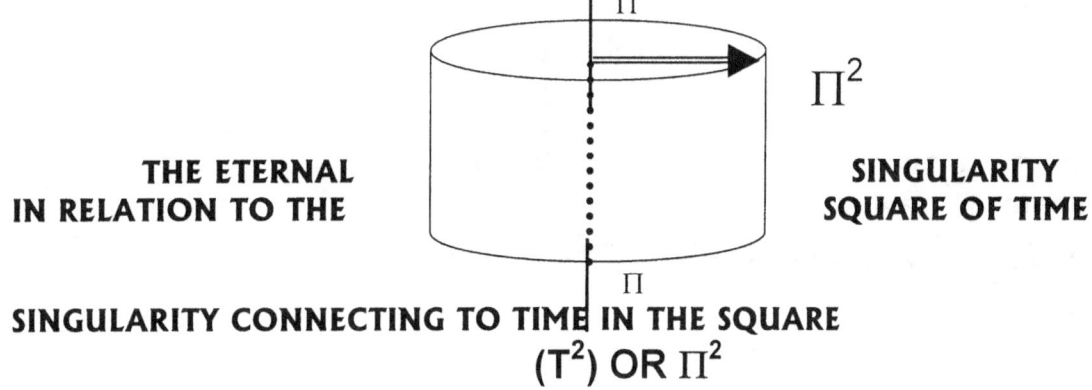

THE ETERNAL Π^2 **SINGULARITY**
IN RELATION TO THE   **SQUARE OF TIME**

SINGULARITY CONNECTING TO TIME IN THE SQUARE
$$(T^2) \text{ OR } \Pi^2$$

To the side of that space which is less, it is the proton Π^2 waiting on its turn to dip into singularity by demising space completely. While waiting its turn, it keeps busy by destroying space to time, demolishing one dimension of space (the neutron), and that leads to space being at that point in a square of time. While this process is going on at the bottom of the ladder, the top also remains busy with the same process in a dimension more than the square of time.

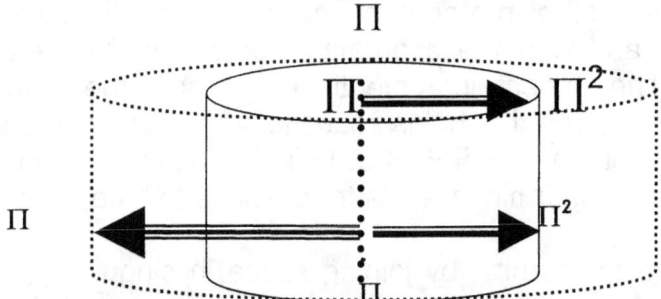

THE ETERNAL SINGULARITY IN RELATION TO
THE SQUARE OF TIME IN THE NEUTRON ALLIGHNMENT

Above the proton dimension destruction, the neutron converts heat from the fourth dimension of heat in space where it reclaims one dimension more than the electron dimension has. To the one end of the neutron, space holds a dimensional value of 3, and to the other side of the neutron the dimensional value becomes Π. The process is much more complicated because there is still the effect of 10 converting to Π^2 and 10Π converting to $\Pi\Pi^2$. I shall not go into all that, because I only need explaining the one side of dimensional alteration or "gravity" to understand the Titius Bode Law. The main consideration is that all over, there is one dimension at loss.

$\Pi^2 \Rightarrow \Pi^3$ while

$\Pi\Pi^2 \Rightarrow \Pi^2$ and

$3 \Rightarrow \Pi\Pi^2$.

This process is more complicated because we only use one relative in heat unoccupied space, but there is still others. I shall shortly explain that very briefly. To summarize there is:

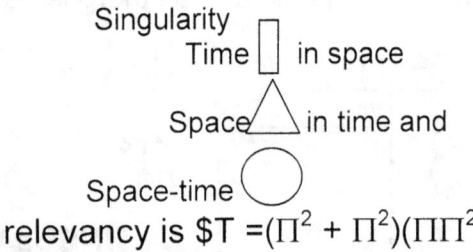

The full relevancy is $\$T = (\Pi^2 + \Pi^2)(\Pi\Pi^2)3$
Where $_1\Pi^2$ relate to Π^3

The full relevancy is $\$T = (\Pi^2 + \Pi^2)(\Pi\Pi^2)3$
Where $_1\Pi^2$ relate to Π^3

$_2\Pi^2$ relate to $_1\Pi^2$

$_{Neutron}\Pi^2$ relate to $_2\Pi^2$

$_{Neutron}\Pi$ relate to $_{Neutron}\Pi^2$ and

$_{Neutron}\Pi$ relate to 3

The one dimension always stands related to the one side as space forming matter 7 / 10 (even heat in space) and to the other side as matter in space forming 10 /7
In reality it is always space to time 10_Π / 7_Π or it is time to space 10_Π/ 7_Π.This

implication brings about that the Universe cannot relate to any point of time duplicating space at two points during the same time. Matter can never be in two places at once. This is very important when I bring proof about the forming of the solar system.

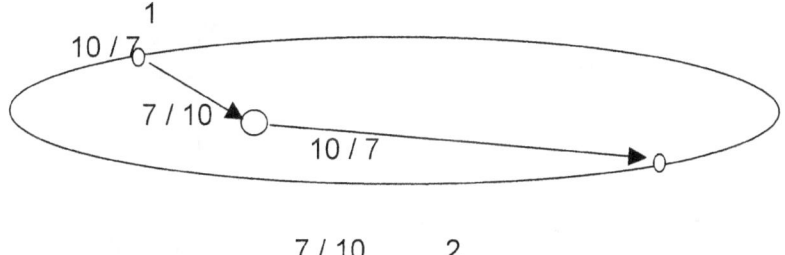

Through this, the equal relevancy of space 2 Π holds the same value as time Π^2. If time is at a value of Π^2 (10 /7) then space will be Π holding Π, at a relativity of space forming at the same token a value of ((7+7) / 10), which converts to 2Π.

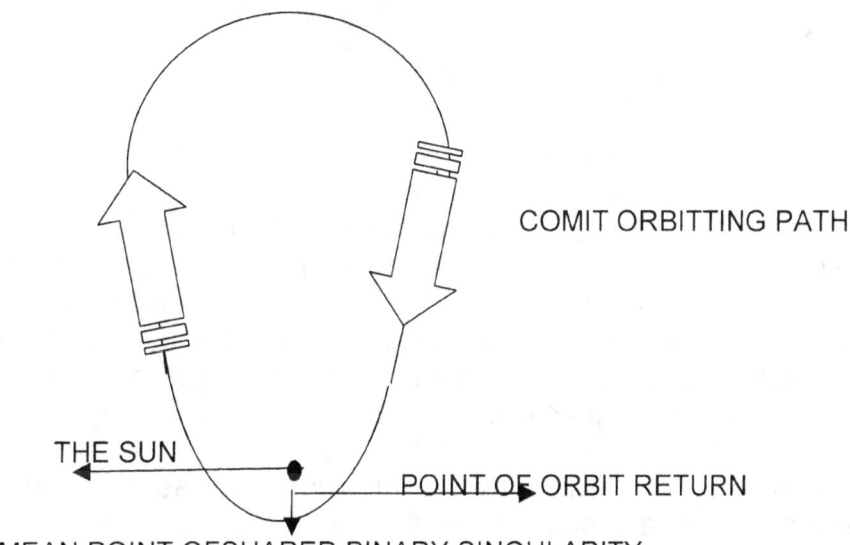

It stands to reason that where there was a binary point holding mutual singularity, the excessive fluctuating of space to time 10/7 and time to space 7/10, will seem more apparent than usual because the space growth and time duration will be more than the Hubble shift will normally allow. There is just a lot more heat to create space in the form of light and unseen heat.

Again we have the principle of Kepler showing what gravity is.

Looking at the entire picture we see the motion of the comet T^2 with **k** relative to singularity k^0 that produces the space a^3 in as much as space moves $a^3 = T^2 k$.

The picture coming from where singularity focuses on space –time in the formula we have $k^0 = a^3 / T^2 k$, where it says that singularity control space-time.

From the sun we have the comet extending the gravity influence of the sun by the measure of **k** where $k = a^3 / T^2$

Finally from the comet we find the comet attaching to the sun by the negative contraction of gravity $k^{-1} = T^2 /a^3$. The Newtonian concept of gravity does not apply only because it does not exist in any form anywhere. It is space flowing towards and space fighting for independence that establish gravity.

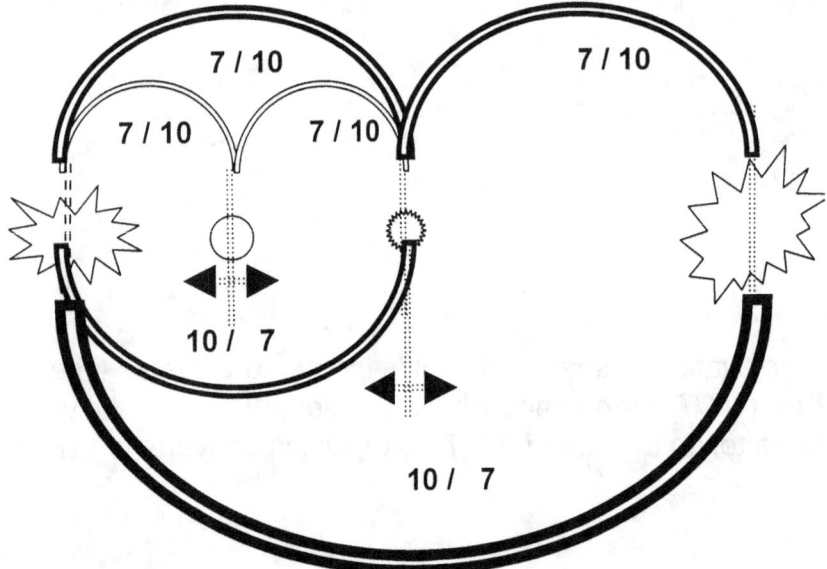

The normal application of the Titius Bode Law provides that a orbital fluctuating of .02 will remain between the perspective the pivotal cosmic atom provide and the orbiting structure. This does not apply to comets because the comet lost its individual singularity and holds on to the original binary-shared point of singularity as if it was its own. The expanding heat turned to space makes the positioning of the original binary hold a small growth, when in relation to the sunspace growth there is virtually no comparison. Heat turning to space has a way of denying us humans to grasp a true perspective.

That brings us back to the dynamics of heat in as far as the speed of light is relevant to the electron and space-time.

Dealing with the speed of light as a cosmic factor with the aim of connecting that to time, has many implications that may well lead to confusion. The photon has no connection to singularity, neither in the wave nor in the proton, but only connects to singularity through single photons within the wave that connects to distant protons outside the wave.

The line of singularity running through the centre of the line of singularity will bring across a double message, one in front of the wave, and one behind the wave. The result from this may lead to the perception that that specific electron holds two positions in space-time, but this cannot be. If such a perception is there, then the mathematical calculation of the specific position of the electron will prove of only one position. This is because the speed of light links to time, but is not time, whereas mathematics do not confuse space and time as the human perception does.

I shall try to explain this in a manner that we see a burning charcoal fire. The smoke is heat, retained by oxygen, the flames is liquid heat, retained by nitrogen and the glowing charcoal is heat transmitted by the coal or carbon. To our perception, it is the very same thing. It is heat, holding space relevant to time where each element holds the relevancy, therefore determines the time of the space and that gives us the perception of three different objects. The truth is that it is the same thing because the fire may lower or increase its intensity but ..... is the same fire.

The value of Π, the value of 3 and the value of 3^3 give us humans an impression of time in different positions, however, it is the very link to singularity and the eccentricity space holds to time in each case that stuns our wits. IT IS OUR WITS BEING STUNNED AND NOT TIME OCCUPYING SPACE IN TWO DIFFERENT POSTIONS AT ONCE.

The Universe is also as much in every atom, as it is in every atom cluster, where that forming a group density where the group density controls the space they occupy, the space they claim through control and the space, which they influence. Newtonians have made very poor judgments leading to theories that miss the target by a mile. For instance, with all Einstein's brilliance and genius, he could redesign the Universe, if the Universe had worked the way he said. Einstein demanded the Universe to be space-time, in singularity and he missed the Universe by the width of the Universe. Not every atom controlling its claimed space and influenced space, loses the space when it is dense to a point it holds the minimum heat in separation. The group or cluster of protons start to combine forces and claim control space, as much as influenced space as a unit, although they perform as a unit, of many individual Universes claiming to be in the centre of the Universe they are in fact. Every atom is its own Universe, therefore every atom claims its part in the Universe and the Universe is that number of atoms cling to that space they claim. The Universe is the atom claiming space to the heat surrounding the claimed space and that heat is the space the atom, atom cluster or atom group seeks to influence to its own benefit.

THE UNIVERSE IS NOT THE SPACE, BUT THE ATOMS CLAIMING THAT SPACE WITH THE SPACE THE ATOMS CONTROL AND THE SPACE IT WISHES TO INFLUENCE.

Without the atoms, the space in whatever form, would not be, because it is the atoms parting singularity from the two ends forming singularity. The singularity is one, there cannot be two, because then there is space to hold in different spaces. That task falls on atoms, forming different spaces, and as much as different time to hold to that space. There the space becomes two, not in the space of singularity, but in the space of the atom, where the atom parts with singularity. That is why the Universe is spinning.

The one side we see but that is maintaining | The supporting sides that we cannot see, the side we see.

Therefore we see one side (1) and half of four other sides (4/2) = 2. In fact we still see three sides even if only one side is in our vision. This is a lot different in the wave. (Let us ignore the pencil line science and Einstein wishes to apply to light). We see Π^2 as the light moves to the nearest proton and if we are able to see the light, that proton will be my eye.

We can only see the light holding the time value of 1 carried by the Π^2 value. Then we can see the light not yet deflected by the 7° outward movement of the sphere. The 7° is reflected in a new wave and top the other side we have another sphere of photons deflecting. Therefore only Π^2 and $(7^0)^2$ remain in our field of vision as the rest of the light goes lost to us in a space occupying (existing) of heat. From the sphere we can only see $7 \times 7 = 49 \times \Pi^2 = 483,61$. This value holds space in the fourth dimension of the triangle in 180° being the compliment of the straight line (180°) and the half circle (180°). The value forms a corner of 90° between space and time, therefore Pythagoras comes into play. That means the square root of 483,61 holds the relevancy, which is 21,991. The relevancy of the light we see holds the value of Π to the space (21,991) and matter (7). The light that reaches us has only the value of Π, with the rest lost to space once more.

That brings the distance between the Sun and the Earth as the factor of Π and Π holds the space relevancy of $10(10^2)$ going in one direction $5(10^2)$. So Π will reduce its value by $5(10^2)$ every eternity parted by an infinity. $T = \Pi$ and $R^3 = 5(10)^2$. That means the distance of the Earth separated by the Sun will be $r/5(10)^2$. The reason why matter forms a square $(7^2) = 49$ is that matter forms the link to time and time is always in a square. From this loss of light to darkness (which is actually light that lost its direct link to time from the position of our vision vantage point) will reduce as the Sun will seem smaller the further it is from the object in its field of vision. From Mercury the Sun will be many times "bigger" than it will be from Pluto. It is due to the loss of density applied by $\Pi^2 = 1$ and the linear factor of Π. This EVERYBODY KNOWS ALL TOO WELL, SO WHAT ABOUT THAT IS NEW THEN, you may ask.

The main issue arising from this, is the following:
The distance between the Sun and the Earth is $149\ 600\ 000/5(10^2) = 299200$ km/sec (the kilometres come from $R^3 = 1m^3$ of water and the time arrive from 360° $\times$ 60 (space) a relative man made). When light was measured, the distance between the Earth and the Sun was larger than average and the distance I use is the average. This means within the heat density in the space-time we hold in time and space at this relative point from the Sun the speed photons displace the heat density is 3×1^{-5} km/sec. That too means that the speed of light at Pluto will be a staggering $11,827 \times 10^6$ km/sec. Fortunately the second will no longer be a second, as will the kilometre no longer be a kilometre. What does this proof? Astro Physics applied by the Newtonians is a joke, and the Universe is immeasurably older, bigger and more massive than what our Super-Educated will ever accept. This makes all their brainpower and calculations irrelevant.

That, once again proves my theory, that the displacement of space-time holds the relevancy and not time neither space. Einstein gave matter in space the constant by searching for the critical density factor and the curvature of space-time. That is a lot of bogus nonsense. Illustrated, this will happen.

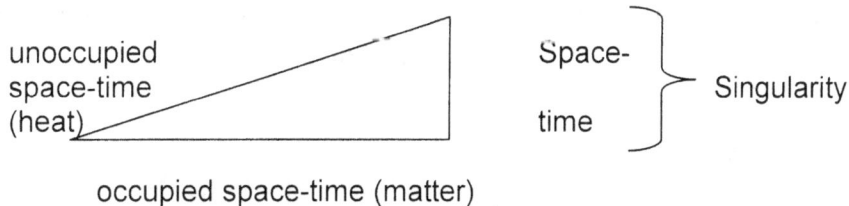

occupied space-time (matter)

When matter (occupied space-time) becomes more time and space will bend. That is nonsense. There is no curvature of space-time and there is no critical density factor because:

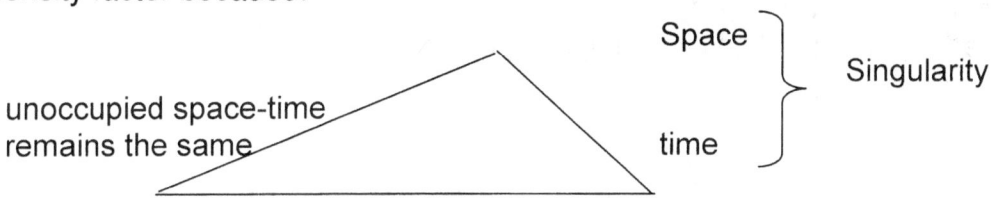

Matter increase

By increasing the matter aspect and maintaining the unoccupied space-time at the same value, then of course will space and time holding singularity bend. But singularity must always be a straight line because it is dimensionless, only parted by heat occupied and heat unoccupied. SPACE AND TIME LINK ETERNITY TO ETERNITY WITH HEAT OCCUPIED AND HEAT UNOCCUPIED PRODUCING INFINITIES. THE ETERNITY LINK REMAINS ETERNAL . IT IS THE NUMBER OF INFINITIES FOLLOWING ONE ANOTHER THAT CHANGES.

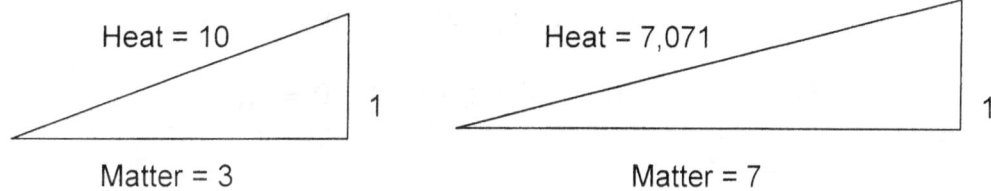

The increase in matter will reduce the density of space-time unoccupied (heat). Throughout this, time linking space will remain 1, eternal.

From the space-time constant the speed of light relates to the cosmos as follows:

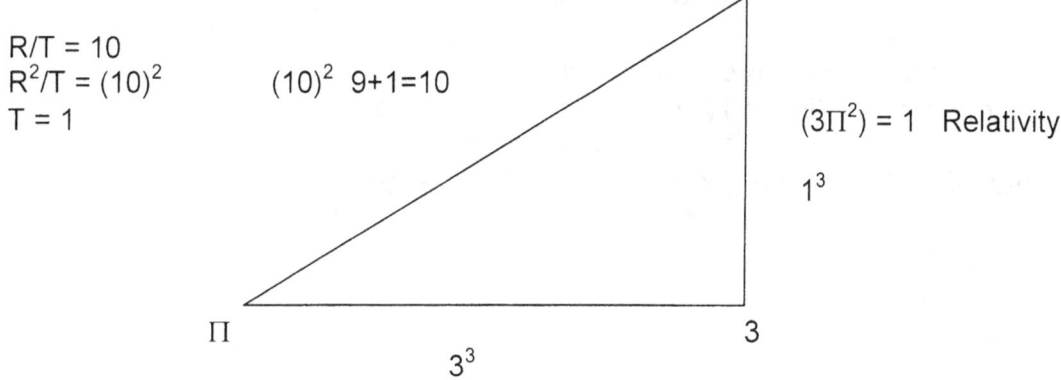

$R/T = 10$
$R^2/T = (10)^2$ $(10)^2$ $9+1=10$
$T = 1$

$(3\Pi^2) = 1$ Relativity

1^3

Π 3

3^3

In this you find the relevancy of space (3) relating the fourth dimension to time (Π^2) through a position matter holds to space and time $(21,99/7) = \Pi$.

But space-time is Π^2 in the space of 10. Π^2 to the fourth dimension are 10 because "gravity reduces 10 to Π^2 and the Π^2 value lies on the space side of the atom.   Therefore the full compliment of "outer space towards time will be the relevancy of $\Pi\Pi^2$ where Π is the Pythagoras value.

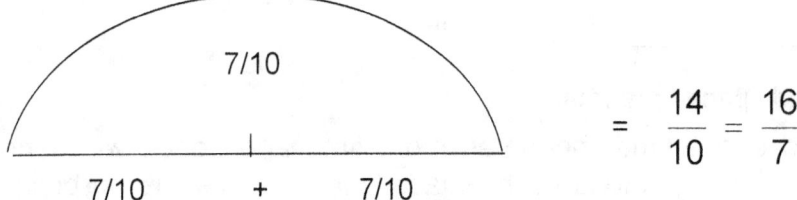

$$= \frac{14}{10} = \frac{16}{7}$$

Therefore $\Pi = 7/10$ and $\Pi^2 = 10/7$
But the matter position relates to the space position.

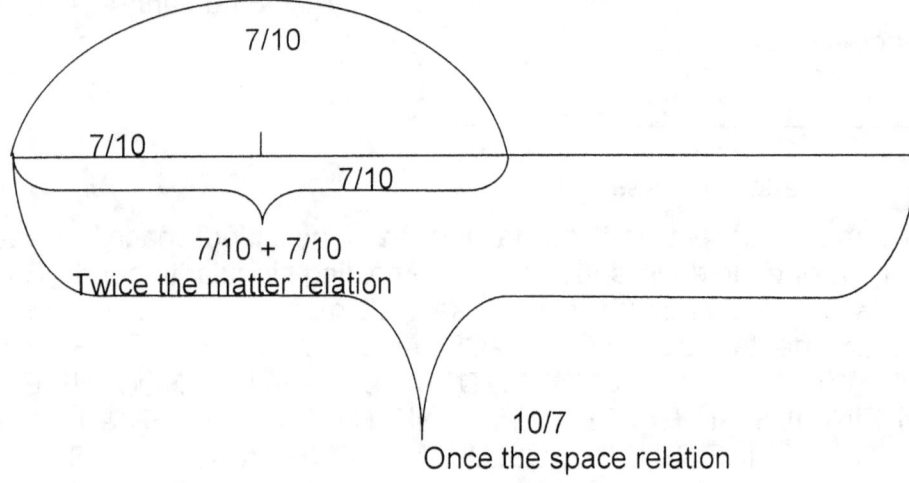

10/7
Once the space relation   Twice

the matter relation   10/7
                              Once the space relation

O   O   AC = 7/10 + 7/10   1,4 + 1,4 = 2,82/Π = 0,9
A   7/10   B
   +10/7

7/10 + 7/10 + 10(7) = 2,8285 / Π = 0,9
Π/(7/10 + 7/10 + 10/7) = Π/28285^2 = 1,10
                        Total   = 2,01

Therefore singularity at Π relating to space-time = 0,9
and space=time relating to Π in singularity   = 1,11

The total of matter to singularity   = 2,01.

Therefore Π^2 to Π = 2Π

This then is 7/10 + 7/10 = 10/7

Therefore space will always hold double to the relevancy of matter.

To that end
.   3. (7/10 + 7/10 + 10/7)
.   3.   6.

Λ   B   C
                 2Π

Then 6 becomes 3 = 2Π(Π=6)=12
A   C   D

Then 12 become 3.   2Π = 24
And that concludes the Titius Bode configuration of 3; 6; 12; 24; 48 etc. by valuing
the triangle and the half circle.

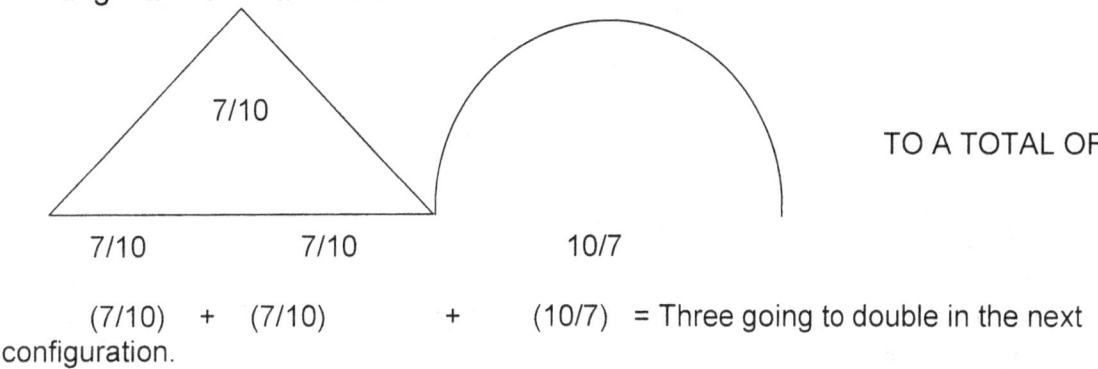

TO A TOTAL OF

7/10

7/10   7/10   10/7

        (7/10)   +   (7/10)   +   (10/7)   = Three going to double in the next
configuration.

The first $\Pi^3 \to \Pi^2$ Π Separating singularity.

This then brought on Π^2 in heat.

Π
Π^2 Π   3Π = 3
Π

From that space developed

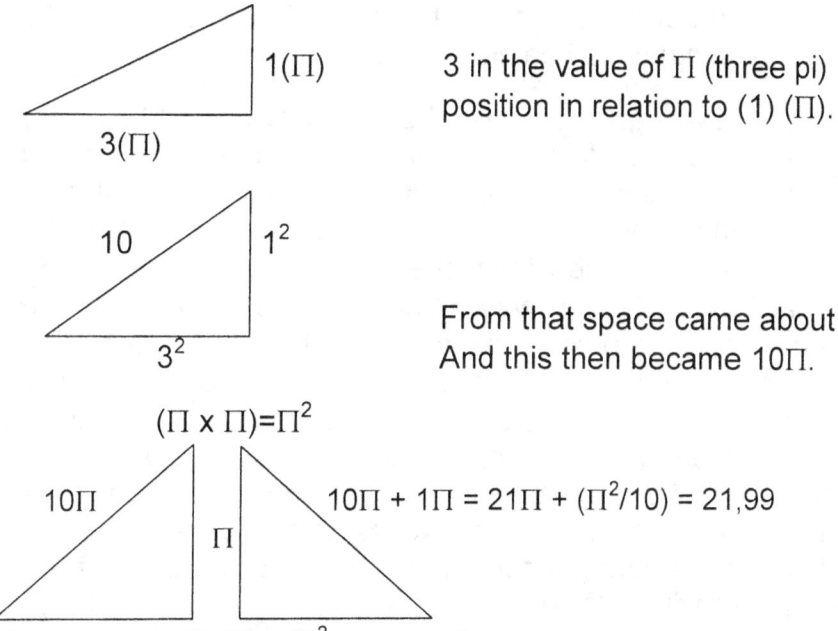

1(Π)   3 in the value of Π (three pi)
              position in relation to (1) (Π).

3(Π)

10   1^2

              From that space came about
3^2 And this then became 10Π.

$(\Pi \times \Pi)=\Pi^2$

10Π   Π   10Π + 1Π = 21Π + (Π^2/10) = 21,99

Π $(\Pi \ \Pi) = \Pi^2$ Π

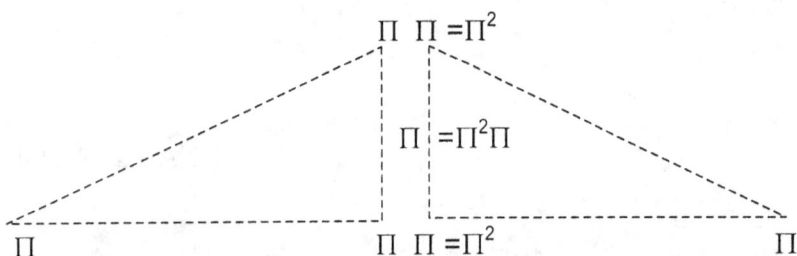

Add that to the seven that holds densified and occupied space-time and there is seven of Π in the triangle of matter adding 3 Π's in the half circle of space ($\Pi^2\Pi$) and the total Are 10Π. With 10Π the value of the total triangle (in a square) and 10Π the total of space-time (matter holding singularity apart, there is a factor of 7 by Π to 10 by Π, with the triangle having two Π - two factors where Π at the bottom formed (Π x Π) = Π^2 and to the top (Π x Π)^2 = Π^2, with all of this constructing space (10Π) to matter Π^7.

$$\Pi \ \ \Pi = \Pi^2$$
$$\Pi = \Pi^2\Pi$$
$$\Pi \qquad \Pi \ \ \Pi = \Pi^2 \qquad \Pi$$

This will bring about 5 ($\Pi^2+\Pi^2$) bringing matter and space to singularity. There are Π to the number of 5. From this comes that 4 (the four Π in the double proton in relation to the proton ($\Pi^2+\Pi^2$) will always form the time value (4($\Pi^2+\Pi^2$). From this comes the fact that 3($\Pi^2+\Pi^2$) will hold the space component as that in fact is half space-time. Space-time being (Π x Π)(Π)(Π x Π) and half of that to any direction is (Π x Π) (Π). This is why space-time in the geodesic sense has the value of 10Π and that 10Π in a sphere are Π^3. It is both triangle relating to one half sphere which is both triangles (7) in the totality of the half sphere (10).

The value of 2($\Pi^2+\Pi^2$) will relate to a position where space passes on time in a dimensional transformation, but not a value transformation. It will be when Π relates to Π^2 as much as $\Pi^2+\Pi^2$ = 19,7 and 6 (the sides available for Π to use in space is 18,84. It will form a border of dimension where space will not apply in the same manner as it did before reading that border. One may say that is the densifying border of heat in space to heat in matter.

The last position of the Lagrangian atom is where only Π (in the triangle) have value and this links space directly to time (matter). At this point concerning the Lagrangian layout, we must view heat, filled space, the stuff we exist in, as an element. Hydrogen becomes liquid at −269 °C and heat (outer space) has a gas value of −273 °C. This is only a dimensional changeover from 10 to Π^2 or from 3 to Π. That was what the "Big Bang" was all about. Creating the Universe in space as we see space was the process where the last natural element formed. The Universe or Cosmos is the last atom, which formed. It was the conclusion of the proton ($\Pi^2+\Pi^2$) in developing. That space can be in gas as we now see it, covering elements to position them in relation to the rest of the Universe as gas, liquid or solid. That is why astro-physics in the ways science apply it at present, is but a good old romantic science-fiction story. Mass holds no value, it is density and that density brings in the heat in space, unoccupied as yet by matter. That density gives a star its "gravity". Where the Sun can only hold heat to a liquid more dense stars will hold heat to a "jelly" and others will take heat all the way to something as hard as tungsten. That is what tungsten is. Tungsten can place heat in a relevancy that the heat relative to tungsten is almost as dense as the neutron within an atom. That is why I refer to the system as the Lagrangian atom, because the Universe holding heat, produces all the relevancy matter can have. The Universe is the final atom.

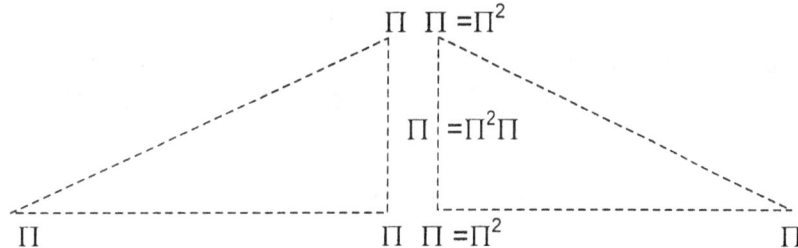

From the illustration it is obvious that the inverse cannot be in a value of $6(\Pi^2+\Pi^2)=118{,}4$. To be in that position an object has to be in two places at once, which frankly in spite of the views our Super Newtonians contribute, is simply illogical. To do that will take space-time to a point where space not yet formed. I do admit at 118 there is something, but we shall never know, because whatever it is, will be even beyond a Proton Star, in the heart of a Black Star.

Because no object can hold dual positions the one proton will link to Π in space-time at the end of the triangle and the other will connect to Π in singularity.

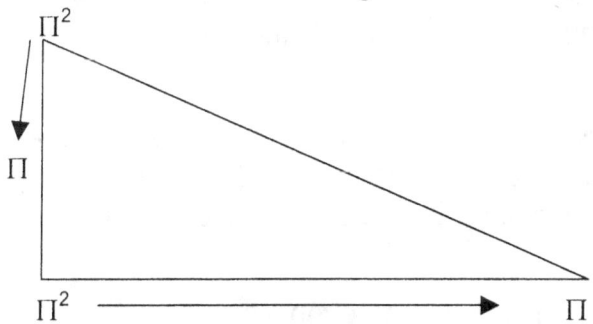

As I mentioned nothing can be in two places at once, not even Π can be in two places at once. Therefore while the one proton connects to the heat establishing space, the other connects to what is space, confirming the circle of dimensions. That is "gravity". It is a dimensional link establishing a transformation from the 6 by Π to the 3 by Π by relating this neutron Π^2 to the neutron Π^2 where the neutron Π^2 links to the " gravity - motion" Π^2 as the " gravity - motion" dips back, unifying Π^2 in singularity to Π in singularity, revaluing Π^3. After establishing this, and with the Newtonians simplistic view on "the gravitational force" I surely would enjoy watching them while they were "creating artificial gravity". What an effort that will be and the means in which they wish to perform, by merely spinning whatever spacecraft, and thereby creating centrifugal force. Newtonians should stop dreaming and start thinking. Dreams, magic and forces are for children, so let the children have them while the grown-ups use their minds for thinking.

The cosmos is the effort that space in singularity will perform energy through which it will produce gravity, to re-unite with time in singularity when time in singularity holds the value as matter Π^2 and heat holds space (Π) in singularity to a position of 6, effectively operating as 3. $T= (\Pi^2+\Pi^2) (\Pi^2\Pi)3=1836$.

With that $ = R^3 = T^2$ and $R^2/T \times R/T = 1$. What that imply is that Π^2 is time as singularity in space Π reverts to both value of Π, be it in singularity or heat.

Connecting Π to both Π^2 is the Roche limit where one object (cosmic atom) holding a position of Π to both ends. It means they are connecting at Π to singularity that means Π holds both one will hold $\Pi/2$.

While one aligns with Π^2 it also holds the other cosmic atom as Π^2 at the point it positions ($\Pi/2$). Therefore the Roche limit is always ($\Pi/2$) with the Titius Bode applying some influence, even if it is holding the Roche limit accountable for service.

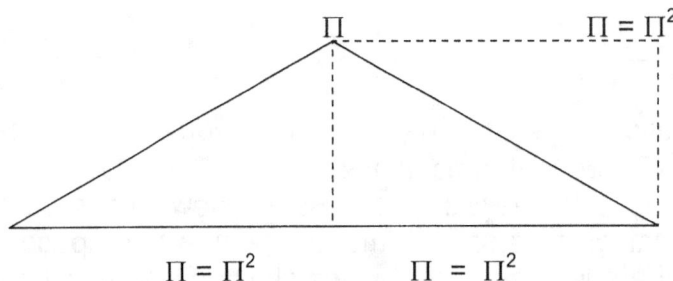

The Roche limit then is the closest any object can be in being in two places at once, because it holds time to one place. It is holding time Π^2 to a quarter of cosmic time 4. That means it is holding position in 4 Π possibilities to time Π^2 $\Pi^2/4$ = Roche limit. This gives iron56 its value of 10 x 21,7 in a relating to 4. The 10 is space with 21,7 the space relevancy to matter (at 7) and iron holds matter, therefore iron holds ($\Pi^2+\Pi^2$) to one.

Space 10(Π) x space (22Π) is 220Π^2 holding time 4Π^2 = 55, the value of iron and iron holds time to singularity through matter $4(\Pi^2\Pi) + 1$ (Π in singularity). Anything less than 55 cannot link with singularity in space, where Π will provide stability.

To compare, let us first view the Newtonian position on events.

The Newtonian approach of gravity is promoting the fact that matter will cling together because the mass sucks like a hovercraft blows, and with the suction all space separating all particles will reduce to nothing. After this reducing of the separation distance parting the particles will still suck, and suck and suck until the particles can withstand the suction no longer. In anger the particles will respond by heating as anybody can feel when one rubs one's hands with force. They see the force we use to pressure our hands together as a similar force that will do some sucking on the one side adding some pushing on the other side and that will produce a burning star. This is total rubbish with no proof whatsoever. If that is gravity, then I reject gravity.

I shall try to explain how I see the diminishing of space comes about in the process we regard as "gravity". By this process I shall also give account of forming of our solar system.

This is what we see as space filled with matter establishing time outside singularity.

What the Universe therefore is, is a box confined to a sphere. In the box three sides are in conduct holding to one side of space.

Even when you only view one side of a box, there are three sides.

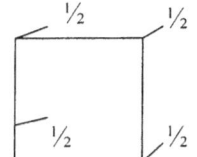

In view there is a square with one side, but in order to remain in a position in space, four halves have to support the position of that one space.

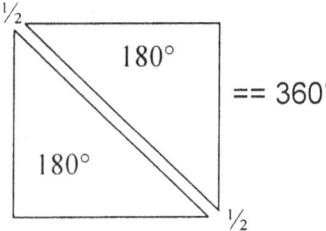

Even looking at the box square on, the same three sides apply.

In the box there is 4 half sides and one full side giving a total of three sides. $4 \times \frac{1}{2} = 2 + 1 = 3$. In cosmic reality this is but a four-sided triangle.

Even with a box holding just one side in full view, three sides still apply.

Convert that to a 'lesser' dimension and it will be $2 \times \frac{1}{2} + 1 = 3$. That is space $3 \times 2 = 6$. Place this in a sphere and there is the Titius-Bode law. The application of the Titius Bode law has a value of 1, 3 , 6, 12 etc. This is a dimensional translation of the neutron value to space-time in the fourth dimension.
 With the Sun in a Lagrangian layout it should be fitting at this point to bring in an explaining concerning the Lagrangian layout

THE LAGRANGIAN ATOM.

  It starts as follows:

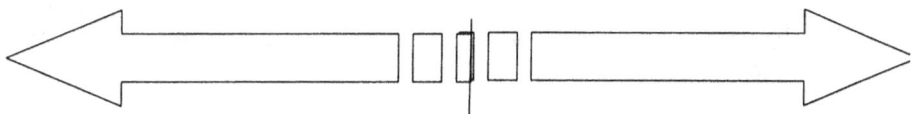

No line can start at zero because having a starting point of zero there is no line (0 X by what ever reduces whatever to zero). The starting point has to be infinity the shortest any line can be leading to eternity the longest any line can ever be. By having infinity there then has to be a VERTUALL ZERO (not zero) and from that point the rest of the line must start running the other way.

This establishes that no line can have a single direction and must have a continuance to both ends. Such a line has to have a value of 360° not 180° as believed.
  This places the single line that is half of the full line at a value of 180°.

That is also the value of the circle and the square, both dimensional components of the cosmos. By reducing the one side of the square to zero (which it cannot be) the square will disappear. Therefore one has to reduce the one line of the square to infinity to produce a square holding a straight line with the line running in both directions from a point of infinity.

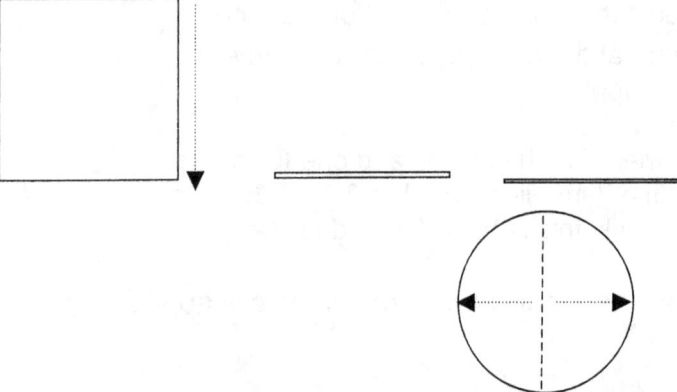

That same principle applies to the circle because the radius of the circle is one side to the round cube or a cube with no corners. That makes the square the circle and the straight line the very same thing of something totally different. The common denominator is the singularity of eternity finding infinity.

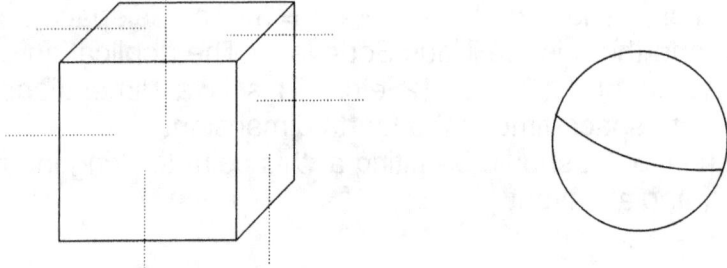

In the cube and the sphere there are a limitless of singularities connecting to one common denominator, the major singularity.

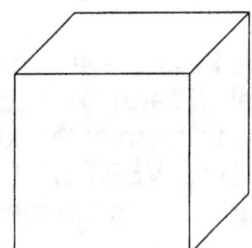

The concept behind the term we use for gravity is the reducing of one dimension.

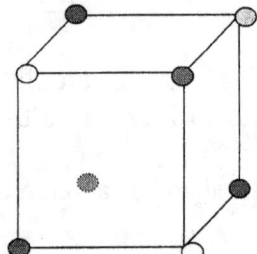

In a cube there are at least eight positions matter can relate with matter through space and therefore gas clouds cannot become structures.

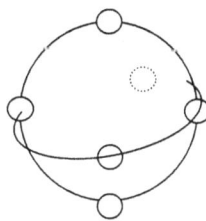

In any sphere six points of singularity positions six possible connecting singularity with that connection connecting to an eternity of singularities When the applying of dimensional reduction sets in by the dominating of one superior singularity controlling the space the other points of singularity occupy, one such a point has to fall away other wise "gravity" (the depletion of a single dimension at a time) cannot apply.

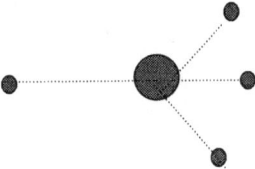

I KNOW THE NEWTONIANS ALL LOOK AT THIS ILLUSTRATION WITH THE EXPRESSION OF " _NO GOOD WELL FINE, SO WHAT_" BUT THEN THEY NEVER ALLOW ME THE TIME TO FULLY EXPLAIN WHERE THIS COMES FROM.

From the line of singularity to both sides forming the edges of singularity starting the border of singularity all aspects of the cosmos holds two halves in four quarters. Each of the halves and all of the quarters are in direct opposition to each other as much as to one another. From any point all space is moving from one side of the Universe to the other side of the Universe in quarterly displacement. The time it takes the movement will be the movement from the starting point at a value of Π to an ending point holding that moment of infinity to a value in eternity therefore bringing the square of Π^2 to the value of time. Because time is the movement bringing about the change in quarters from any given point to any other given point nothing can be in two places of the Universe simultaneously as much as no one can freeze time to single instant to calculate time.

When I came to this point I realised that there was a time before space. There was a time before material and there was a time when time was form. The Roche limit splits the Universe by 2.4674.

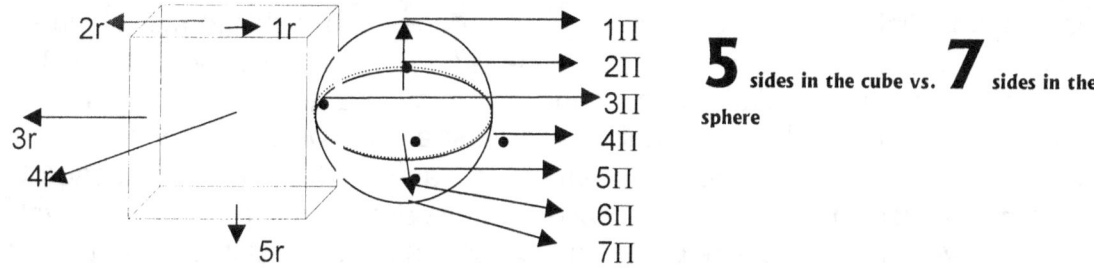

5 sides in the cube vs. 7 sides in the sphere

However if form was in place at the start of time then by the square of space 5 should divide the line forming the Roche because it is the five that should split the difference becoming 2.5 and not 2.4674. From that I made my first deduction that there was a point before space when time was the only consideration because time is the flow of heat in space and that is Π^2.

LAGRANGIAN POINT:

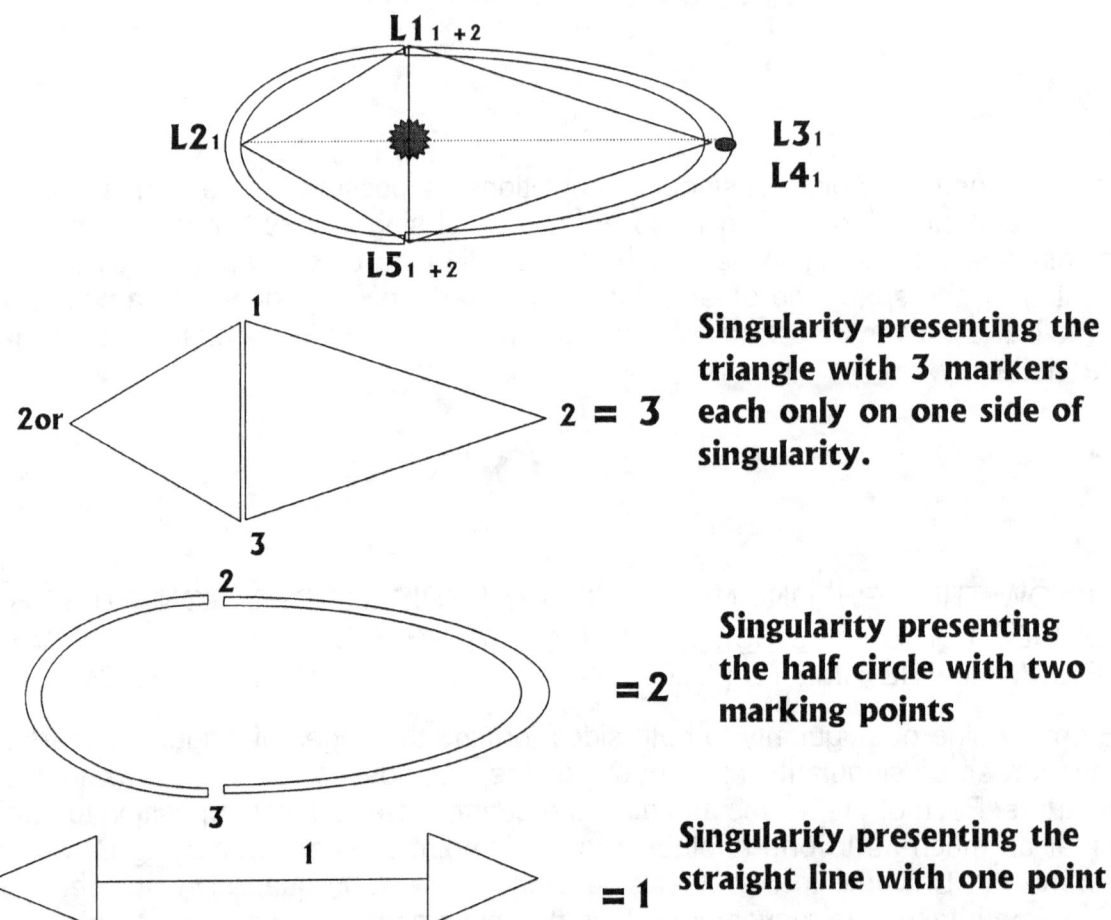

Singularity presenting the triangle with 3 markers each only on one side of singularity.

Singularity presenting the half circle with two marking points

Singularity presenting the straight line with one point

Triangle holds three, the circle holds two and the line holds one point away from singularity infinite.

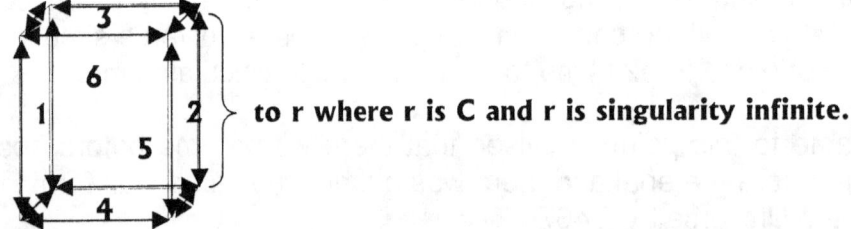

to r where r is C and r is singularity infinite.

That makes light the seventh link to the cube and light forms singularity connecting the Universe in unoccupied matter whereas the sphere holds Π to seven positions of singularity connecting occupied heat. In the factor of space unoccupied where r holds singularity there must always be space between r and the next r to prevent r from being Π as in the case of Π not allowing space but only time between Π and Π^2. By disallowing r space Π is the connecting value and whenever matter unoccupied is a factor, space always must form part of time to distinguish r from Π. With 6 positions away from singularity forming Π and matter r has to withhold one position in space to allow space and therefore distinguish r from Π. The maximum space can become is six position and r. However in the event of space connecting to Π and space holds more than five positions r will disappear in becoming Π. That is not possible as r is in clear distinction of Π. That is the basis for the Lagrangian system as well as the basis for the Lagrangian atom and that is the reasons all stars have space collapsing after 5^2 x r + 1 = iron $_{56}$ in the proton value but in the last remark there are other factors also involved.

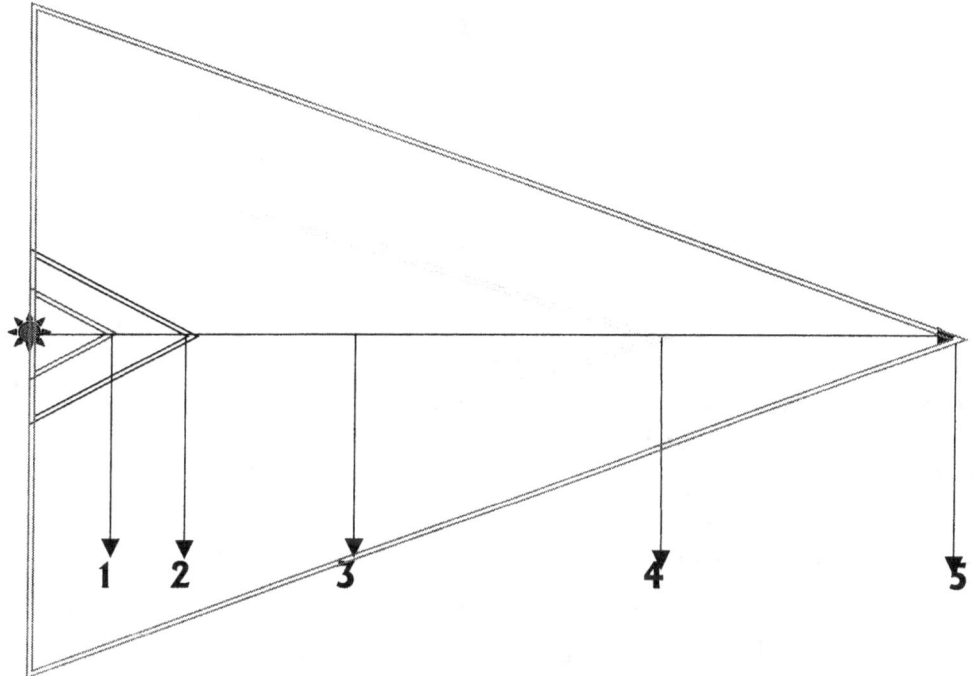

It all forms part of singularity

As a unit by three, two and one holding five in total and with space being the result from matter dismissing Π to favour r, space must either join matter by becoming Π and dismissing r or maintain r and hold a maximum of five points to singularity at the greatest value. With the Universe always in division by singularity the singularity holding seven position to Π will relate to the two singularities affecting the position of a cosmic atom. That will form as double points to space where five then multiplies with the two aspects of singularity divide and form the value of 10

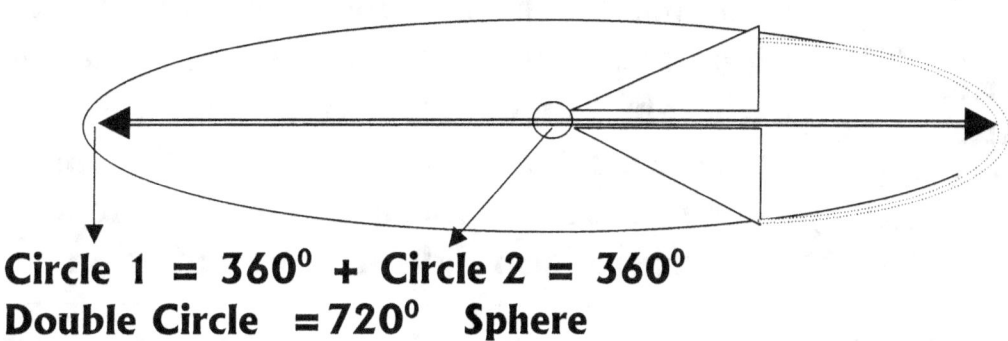

Circle 1 = 360⁰ + Circle 2 = 360⁰
Double Circle = 720⁰   Sphere

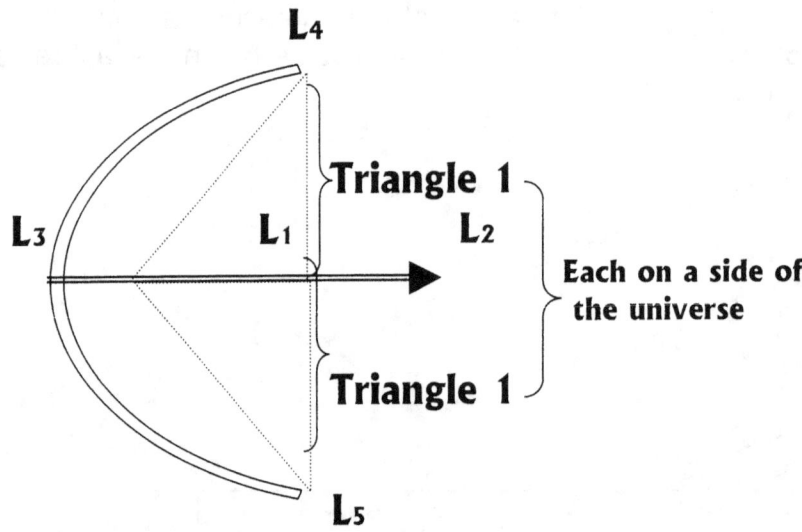

1 Half circle	$= 180^{0}$	L_3 L_4 L_5
2 Triangle 1	$= 180^{0}$	L_3 L_4 L_5
3 Triangle 2	$= 180^{0}$	L_3 L_4 L_5
4 Straight Line	$= 180^{0}$	Singularity
5 Double Circle	$= 720^{0}$	Sphere

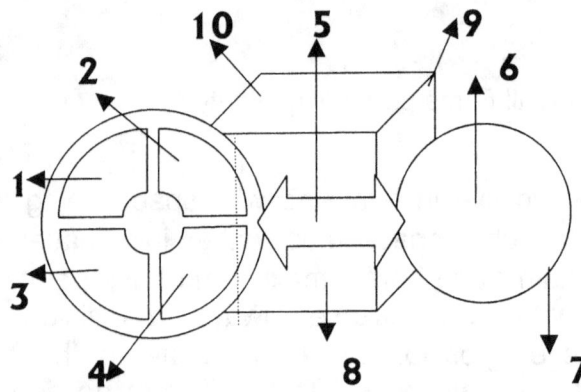

1 Singularity $X\frac{1}{4}$

2 Singularity $X\frac{1}{4}$

3 Singularity $X\frac{1}{4}$

4 Singularity $X\frac{1}{4}$

5 Singularity Π Extend

6 Matter

7 Matter to space

8 Dimension (1)

9 Dimension (2)

10 Dimension (3)

From this positioning dimensions come all the cosmic principles where singularity holds five positions and duplicate five positions as matter whereby singularity extends matter and past matter to space and three dimensions as space being fragmented singularity also light heat or radiation

From the line of singularity to both sides forming the edges of singularity starting the border of singularity all aspects of the cosmos holds two halves in four quarters. Each of the halves and all of the quarters are in direct opposition to each other as much as to one another. From any point all space is moving from one side of the Universe to the other side of the Universe in quarterly displacement. The time it takes the movement will be the movement from the starting point at a value of Π to an ending point holding that moment of infinity to a value in eternity therefore bringing the square of Π^2 to the value of time. Because time is the movement bringing about the change in quarters from any given point to any other given point nothing can be in two places of the Universe simultaneously

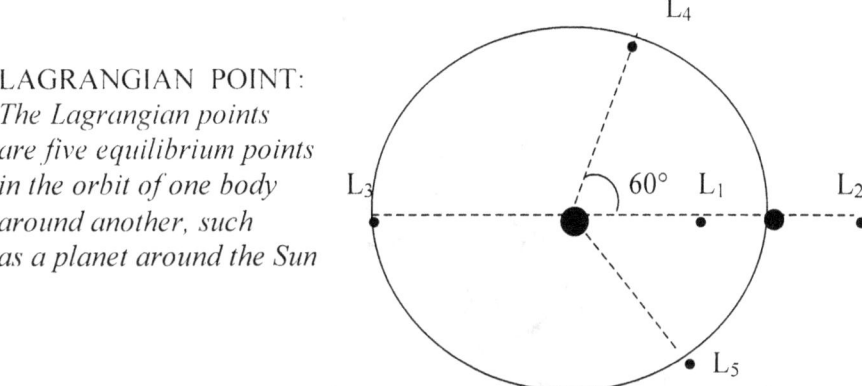

LAGRANGIAN POINT:
The Lagrangian points
are five equilibrium points
in the orbit of one body
around another, such
as a planet around the Sun

From the line of singularity

Three values each holding 180^0 comes from singularity and this fact science is familiar with. The straight line is always a potential triangle with on side apparent and the other side in infinity.

There is no zero from where a line can start and because of the absence of such a point mathematics brought about a diversion to escape the zero mark not existing. If the straight line did cross zero in would not be one line but no line since the one line will discontinue cancelling the line at one point. All lines have to have a start and an end therefore no line can be half a line.

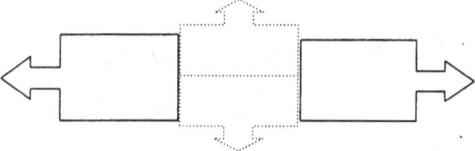

In order to overcome such a problem the straight line holds another line as a point in infinity to half the line as to enable the line diverting from zero.

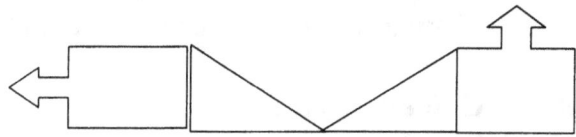

Because each line represents the other side of singularity dividing singularity by half the square of such a halves represent singularity two the half of a circle thus bringing total of the two halves would match the other half of singularity in half the circle.

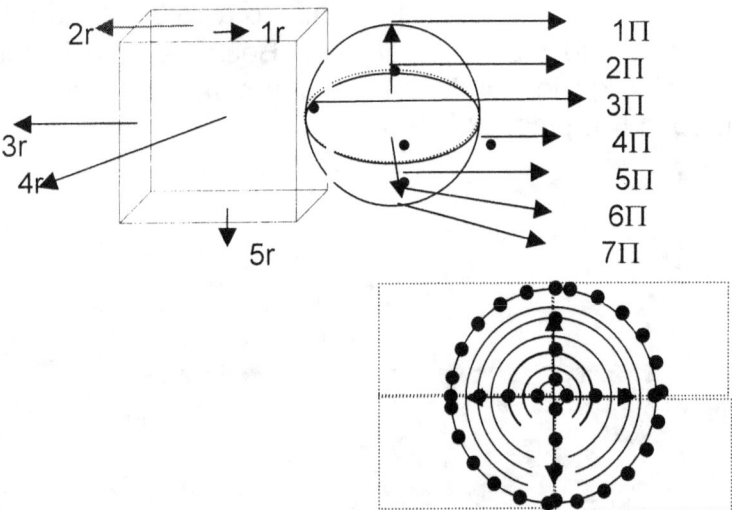

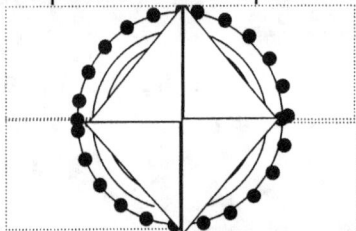

From space the cube holds the value of 5 times four quarters in relation to singularity forming the four five points of the square in the cube.

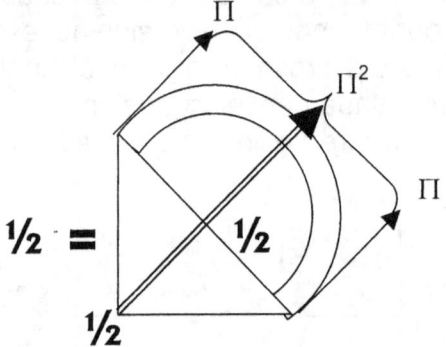

From singularity holding the relevancy the five sides in the cube as a square holds four triangles to two circles.

$\frac{1}{2} + \frac{1}{2} = \sqrt{1}$

$360^0 / 5 = 72 / \prod = 22.91$

.91

1 from singularity on the one side of the Universe and 1 from singularity on the other side...

10 from singularity on the one side of the Universe and....

10 from singularity on the other side bringing about the...

$\prod$ **that holds 21.91 to 7.**

Since the sphere is double the circle and half the circle represents singularity by the square, half the square of the triangle is a straight line diverting singularity the law of Pythagoras is valid.

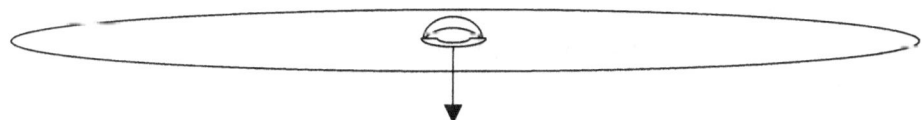

The divide bringing about the two sides of the Universe where the one (1) to singularity depicts the one side of the Universe and the other side (1) depicts the other side holding space from singularity (.91) bringing about the singularity value of 2.91

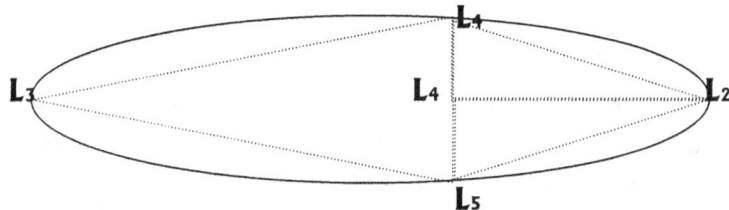

Dividing the four fives, singularity holds a centre line (.91) with one on either side but since it is a space relating to a sphere only one of the quarters on either side of the divide relates to a specific. Therefore unlike the sphere where the full value of Π relates to four fives, bringing about Π as the dominant the space separating the sphere from the points in space holds a combined value of one cube in line with the divide singularity supplies having five points.

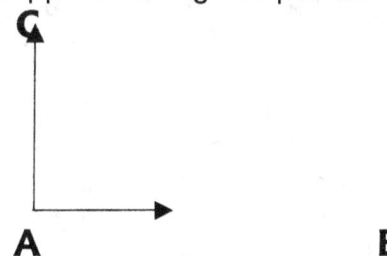

It takes any line time to relate to space and only nothing is instantly and nothing is what it is, the moving of the line through the space it covers are the square of time. The time factor is in all cases in relation to the space factor in the square while the space is in the cube and that is a law of cosmology. To that reason I reintroduced Kepler's formula to the formula $R/T \times R^2/T = 1$ and this puts a relevancy between time in the line and space to the line.

Since the triangle in singularity are on both sides of the divide of singularity and the circle holds the time aspect relating to space in the square the triangle therefore must relate in a square in order not to duplex singularity in the divide.

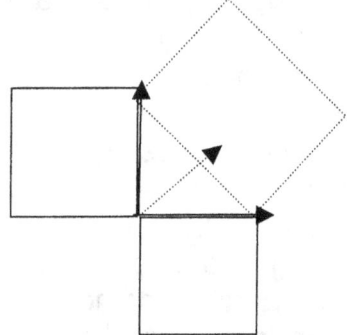

The time affecting the space of AC will relate equal to the time the line AB relates to time and space and where time is always in the square the lines will be the square of the triangle forming in relation to the square existing in the total of the time to the space relation forming between the lines.

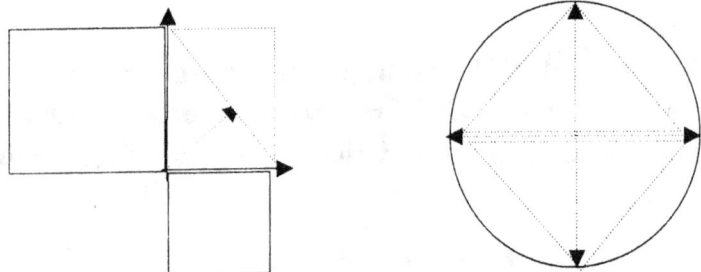

Since the triangle forms on both sides of the divide and all things concerning singularity is in duplication the double triangle will be the square. In the same way the circle represents both sides of the divide singularity forms and then also has the value of 2×180^0 as does the double triangles. That brings about that Π relate to the square and r to space and from this fact mathematics can substitute Π by using r. However that is not the case in the dimensional aspect and as mathematics gets its queue from the cosmos and not the other way around the substation may apply to singularity as space in the sphere but not as space as the sphere. Heat is concentrated space and space is expanded heat. Gravity and electricity is the very same thing where electricity is a concentration of heat demolishing space in a very specific location and gravity is the concentration of heat in a less dynamic form but acting in a much broader space. In both instances it is the polarized motion of iron $_{56}$ linking time directly to singularity through the conducting of heat surrounding elements. To this effect I wish to point out that no element is either a liquid, a gas or a solid as all elements are all three forms and it is only the state of the relation that apply to an element at a very specific position in space-time that will allow the element to act in either of the conditions that the heat or space which is the same thing will allow. Gravity is the dimensional destructing of space to the concentrating of heat in that space by increasing the time duration through the Titius Bode principle or when matter holds less space to the normal allowing of matter occupying space, matter will produce the Roche principle in guarding its individual singularity of the mutual singularity between objects.

This principle also is the only difference of notoriety between electricity on Earth and gravity on Earth. On Earth Π has a very slightly higher value than space in as much as space is 3 and the atmosphere is Π. In a cosmic midget as the Sun are the relevancy changes are considerable and the space to atmosphere can be 10 Π. There is no chance of generating electricity in the atmosphere of the Sun because the atmosphere of the Sun is electricity in as much as the gravity being 10 Π to the Π of the Earth. That explains the fact that the Sun liquefies heat to a watery substance. With the heat in a liquid the Sun becomes a sea of heat. By matter applying Π as the reference, there is little man can do to change that. In the case of electricity using r to form the value C we can change that because as r relate to space and we are part of space as space above the Earth in the neutron zone we can change the space holding r as value. When looking at the Sun applying gravity and relating what we see to what we find in electricity there is hardly anything to recognise a similarity by except that one can change and inter change the two aspects of heat. But when in view of dimensional dynamics it is the same thing because in the Sun even r holding heat becomes Π holding matter as heat becomes liquid and that stands in between matter and gas. With this view in mind it would be worthwhile to have another look at the way we see how creation started and bring heat in as related to r and matter being related to Π.

Gravity is the transformation of space to heat in one specific dimension changing that particular dimension in relation to the other five dimensions. That brings the reason why the Lagrangian system can only allow five positions and allowing any more will destroy any form of dimensional implication between object relating to one another while sharing space occupying in time duration. The Big Bang had its massive motion brought on by first implementing the Roche factor of $(\Pi / 2)^2$ after which when matter had a larger claim to space and space broadened the Roche factor adjusted to Π2 / 2 and then implementing the Titius Bode principle very much later on toΠ.

It starts as follows:
ALL OF CREATION STARTED WITH Π = TIME AND AS TIME WAS ETERNAL,

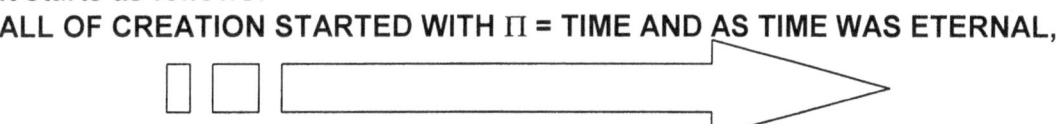

The spot grew into a dot by the value of 1^0 going onto 1^1. However this also was Π^0 going onto Π. However with no space yet available and the spot forming being so small (as we still can see when observing the centre of the spinning top because the line inside the line that is inside the line we cannot see is the line we are referring too.) The line is there as it was, but our inability to go that far into singularity places us at the disadvantage not to recognise the line as it is. Because it is spinning it has to be Π and because it is spinning it also has to be a line and all of that is there to witness for all those disbelievers.

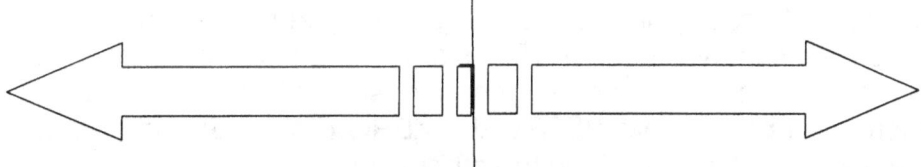

The line formed a spiralling line that was so small it was continuous and never broken due to the lack of space and those not believing me I advise you to inspect the spinning top. The line formed next to the running line and that too is in plane view of all but we cannot see. The line has to form a line on this side of the divide because it forms a divides as much as there has to be a line on that side of the divide for the very same reason

This places the single line that is half of the full line at a value of 180°.

That is also the value of the circle and the square both dimensional components of the cosmos. By reducing the one side of the square to zero (which it cannot be) the square will disappear. Therefore one has to reduce the one line of the square

to infinity to produce a square holding a straight line with the line running in both directions from a point of infinity.

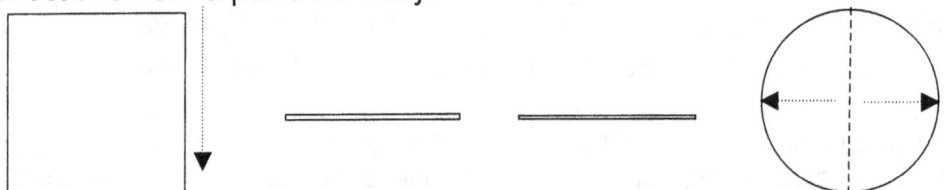

That same principle applies to the circle because the radius of the circle is one side to the round cube. That makes the square the circle and the straight line the very same thing of something totally different. The common denominator is the singularity of eternity finding infinity.

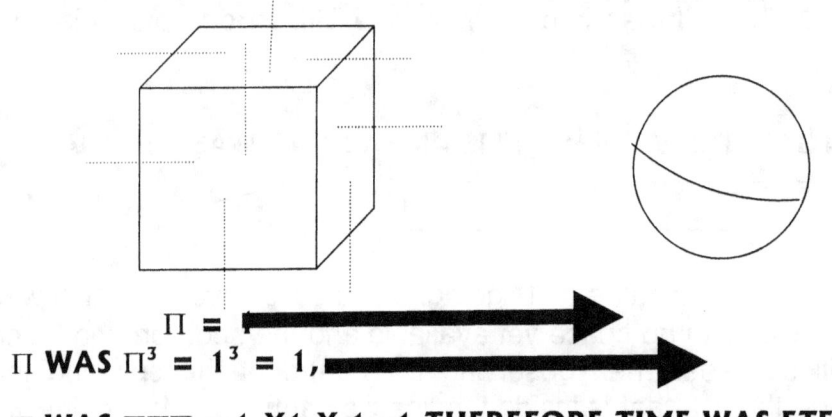

$\Pi =$

Π WAS $\Pi^3 = 1^3 = 1,$

Π WAS $\Pi\Pi\Pi = 1$ X1 X 1 = 1 THEREFORE TIME WAS ETERNAL:
E $=\Pi = 1$

Let us have another look at the straight line

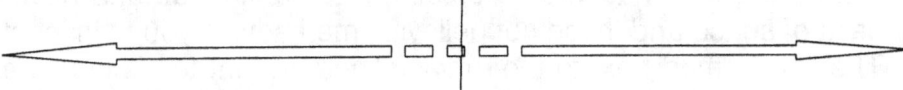

No line can start at zero because having a starting point of zero there is no line (0 X by what ever reduces whatever to zero). The starting point has to be infinity the shortest any line can be leading to eternity the longest any line can ever be. By having infinity there then has to be a VERTUALL ZERO (not zero) and from that point the rest of the line must start running the other way.

This establishes that no line can have a single direction and must have a continuance to both ends. Such a line has to have a value of 360° not 180° as believed.

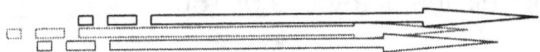

From infinity came the straight-line borders bordering the straight line on both sides of infinity

MATTER EVOLVED FROM TIME AS TIME = 7 ($\Pi^2 + \Pi^2$)
THEN TIME, BECAUSE IT WAS ONE, MOVED A DIMENSION UP AND DOWN, WITH BOTH DIMENSIONS BEING EQUAL, THE SAME IN EVERY WAY.

Let the lines be somewhat bigger than infinity where we can see the lines more clearly

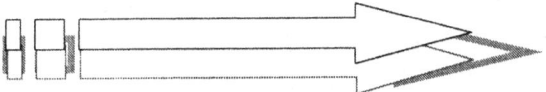

MATTER RECEIVED THE VALUE OF Π^2 AS FOLLOWS:

THEN

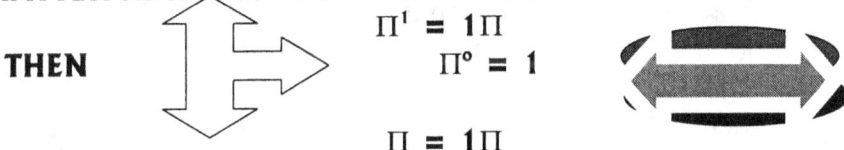

$$\Pi^1 = 1\Pi$$
$$\Pi^0 = 1$$

$$\Pi = 1\Pi$$

THEN TIME HAD A DIMENSIONAL VALUE OF Π = 1; Π =1 AND Π =1, BUT AT THE SAME Π DUE TO THE PROXIMITY OF SINGULARITY

POINT OF SIGULARITY

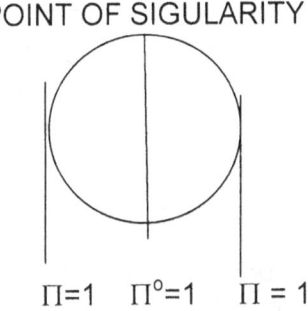

$$\Pi=1 \quad \Pi^0=1 \quad \Pi = 1$$

The Universe formed singularity in the straight line with two points of Π to both sides. There was no radius because the radius was infinite small. Then time formed as part of singularity in the value of Π going on to $\Pi\Pi\Pi$ going on to Π^3. The one Π remained in singularity Π while the other $\Pi\Pi$ became Π^2.

TIME Π

$$\Pi = \Pi\Pi\Pi = \textbf{SPACE} = 31 = 10\Pi$$
$$= \Pi\Pi\Pi = \textbf{MATTER} = 31 = \Pi\Pi^2$$
$$\Pi = \Pi\Pi\Pi\Pi = \textbf{TIME} = 31 = \Pi^3$$

While on the other side of the universal atom Π^3 formed space in time in the value of $\Pi\Pi^2$. Between the two universal atoms the line to singularity was the combination of Π^3 holding Π to Π^2 and forming space in matter to the value of $\Pi\Pi^2$.

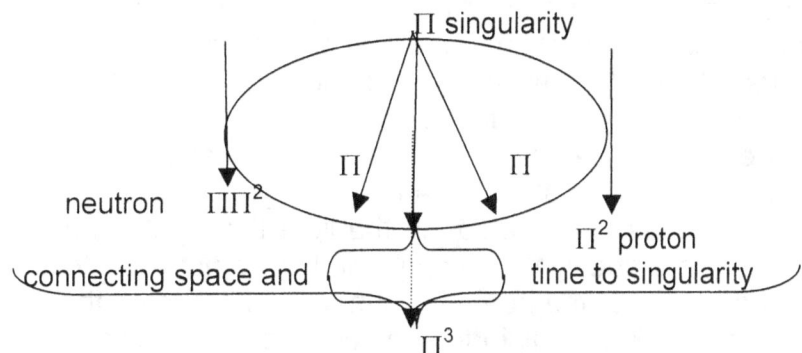

In all it became 10 Π in all being in singularity

Whatever there is today started at a point we cannot trace. It started at a point we are able to envisage and locate but that is only with a mind that can accept what is not there to view. It is acceptable through intelligence because through intelligence we can detect it albeit outside the Universe we have. It is just like religion and worst is that it is generating what there is because what there is are not if not generated. There was Π^0, which was α^0 or if you would rather have it Ω^0 or it maybe was 1^0, but more correctly it was all the above and the beyond because multiplying what ever constitute the mentioned will bring about what is mentioned to a precise equality. It was a spot that was not. The spot is still there because the spot is still not there. It was a line that ran eternal but because it ran eternal and kept repeating exactly what was before to the precise what came afterwards, the line was there and was eternally running, while never changing in the least or growing by any measure. It was not one because before it could reach one, it returned to what was repeated and the process cycled back to before one was reached and even before one could be accomplished. It was such a continuing of the monotony, no change occurred and therefore never did the running produce progress because the progress was in the perfect repeat of what was before. The duplication brought contraction to the minutes detail. That is where our atheists get one hiccup. The repeat brought eternity and the repeat was so perfect that the repeat continued. The repeat still is with us as much as we are within the repeat. To bring change to the eternal repeating of the monotonous there had something beyond the Universe that institutes change. There was something that brought a difference and we are within that difference. That difference was time and that time is what we move through as much as what we see at night. Oh, how stupid and how thoughtless the minds of atheist and other atheistic animals are. Baboons do not recognize the light we are within because they cannot think and are therefore atheists. Spiders cannot think and therefore they are atheists, as they do not think where the line is that is not. Reptiles cannot think and without thought they are incapable to see what time is, how time that is not generate space that is in time that is not. Mammals cannot envisage what space is, what light is and what makes us see the darkness cannot be. All the animals I have mentioned are mindless atheists because they fail to see beyond the visible into the realms of the thinkable. All that I have mentioned passes them by including religion and the accepting of God because through being incapable of thought and reason they cannot envisage what only intellect can bring to mind. Because of the incapacity to think the animals are both mindless and they are atheists. Therefore atheists are mindless. The night sky is such a bright light our evolution protected our vision from the brightness in order to give as much better vision. Through evolution development our eyes are protected by how we remove the qualities from the light. However animals do use that light and not our light to see by. You can shine a bright hunting spotlight onto an animal at night and the animal will not be able to see the light on it. The animal does not use the light to see better as the animal is totally unaware of the light. Then a prowling cat comes from the night and sees the antelope in the light the night provides. It does not use the light, which the spotlight casts and the light is not even traceable to either animal being the hunter or the hunted. From there we accept that during the day the animals must be using our light to see because the nightlight is inferior to see by. Who says they use the daylight much different from the nightlight because all evidence is there that they cannot recognize our light as light. It is very evident in the manner they go on hunting and grazing while being totally unaffected by our form of light. That which you see at night because you cannot see darkness and you cannot see black is the light the Universe is painted in just like the Bible says.

That is the light that started all because that is the light holding us away from the eternal darkness. This is not religion and it is not a sermon, it is hard-core and brutal basic science and it the most fundamental basic physics there is. It is the start of the mathematical Universe portraying the only physical way it could ever be. The light that came from the Command is the light allowing material to move.

My atheistic idiots, your mindless caught up with you!

Then came this light that the Bible refers to as the first of what ever was and what our stupidity tells us is darkness. This was moment-Alfa. The darkness was there and from the darkness heat came about. Only heat expands and it interrupted the true invisible darkness, the blackness of a Black Hole, the invisibility coming from within the Black Hole. Eternity tore from infinity. Darkness broke from light. Heat broke from cold. Relevancies parted by 1^0 going 1^1. There was one but also there was two too because one cannot be without two being there to ensure one is one. The marks are still with us but to see the marks requires a great deal of intellect.

$\Pi^0 \Rightarrow \Pi$. In this there was only space for one being one in the two forming one. It was $\Pi^0 \Rightarrow \Pi$ however there was no space to be $\Pi^0 \Rightarrow \Pi$ and there fore because of the lack of space to be which is the infinity of time braking the eternity of time the true measure was $\Pi^0 \Rightarrow \Pi$ but realized only 1^0 going 1^1. Π was to the future because of the motion of time involved and the space less ness of space at the time. By inclining to move the process crossed the Universe but also it took one eternity to accomplish the feat.

> •• ▶ 2

The fact that 1^0 going 1^1 brought movement can only become a reality as a result of light. Light is heat and the heat is expanding.

1^0 going 1^1 where $1^0 \Rightarrow 1^1$

1^0 going 1^1 1^0 ▶ 1^1 had to bring about 1^0 going 1^1 1^0 ▶ 1^1, because the eternal repeat of duplicating while contracting was not relieved from the Universe. Before the contracting was equal to the duplicating because by measure the heat was identical to the cold. It was eternity that was interrupted by one cycle of infinity and was in repeat of eternity. Once something is part of the Universe there is nowhere else to take it so it has to remain as a part of the Universe.

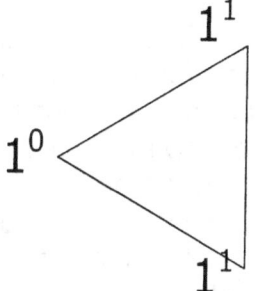

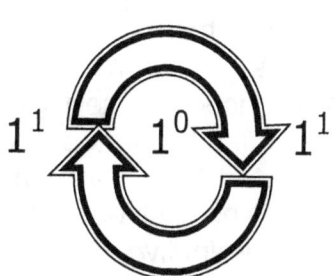

Then came three because motion was so limited that the least inclination to move threw what wished to move to the other side of the Universe, As it moves it also moved across singularity. It crossed the entire Universe as it moved because it moved and finding nowhere to move too. It crossed the entire Universe and it took one eternity less the measure of one period lasting infinity to achieve that. That brought to relevance three points where each was in measuring quantity exactly equal but also one Universe apart.

In the reality there was now two points holding singularity on both sides of the Universe because by crossing the divide that crossing set in place the two sides relevant of singularity governing. However infinity was bridges at two points holding infinity with which process eternity repeated the past into the future.

$$\cdot\cdot\blacktriangleright\ _2\cdot\cdot\cdot\blacktriangleright\ _3\cdot\cdot\cdot\cdot\blacktriangleright\ _4$$

This then is the occasion where Pythagoras stepped in. Since it as a crossing of the divide the crossing involved a line that formed a half circle connecting a triangle. But the crossing was done in the space of half the Universe and since the Universe was 180^0 half of the Universe was 90^0. That involved Pythagoras as mathematics was born. Up to this point it was arithmetic with adding but now mathematics came into place. Remember we are a few eternities in side the development of the Universe.

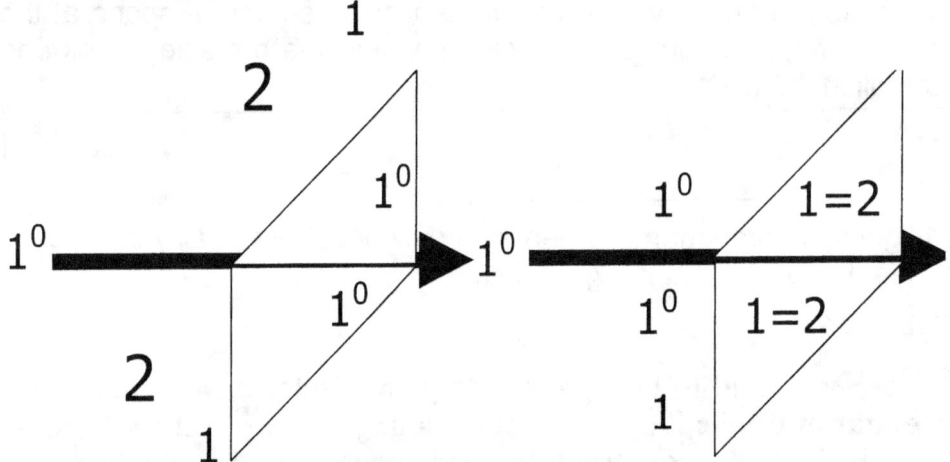

In the three came four that brought along five.

$$\cdot\cdot\blacktriangleright\ _2\cdot\cdot\cdot\blacktriangleright\ _3\cdot\cdot\cdot\cdot\blacktriangleright\ _4\cdot\cdot\cdot\cdot\cdot\blacktriangleright\ _5$$

Reaching five is a benchmark because at that point half the Universe was finalised. From the five that formed the Universe continued and formed space-time. All progress balanced on the five that formed where two parts of equality parted eternally as liquid separated from solid, motion moved apart from the motionless, heat diverted from cold. The four in time grew away from the one point space formed being just outside the control of time. It was the start of material since from the point five formed just outside the four of time. There was a lagging behind of heat building and not being in the range of the immediate control of time. Time could generate a point in heat without bringing immediate demise to the point through cooling.

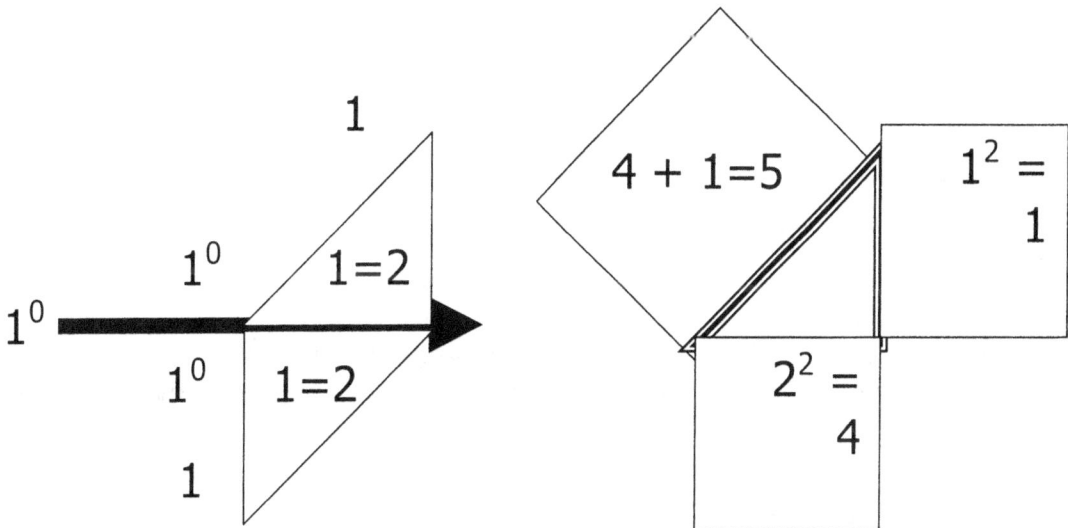

Then five filled the one half of the Universe that was able to contract and cool while the Universe divided the other half into sectors of what was (5) and what will be (5), which put material (7) in relation to the half of the Universe $\Pi^0 \Rightarrow \Pi$ in which the material was at that specific point (five relating to seven) in time.

$$1^0 \quad \blacktriangleright \quad 5$$
$$1^0 \qquad = 1 + 5 = 6$$

The motion consisted of Π moving to Π and thereby duplicating Π to relieve Π^0 of the burden of overheating. On the one side of The Universe there was Π^2 being relevant to Π which was forming on the other side of the Universe. The entire Universe had the combined value of Π^2 on the one side in addition of Π forming on the other side. I wish to remind the reader that any and all points formed by singularity was as much representing the Universe as it was the Universe at all times because $\Pi^0 = 1^0 = 1$. That made the entire Universe being any point affirming singularity by forming about singularity.

But that meant that the Universe was a total of $\Pi^2 + \Pi$ which when added was also $\Pi^2 + \Pi = 13.0$

$$\bullet \; \bullet \blacktriangleright_2 \bullet \bullet \bullet \blacktriangleright_3 \bullet \bullet \bullet \bullet \blacktriangleright_4 \bullet \bullet \bullet \bullet \bullet \blacktriangleright_5 \bullet \bullet \bullet \bullet \bullet \blacktriangleright_6$$

$13.0 - \Pi^0 = 12$ because singularity cannot be part of space-time developing as the space, which later was filled with the material that formed, filled this part.

$12 / 2 = 6$ Material formed at the point where six was located.

$$6 \; + 6 = 13 - (\Pi^0) = 12$$

Because singularity is a divide and is not part of space-time singularity as a factor removes from space-time. Why it adds with five to form six is because to the one side only singularity is in the other side of the divide. Only nothing can be in two places at the same time therefore on any one side was the half of twelve, which divided 12 in two parts. That then was $12 / 2 = 6$

$$6 + 6 = 12 + (\Pi^0) = 13$$

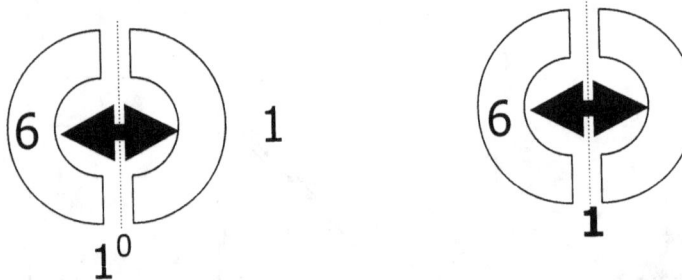

Developing six was an addition to the square as the line flowed and did not involve the crossing of the divide. Therefore Pythagoras was not involved by the forming of six.

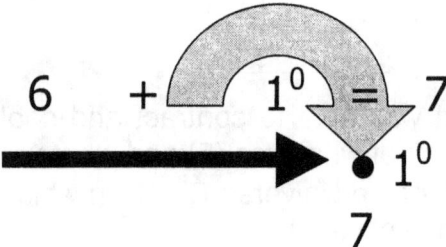

At the point where the space filled with heat meets the point in time representing singularity the end of material (6) confirmed the following spot (+1) at 7.

Forming seven very much involved singularity because it confirms appoint where space ends and space (8) begins.

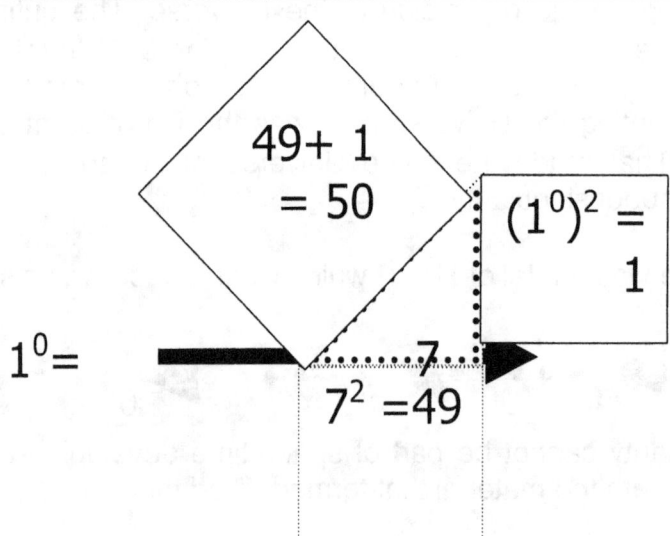

By taking singularity into Pythagoras and filling the Universe by halving the square of space seven completed the required circle within one half of the Universe in order to relate to half the time it takes material to fill time by duplicating. To find the necessary cooling required for control material has to use five points to be within because of the square involved. The there has to be another double five amounting to ten to fill the void from time in the past (position of five) and time in the future (another position of five).

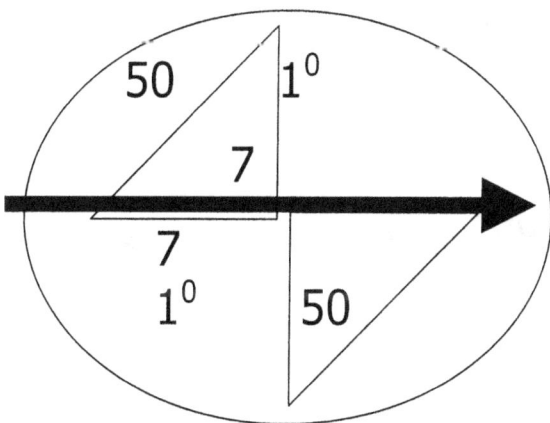

$$50 + 50 = 100; \; (100)^{1/2} = 10$$

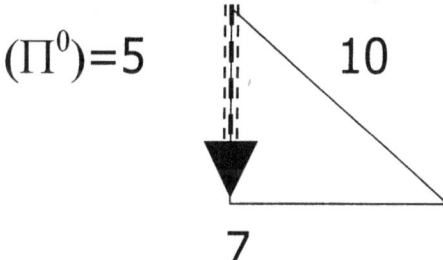

The Titius Bode require meant is seven holding relevance to ten and ten being relevant of seven while being in half the Universe $\Pi^0 = 5$

Then come eight causing a line of material to break.

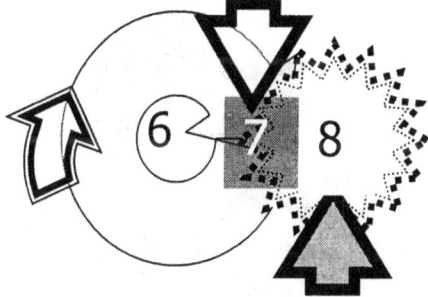

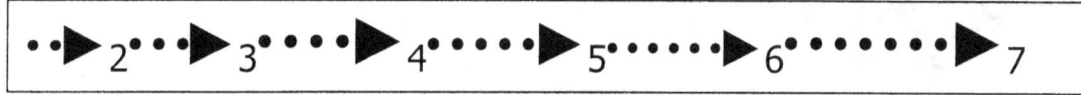

At seven the line completes at a point distinguishing material with in space from space without material

The circle of development has finalized a point. Seven has gone square 7^2 and realized with singularity half of the final of space in the absolute square.

It is this eight to ten science does not recognize and do not distinct as one other part of time. This in relation with the finality that came about at the point seven marked by using Pythagoras that another space, this time in time was developed to compromise for the lagging of time within space-time.

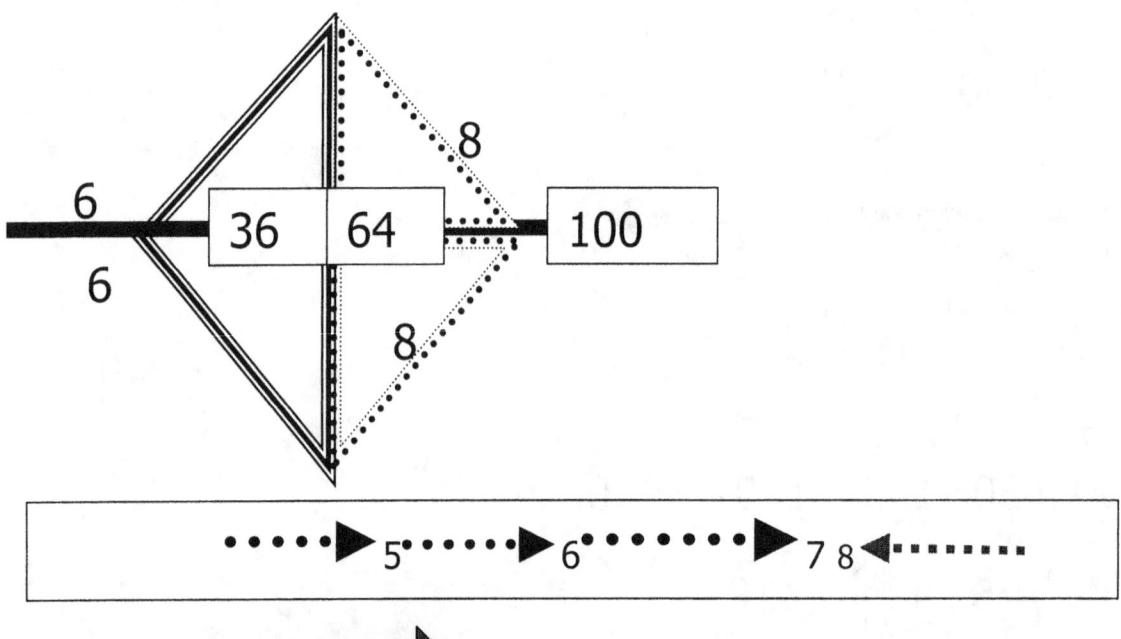

$$3^2 + 1^2 = 10$$

$$\Pi^0 = 1$$

The cycle of eternity could then complete one more time by forming singularity once more

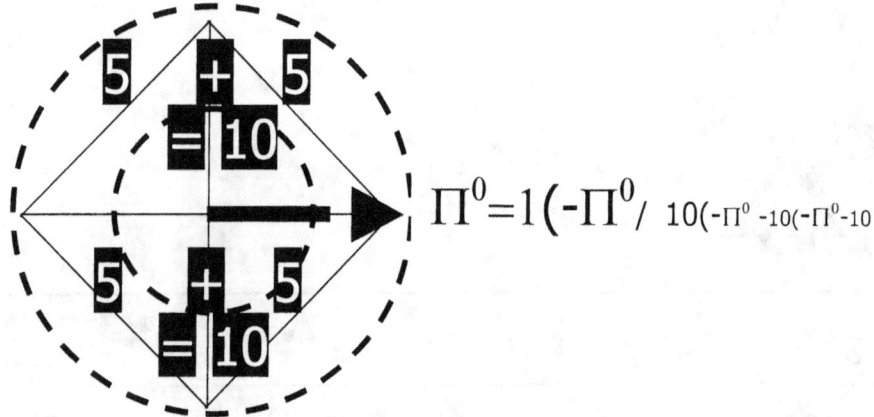

$$\Pi^0 = 1 (-\Pi^0 /\ {}_{10(-\Pi^0\ -10(-\Pi^0-10}$$

With the Universe established at ten crossing the divide meant that Π^2 at four was a half and five was completing the one half.

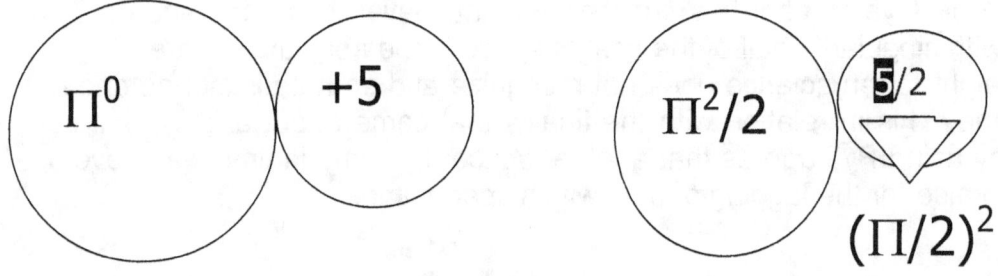

The Roche limit shows that singularity needs at least more than half the Universe (5/2) to share and…the Lagrangian system is at least half the Universe.

Therefore the circle extended to the point outside singularity where matter was holding a space value of $10\ \Pi^3$ and space holding matter was in $10\ \Pi$. Then there came the proton connecting space in time to singularity ON THE OTHER SIDE OF THE UNIVERSE.

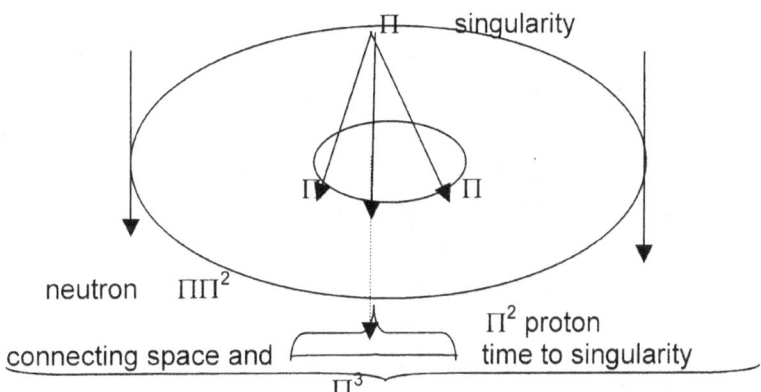

In all it became $10\ \Pi$ in all being in singularity

To connect time to space the $10\ \Pi$ divided into the $10\ \Pi^3$ forming the other proton of Π^2. THE PROTON THE NEUTRON ELECTRON

$$(\Pi^2 + \Pi^2)\ X\ (\Pi\ \Pi^2)\ X\qquad 3\qquad = 1836$$

MATTER AT $7\Pi + \Pi3$ SPACE OUTSIGE SINGULARITY IS 10 THE ATOM

We can thank that same process forming our solar system in the small manner as I described. After running through the cosmic formation principle I wish to show in what way one might be able to use the relevancies to further one's understanding of the cosmos and in this case in particular one's understanding of the solar system and the manner in which it formed. The principles I apply is real and does not reflect on the Newtonian Mede Avail mind boggling idea that dust can fall on dust and forms layers that will eventually turn as hard as rock by the mere wonderful magical medicinal forces unleashed by the magic of gravity.

We take the approach that the cosmos does prefer to use science instead of the wonders of magic and that we are able to explain everything there might be in the cosmos through the medium science renders us.

When I was very young I had a good look at how I was told in books written by grown ups how planets is formed from rings of dust they accumulate and then compress to form stars and in the case of the smaller stars it became planets since the material was insufficient to form stars. Then there was the fact about comets being small pieces of rock and so meteors were too and with those small pieces of rock getting that solid I was starting to smell rotting fish in some stories I was told. A comet and a meteor were small, yet they were big enough to start to form as solid blots so hard as what any rock could be.

Hey, for a child that still imitated Superman on the radio show, it sounded fair accept for one small detail. The sun was big. The earth was small. The sun was gas and the Earth was rock. How much more material was accumulated by the mighty gravity of the sun and how little material did the Earth use to accumulate and what force had the earth availbele to compress all of that in comparrison with the Sun. Yet the Earth and all its smaller planet sisters could compress the little it had to rock but the bigger they got and with the more gravity there was in each one the more the object was made of gas. Gas can't bring about the amount of mass per inch going square (we still had inches then) than that what rock can however, the bigger they get the more gas they collect. That made me realize that Superman was a possibility, but this was a lot of rubbish.

If it was the case that dust could collect and lie one on top of the other in layers and form from weight to become dense while becoming so dense it formed stars with the aid of gravity, and gravity was the pull of material making them dense, and gravity was about the more there are of what there is the stronger the pulling is and on top of that the Sun was Big and the Earth was small, yet the Sun was a gas and the Earth was rock, then someone painting me this picture in monochrome got black and white confused and a lot of disciplines mixed up.

If gravity was about a lot pulling more than the little was pulling less because of the lack of weight and the lot compressed, what was available to compress so hard it became solid hard rock by pulling, then it had to be the other way around. The Sun must be awfully hard while the Earth is still trying to sort out the rings of flying dust that the Earth was unable to control with the little gravity the Earth had available to use. As I said before, back then I was a child and was strongly influenced by the opinion of my superiors and Superman made a lot of sense.

Nevertheless… that lot got me suspicious and when I had a good look at the tale of the illusive comet that refused to be caught by the big bullying Sun and how the comet gets away every time as Robin Hood always did in the light of what gravity was available to the Sun and in the amount the Sun had to its disposal while the Sun in all its splendor was not doing to the comet versus what gravity was suppose to do with the comet, made me ask questions.

After a long time passing by with no one being able or in mind to answer my asking questions I came to the conclusion I had to provide some sensibility myself if I was reluctant to accept the insanity of what cosmology in theory had to offer. That what cosmology offered was harder to believe that Superman and by then I was getting a lot of doubt about the fanciful abilities of a man called Clark Kent. To go into more detailed suspicions I had and how I turned that around to fit my theory I eventually devised will be much too roomy to have in a compressed book such as this letter is supposed to be.

The Sun was one part of a binary. The binary was part of a Lagrangian layout.

O   Saturn

Binary
O   Neptune   oo   O   Jupiter

O   Uranus

This was all frozen before the Big Bang in the same way as the structures in the inner core of galactica still are at present. The binary started spinning around their common or joint axis at $(\Pi/2)^2$. The Sun eventually overpowered the one structure in the binary and that star, (which I named Unknown Star) overheated as it could not apply enough cooling to its inner-core.

As it broke into bits, those bits of inner core formed the solid planets, the asteroids, the comets and even the Pluto double rock structure. All the moons and satellites are also remaining debris from this event. The outburst caused by the Unknown Star destruction led to the release of liquid heat and this liquid heat is what the Authentic Author saw as water. The winds he refer to is nuclear winds, established as the heat formed gas as space from its position of being liquid heat, developed into a gas heat we call space.

Since the First movement of time, the Roche factor was present and from the Roche factor came the Titius Bode principle. Each object seen, as well as not seen, represents a different period contained by a different specific value of space. All matter is time in a different frozen state in space and the time gives the space its particular value.

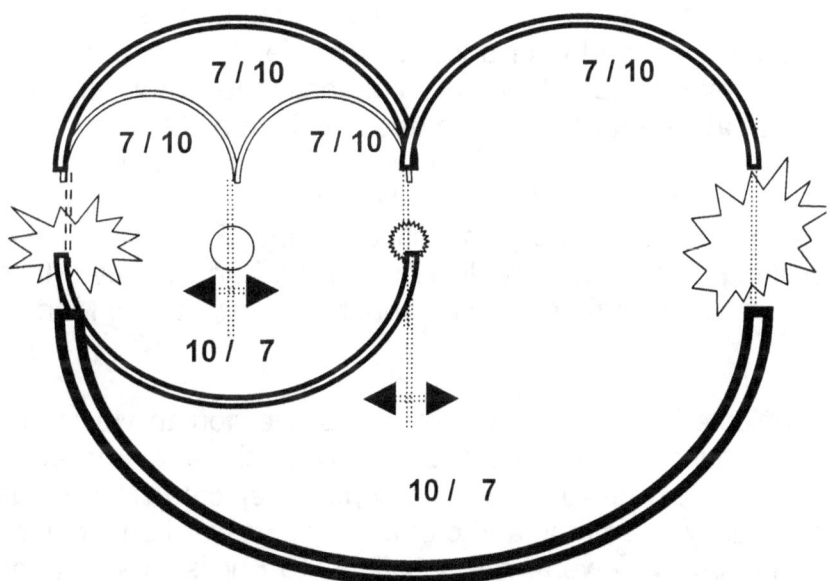

We now can use what is cosmic law to interpret how development actually took place by cosmic construction. The form the cosmos adapted remained d present to this day because what once is part of the cosmos does not go away and the cosmos use the same there is in so many ways the many ways make what there is confusing to us.

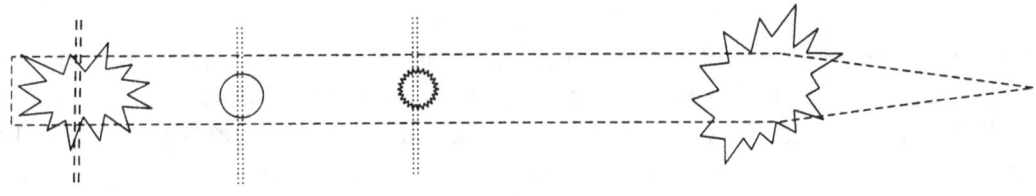

In view of the Titius Bode time depletion, time in flow creating space has a far more complicated arrangement than Xepted science can even produce on a chart. To be honest every person knows that Xepted science cannot even place the planets on a chart, depicting true distance to size, but they WILLFULLY never mention that information when the chart they show is as false as a three-dollar bill. In a sense it does no harm leading people down the ally in such a way, because others in my class of mental insignificance in society is far to un-intelligent to realize the correct way and will therefore not understand the correct way in any event. That is a mistake with a stinging tale. It is as dangerous as a scorpion to Xepted science. In an introducing article, I named Anglo-American Mythology, I pointed out how misconception feeds society, favouring the lies and untruths and blatantly ignoring the truth. In the past, since time began the powerful used this on the brainless masses, and for a period where that civilization lasted, got away with that strategy. The next civilization that came to power, followed the same methods applying the same dogma, and in the end paid the same penalty because their greed, lust for power, and sublimations gave them control over the masses for a while. The misconception those in favourable positions forced onto the masses, made the very people in power so shortsighted, their course on vanity lasted but a few generations. This is achieved because our Earth environment is tolerant and can buffer a lot, to save life in the end for life's contamination on Earth. They wish to extend life's connection to Earth, as being a connection to the cosmos at large and will be in effect as long as life remains in the cosmos. When we are "going abroad" to our "next door planet" that apparently holds all the supporting evidence of life carrying organisms, the connection to the cosmos remains and connecting to the Earth is of little consequence. That bluffing must stop. I realize no one on Earth will ever take note of what one sod (like me) on Earth is shouting, but misery awaits our Martian Colonists.

I started using the term Xepted because I do not accept Newtonian views and therefore I except Newtonian views but this bloody brainy machine tell me every time what I can and can't write when I refer to excepted (rejected) Newtonian science. Then I got brainy all by myself too and created my own word and told the machine to accept and shut up. Now we all are satisfied and English just got one word richer by my inventing Xepted science which is what I use when I refer to Newtonian science that science wilfully accept while they very well know it is Xepted science.

Suffering will be the reward for the fools attempting to catch the bounty of "fame, riches and glory" on behalf of the All Powerful Dollar and the dollars absolute true benefactors. Those with eyes, let them see, those with ears let them hear and let the rest self demolish.

Binary stars, spinning to self-destruction will produce significant heat. Heat create space, space forms winds. That is facts that the Bible present and is indisputable. Where the Earth was, was still a void, containing a sphere of circular displacement and this will reduce linear displacement to zero. Linear displacement is space and circular displacement is containing heat for matter survival.

Binary Star Minor overheated. That is why the core brittle and fragmented. This action will release tremendous contained heal, the heat will produce magma flowing in space like water in space and this eruption of heat space that created winds. Once again the recollection fits the scenario. Releasing the heat and producing space will establish space-time and fill the void where the Earth should fit. This is fact and if anybody even tries to dismiss this will be because of abstinence on his or her part. I did not prove the Bible correct. The Bible told the truth and in such correct detail, it is beyond human comprehension, but sublimation on the part of Newtonians and science before them, disallowed their ability seeing it.

That is what an insignificant formula $R^3/T^2 = 1$ where $R^2/T \times R/T = 1$ represents space-time in singularity as well as space-time in densified, occupied and unoccupied format. That means the everything of the whole lot, or as we say in Afrikaans, the "Heelal" meaning Universe. It refers to space-time for the first time while nuclear explosions are the epitome of $R^3/T^2 = 1$ where $R^2/T \times R/T = 1$ and how long has nuclear explosions been with us?

Now comes the proof: In the electron dimension the value is Π^2 Π in relation to 3. In the cosmos there are always at least six sides to any object.

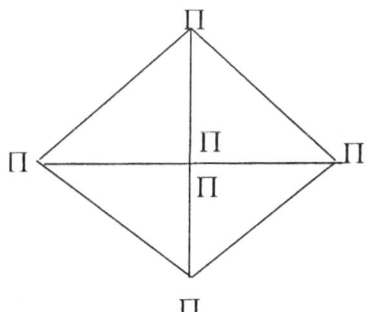

This gives the value of Π^6 in relation to the Titius Bode value of 7/10 in relation to the 6 sides in space (10).

$$\$T = 7/10\ (\Pi^6) / (6 \times 10) = 112.$$

From this comes the Newton formula holding time in the fourth dimension to space in the fourth dimension. At all times the dimension application will be the value of time in singularity (Π) to the time of matter (Π^2). Separating the illustration will be as follows:

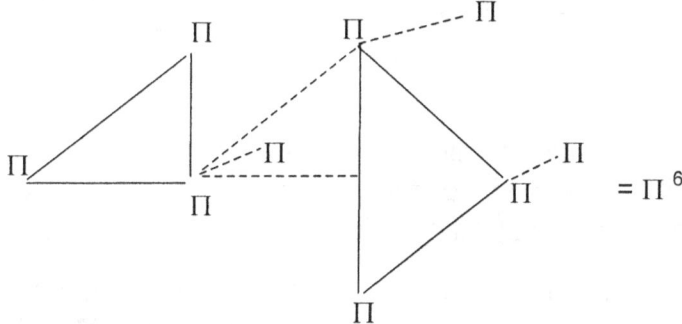

Through this "gravity" has the ability to produce a dimensional change from 3 to Π. Heat in a liquid form will be $3\Pi^2$ while one step more concentrated (the neutron state) heat will be $\Pi\Pi^2$. This reduces the triangle in space-time to the half circle in space-time.

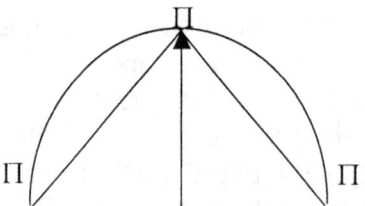

The one side is connecting the 3 x Π and with the change in dimension cross over falls away changing the 3 (number of Π) Π^2 removing of the dimensional value of a half cube (6/2) to that of a half circle. ($\Pi\Pi^2$).

Both remain 180° contact with time in singularity Π^3, which is a straight line (180°). But there can never be a dual application. The one holds the other in support. In a star this fact becomes irrelevant but a star is the cosmos, applied in reverse. In the galactica (of which the solar system is a fragment) the relevancy of support will always apply at the outer circles and not apply within the inner layers. However, this is beside the point as we are dealing with the Titius Bode configuration of 3 ; 6 ; 12 ; 24 etc. that apply to time in space formed by heat.

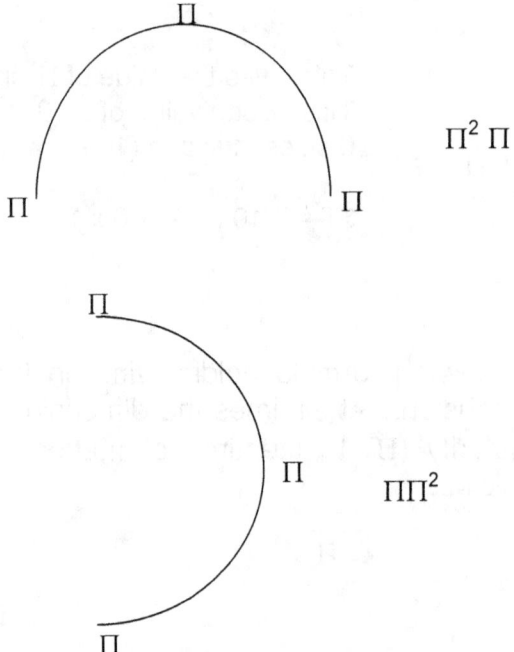

The relevancy applying in a cosmic cluster will always be that of a half circle applying as a triangle because it is in support of the triangle. The single Π will always support the Π^2 of time and this configuration will be a half circle.

Only the half circle can apply at any given point in any given moment. The value of Π^2 extends as the value of Π because Π^2 is the circular of "gravity" and Π is the linear of "gravity".

The rest is everyday mathematics. The invert square law applies just as well to a half circle than it does to a full circle. The value of Π in the next circle will be that of Π^2 in the previous circle. Well in a way presenting it as the invert square law does apply, but it has a cosmic sting to it.

There is a much more substantial explanation about how the Titius-Bode law arrive at the configuration of 1, 3, 6, 12, 24 doubling every time. However, such an explanation covers a great volume of facts because we have to cover the neutron's calculation from every angle available.

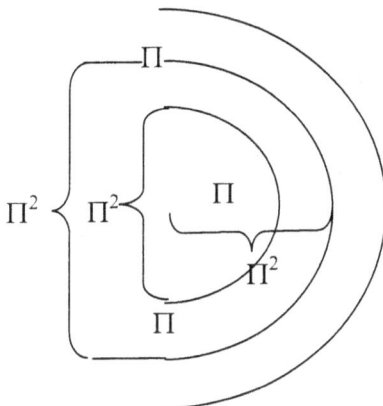

In the first configuration matter lends space all value at $\Pi^2\Pi$ configuring to the dimensional influence change of 3. This will be as such.

Matter

$\Pi^2\,\Pi$ $\Pi = 7/10$ $3 \times (\Pi$ and $\Pi = 7/10)$ becomes 3

                 Π

          7/10

7/10 7/10 $\Pi = 7/10$ ◄──► $7/10 = \Pi$

Π Π

Every time another body develops to the outside of this the inner space will be three dimensional space $3^2 + 1^2 = 10$ and this apply to the sevens, therefore the space to the inside becomes 10/7 and from the space the next applies a matter value of $\Pi^2 = 2\Pi = 7/10 + 7/10$ with Π at 7/10. Only Π^2 determines time therefore $2\Pi = 2(7/10) = 14/10 = 1{,}4$.

Space holds the value to the already developed part as $10/7 = 1{,}42$. Therefore space will be 1,42 followed by matter, 1,4 and that leaves the Titius Bode law to double its distance. $3 \rightarrow 6 \rightarrow 12 \rightarrow 24 \rightarrow 48$.

Where Π^2 are the radius of the one circle it will double to become Π in the next circle. It is the manifestation of the neutron dimension applied in the electron dimension where all 3 of Π holds equal value therefore Π^2 will become $2 \times \Pi$. This relevancy will apply wherever heat and matter produce space-time. This is a given, standing as firmly as the Hubble constant, the Roche limit and any other law application.

Every person in the past sought a relevancy in the application of this phenomenon in as much as it applies to the solar system. The proof there is, is not in applying, but in the way it does not apply and the reasons why it does not apply.

When testing for proof in the application of the Roche limit as far as it stands in figuration of the solar system, we will find it does not apply at all. That means the fact that is NOT PRESENT, PRESENTS THE PROBLEM. The absolute importance of the Roche principle not only reflects on the influence of the Roche principle alone but the Titius Bode space-time growth and the Titius Bode configuration that is an extension of the Roche principle.

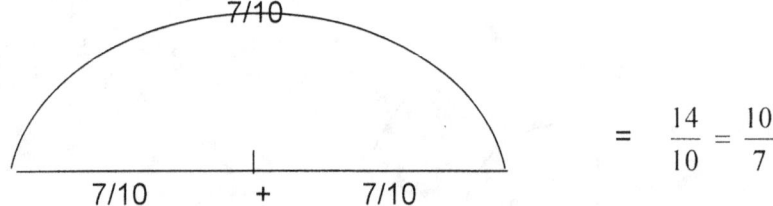

$$= \frac{14}{10} = \frac{10}{7}$$

But the matter position relate to the space position.

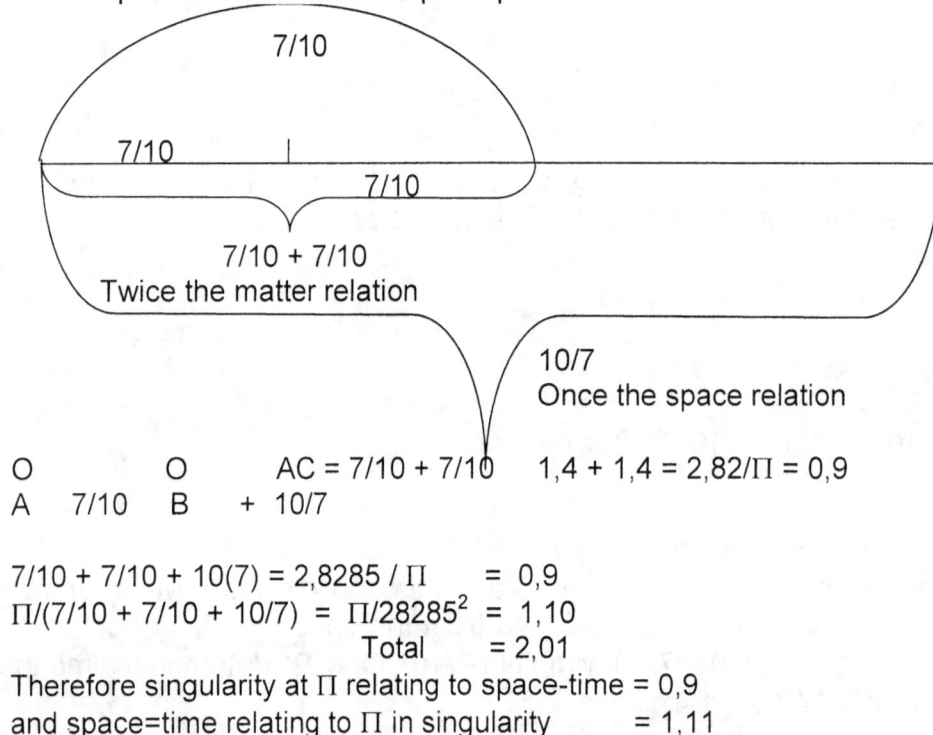

7/10 + 7/10
Twice the matter relation

10/7
Once the space relation

O   O   AC = 7/10 + 7/10   1,4 + 1,4 = 2,82/Π = 0,9
A 7/10 B   + 10/7

7/10 + 7/10 + 10(7) = 2,8285 / Π   = 0,9
Π/(7/10 + 7/10 + 10/7) = Π/28285² = 1,10
                Total   = 2,01
Therefore singularity at Π relating to space-time = 0,9
and space=time relating to Π in singularity   = 1,11
The total of matter to singularity   = 2,01.

Therefore Π² to Π = 2Π . This then is 7/10 + 7/10 = 10/7
Therefore space will always hold double to the relevancy of matter.
To that end
.   3   . (7/10 + 7/10 + 10/7)
.   3   6   .
A   B   C
        2Π
Then 6 becomes 3 = 2Π(Π=6)=12
A   C   D
Then 12 become 3   .   2Π = 24

And that concludes the Titius Bode configuration of 3; 6; 12; 24; 48 etc. by valuing the triangle and the half circle.

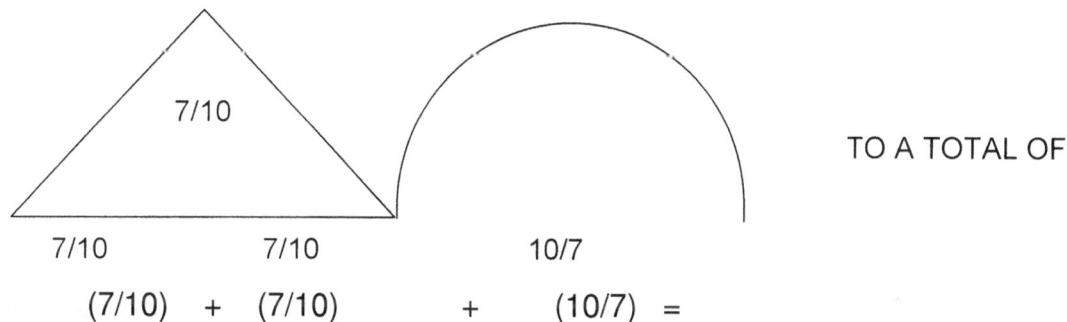

TO A TOTAL OF

(7/10) + (7/10)   +   (10/7) =

Three going to double in the next configuration.

The first $\Pi^3 \rightarrow \Pi^2$ Π Separating singularity.

This then brought on Π^2 in heat.

$$\Pi$$
$$\Pi^2 \quad \Pi \qquad\qquad 3\Pi = 3$$
$$\Pi$$

From that space developed

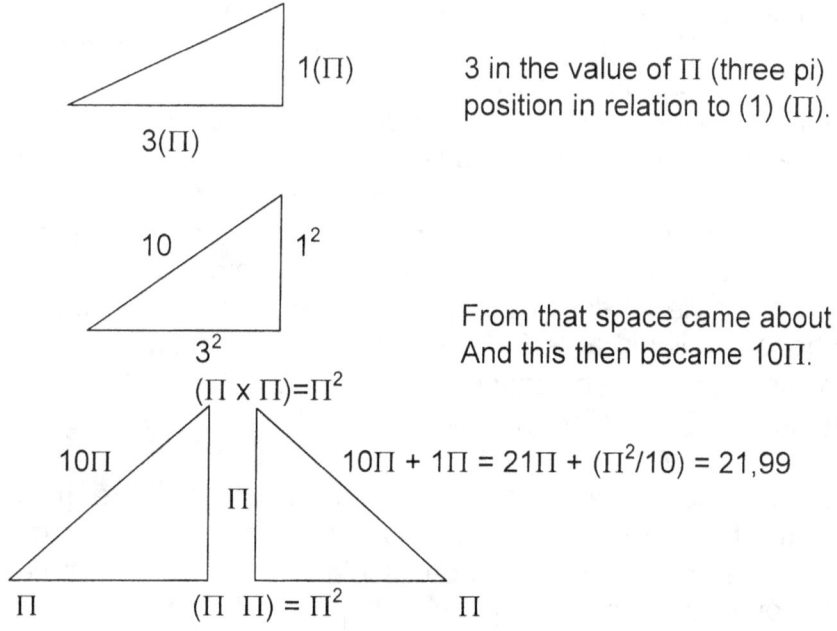

3 in the value of Π (three pi)
position in relation to (1) (Π).

From that space came about
And this then became 10Π.

$10\Pi + 1\Pi = 21\Pi + (\Pi^2/10) = 21{,}99$

On top of that the sphere established.

Add that to the seven that holds densified and occupied space-time and there is seven of Π in the triangle of matter adding 3 Π's in the half circle of space ($\Pi^2\Pi$) and the total is 10Π. With 10Π the value of the total triangle (in a square) and 10Π the total of space-time (matter holding singularity apart, there is a factor of 7 by Π to 10 by Π, with the triangle having two Π - two factors where Π at the bottom formed ($\Pi \times \Pi$) = Π^2 and to the top ($\Pi \times \Pi)^2$ = Π^2, with all of this constructing space (10Π) to matter Π^7.

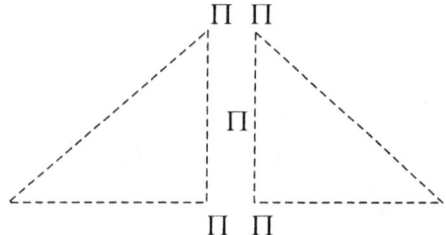

This will bring about 5 $(\Pi^2+\Pi^2)$ bringing matter and space to singularity. There are Π to the number of 5. From this comes that 4 (the four Π in the double proton in relation to the proton $(\Pi^2+\Pi^2)$ will always form the time value $(4(\Pi^2+\Pi^2)$. From this comes the fact that $3(\Pi^2+\Pi^2)$ will hold the space component as that in fact is half space-time.

Space-time being $(\Pi \times \Pi)(\Pi)(\Pi \times \Pi)$ and half of that to any direction is $(\Pi \times \Pi)$ (Π). This is why space-time in the geodesic sense has the value of 10Π and that 10Π in a sphere are Π^3. It is both triangle relating to one half sphere which is both triangles (7) in the totality of the half sphere (10).

The value of $2(\Pi^2+\Pi^2)$ will relate to a position where space passes on time in a dimensional transformation, but not a value transformation. It will be when Π relate to Π^2 as much as $\Pi^2+\Pi^2 = 19,7$ and 6 (the sides available for Π to use in space is 18,84. It will form a border of dimension where space will not apply in the same manner as it did before reading that border. One may say that is the densifying border of heat in space to heat in matter.

The last position of the Lagrangian atom is where only Π (in the triangle) have value and this links space directly to time (matter). At this point concerning the Lagrangian layout, we must view heat, filled space, the stuff we exist in, as an element. Hydrogen becomes liquid at $-269°C$ and heat (outer space) has a gas value of $-273°C$. This is only a dimensional changeover from 10 to Π^2 or from 3 to Π. That was what the "Big Bang" was all about.

Creating the Universe in space as we see space, was the process where the last natural element formed. The Universe or Cosmos is the last atom, which formed. It was the conclusion of the proton $(\Pi^2+\Pi^2)$ in developing. That space can be in gas as we now see it, covering elements to position them in relation to the rest of the Universe as gas, liquid or solid. That is why astro-physics in the ways science apply it at present, is but a good old romantic science-fiction story.

Mass is the frustration of motion while motion in duplication as well as contraction representing gravity. Even with our experiencing of mass it is the tendency we experience to move that is the gravity and the mass part is the blocking of the space we with our bodies wish to claim. While the earth is blocking our claiming of the space we wish to occupy the earth as well as us are both in mass but the mass has precious little and a lot of nothing to do with the fact that we are moving through the application of gravity. To quote Kepler space a^3 filled with material has to move in line with a centre that is controlling the moving of space through time and in time as well. Therefore I repeat what I said before. Mass holds no value, it is density and that density brings in the heat in space, unoccupied as yet by matter. The motion represents density of time in space. That density gives a star its "gravity". The space has to move through time by duplicating and the more dense time is the more effort such duplicating requires. Where the Sun can only hold heat to liquid more dense stars will hold heat to a "jelly" and others will take heat all the way to something as hoard as tungsten. That is what tungsten is. Tungsten can place heat in a relevancy that the heat relative to tungsten is almost as dense as the neutron within an atom. That is why I refer to the system as the Lagrangian atom, because the Universe holding heat, produces all relevancy matter can have. The Universe is the final atom.

The diameter of the Sun is 1391,980 km. Bring this radius in relevance to the Roche factor and the first orbiting structure will be Π^2 relating to $(\Pi/2)^2$. With Π^2 at a value of 1391980 x $(\Pi/2)^2$ the position of Mercury must therefore be 3 4345 73 km making it approximately $3,5 \times 10^6$ km. With the effect of the Titius Bode configuration the next position must be 7×10^6 km and the third at 14×10^6 km. If the orbiting structure were that close, as it should be under all normal conditions, we would have been roasted toast.

1. Mercury
2. Venus
3. Earth
4. Mars
5. Ceres
6. Jupiter
7. Saturn
8. Uranus
9. Neptune

Roche Limit according to actual dist variation	1	2	3	4	5	6	7	8	9
	3.5	7	14	28	56	112	224	448	896
	57.9	108.2	149.6	227	414	778	1427	2871	4497
	16.5	15,4	10	8	7,39	6,5	6.3	6.4	(5)

If there were only the Sun that affected the positioning of the gas, "planets" the diameter of the Sun will be Π^2 placing Π at a position where dimensional implication becomes valid. Since the structures are still in space-time positioning, the effect of the Roche limit will still be in place. Therefore the Sun would be Π^2 arranging $(\Pi/2)$ accordingly.

Sun	1	2	3	4
Π^2 o	10/7 o	10/7 O	10/7 O	10/7
2(7/10)	2(7/10)	2(7/10)	2(7/10)	

I shall explain the layout as follows: To every structure the value of space-time in the electron dimension is

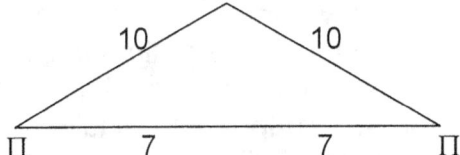

Therefore the matter positioning will be 2Π and that in terms of matter relates to (2x7/10)=1,4. The space-time to space will therefore be 10/7-1,42 because matter is two parts while space remains in singularity and singularity is always one. Therefore 10 cannot double it is one in relation to any one matter part at any given point. This then means Π² means in real terms (2 x 7) /10 as one Π and 10/7 as the other Π. Because matter relate to the dome in the half circle and space to the triangle, in relation to space it is Π+Π and to matter it is Π x Π. In relevancy to matter as we apply our attention to the two structures, the correct connection is (Π/2)². However, in the space factor it will be (Π + 10Π) / 7 x (2²) therefore it will be on the one side

(7 + 7) / 10 (matter plus matter) = 1,4.

On the other side though it is (10/7) = 1,42.

This is 1,4 + 1,4 = 2,82 in the space where Π relate to 10 and in this instance Π is 2,82. This means that bringing the relation back to matter will effectively mean it is (2,82) x 10 relating to 7 in conjunction with 2.

The Roche limit is Π which in this case is in space, therefore cannot be a square since the ten already apply as a square.

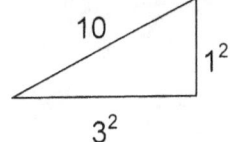

This all relate to the fourth dimension in space, but matter relate to the third or neutron dimension that holds time to a square. The matter as such remains in singularity (7) but time stands in regard to the square of half of Π. This then means matter (7) holds a relative to half the time effects matter (7x 2).

Space-time outside singularity then is

Space-time outside singularity then is

$T = \dfrac{28,2}{7 \times 2}$ which is (1,4 + 1,42) x 10 , which is matter (7) to time 2

$T = \dfrac{28,2}{14}$ = 2

Therefore every object holds the inner structure as 1 and in accordance to its own position of 2. This then is $\dfrac{10}{7} + \dfrac{2(7)}{10}$. That is the Titius Bode implication, however, I explain this better when dealing with the Titius Bode just before this part. Therefore the first portion Mercury must have is Π² (the sun with the diameter of 1 391 980) with the Roche factor implicating (Π/2)². According to this the positions are as follows:

○ ———— $3,5 \times 10^6$

○———— (1) ———— $2 \times 3,5 \times 10^6 = 7$

○————————— (1)———— $(2 \times 7) = 14.$

○——————————— (1)————$(14 \times 2) = 28$ and so on

We find that the gas planets are on average about 2 Π overshooting the development that would apply in the case where the Roche limit would result in positioning orbiting structures. As explained a few paragraphs ago, the application of Π^2 in terms of the Titius Bode configuration will be 2Π, a dimensional factor change. As I have indicated the fact that it shows as 2Π, in the electron dimension, become Π^2 through putting matter in space.

Therefore the two Π you see, is the Π^2 you get. Having Π^2 means one thing: there was another object (Π^2) that related to 2Π and the two Π can only come from one more object that filled that space during some duration of time in the past.

The Titius-Bode principle relating space-to-matter at a value of $R^0 / T^2 = 1$ where space holds the square of 10 and matter is 7

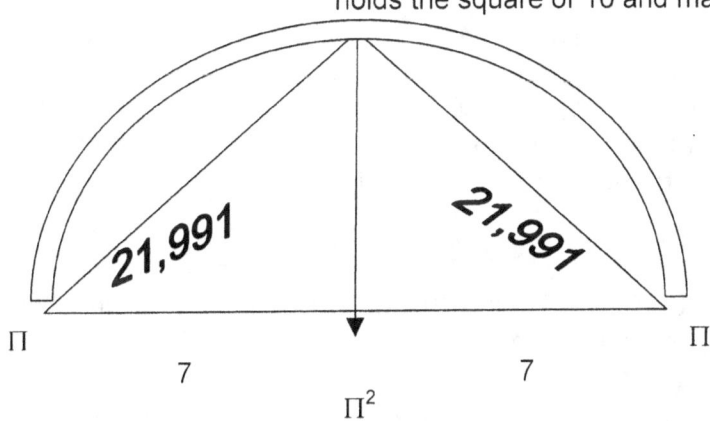

Therefore we may correctly surmise that something to the value of a relevancy of 2Π becoming Π^2 played a part in the positioning of the outer "planets" As shown repeatedly a double star would apply as $\Pi^2 + \Pi^2$ with $(\Pi/2)^2$ separating the stars where the Sun holds the position of Π^2, and therefore another object was present during space development in the time the Sun released from eternity. In other words THERE WAS ANOTHER STAR IN BINARY TO THE SUN.

In the half circle the centre is 2Π's and in the triangle the corners form 2Π.

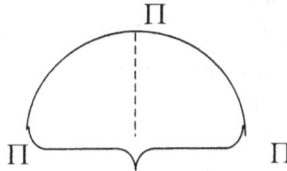

The dimension "gravity" removes as it replaces it with Π^2

In the half circle the fourth dimension holds a triangle.

That is singularity in time to the value of Π^3. The aim we have is determining the value of Π^3. The aim we have is determining the value of $\Pi2$. Known to all at this stage is that there is 7/10 in the Titius Bode law and in all spheres, including the Earth, we have a space dome of 21,991 holding space to the 7 holding matter. When a spacecraft re-enters the atmosphere, the angle of entry must be not less than seven and not more than 21,991. Therefore there are 7 holding 21,991 to the value of Π. This is the dimensional equal of the Titius-Bode law of 7/10. In order to determine Π^2 you therefore have to translate that value to the fourth dimension, giving it a value from singularity (linking time and space to a figure of 1) therefore 1, to its time position of Π^2.

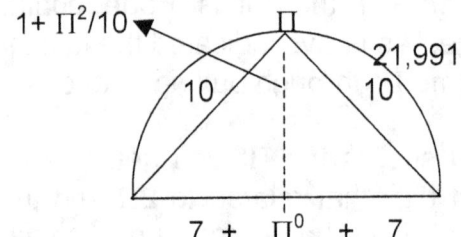

$$1+ \Pi^2/10 \qquad \Pi \qquad 21{,}991$$
$$10 \qquad 10$$
$$7 + \Pi^0 + 7$$

First of all, all Newtonians are educated in mathematics, therefore they will know that the triangle holds 180° the equal to the next dimension of the half circle at 180°, also equal to the straight line of 180°.

Secondly, Newtonians are aware that multiplying in the one dimension translates to the next dimension in the form of adding i.e. 4 = 2 x 2 and 4 x 2 = 8. Therefore $2^{1+1}=4$ and $2^{1+1+1}=8$. In the one dimension adding is the same as multiplying in the other dimension. However, in astro- physics one do not merely transfer dimensions, you work with dimensions running concurrent in value. Therefore to the top you add or subtract, and to the bottom you multiply.

First we subtract the top from 21,991.

Titius Bode 10 + Titius Bode 10 is 20. Adding the one, the link between space and time in singularity, holds a position of one.
This leaves us 21.

Then determining the point where gravity (Π^2) will be at space (10) and will therefore be $\Pi^2/10 = 0{,}99$.

In order to get Pythagoras, you add 10(T.B.) plus 10 (T.B) plus 1 (singularity) plus $\Pi^2/10$ ("gravity" ending in space) and you. Square this total of 21,991 as well as divide it with the value of 7 x 7. (The top you add, while the bottom you multiply).

Therefore the top is 483,6 and the bottom is 49. To get to the "gravity" part in another dimension you divide (not subtract) the square of space (483,6) with the square of matter (49) and the value will then be 9,869467, the value of Π^2 relating to singularity. This means that the 2Π space holds, were filled with matter (Π^2) . There can be no argument about that fact and we can at present see the other Π^2 being the sun.

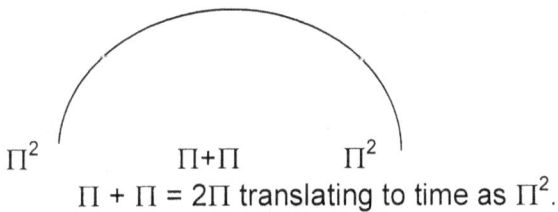

$\Pi + \Pi = 2\Pi$ translating to time as Π^2.

That is what the Roche limit is all about. It is the point where a cosmic proton (Π^2) shares (that means halved) a space relating position of Π in neutron dimension of time (Π^2) $\Pi \rightarrow (\Pi/2)^2$.

$\{(\Pi = 7) + (\Pi = 7) + (\Pi = 7) + (\Pi^2 / 10 = .991)\} = 21.991 / \{(7 / 2) + (7 / 2)\} = 7$

$R^2 / T (S\$^3 S\$^2) R / T(+ S\$^2) = (7 \times 7) / T^2 = (7 / 2) + (7 / 2) = 7$

$\Pi = \$_T = 7$

$$(S\$^3 S\$^2) + S\$^2 = \quad = 483.74$$
$$= \sqrt{483.74} = 21.991$$

$+ \quad \Pi = 7$

$\Pi \quad = 7$ **21,991** **21,991** $\Pi = 7$

Π

$\Pi / 2 = 3.5 \qquad + \quad \Pi / 2 = 3.5$
$\Pi = \$_T = 7 \qquad (\Pi^2 + \Pi^2) \qquad \Pi = \$_T = 7$

 If any person wishes to cling to Einstein's view that the speed of light is the fastest that matter can apply velocity, explain how the Black Hole works. Inside the Black Hole must be matter, because there is no space, yet time does apply because it takes the particles spiralling inwards to the centre time to move from point to point. Matter in motion is time. However, no light can return to the surface, therefore the light is slower than the moving particles within the star. The only thing about the star, is that it maintains a higher relevancy than the relevancy the speed of light can apply. By accepting the existence of a Black Hole, the Einstein claim about the speed of light being one, becomes zero.

Another place where the speed of light becomes obsolete is within the centre of galactica, where the accumulative movement of matter exceeds the speed of light. That is where doctor Hawkins saw a Black Hole that is not a Black Hole, but the precise opposite. Light matter and heat, moves inward in an effort to maintain cooling as the group of proto, proto stars two era to the future) still claims their share of heat maintenance. Those particles in such close proximity, establishes a time well above that of the speed of light.

Everything in the cosmos is all about relevancies. Time started at such a high velocity, it had to be eternal. Nothing diverting from eternal can become more than eternal so it has to be less than eternal. It is fragmenting eternity into parts making eternity smaller. The smaller eternity becomes, the lesser eternity will be. That means that time started at eternity and became shorter with the introduction of infinities that broke the monotony of eternity. The more inanities there are affecting eternities, the shorter will eternities be.

I prove in "Matter's Time in Space – The Thesis" where Einstein went wrong in his theory about "The curvature of space-time". There is the space-time complying with singularity and filling the space-time in singularity is heat and matter valuing space-time. Space-time (Π^3 to Π) cannot bend, cannot curve, forms a straight line, but what fill space-forming time is matter (Π^2) and heat (3).

That part changes. The atom cannot be gas, or liquid, or a solid, because the atom is densified in occupation of space-time. It is the heat in unoccupied space-time that produces the gas, liquid, or solid that all substances can form. It is THE HEAT in SPACE that produces TIME, that can and does curve, bend or whatever. That HEAT in SPACE forming TIME that forms the relevancy of space-time and that does bend. If, by applying the forming of gas, or solid or liquid to the element instead of the space between the elements, of course you will get the incorrect vision Π, where the space-time (matter holding singularity form singularity) is doing all the binding that apply to the curvature of space (validating time) and time in singularity (a straight line) will be solid. Einstein placed the relevancy incorrectly on singularity, instead of heat.

Once again I do admit, IT IS A LOT MORE COMPLICATED THAT WHAT I MAKE IT TO BE AT THIS POINT, but the motto is, Keep it simple.

If you wish to keep time in space constant, everything in the Universe will be oblong. That is why the Newtonians have an absurd view of the cosmos, and they present facts in the cosmos in a way nobody (least of all the Newtonians) can understand.

Time was slow, time became faster and faster because by extending the position of Π, Π^2 will produce speed.

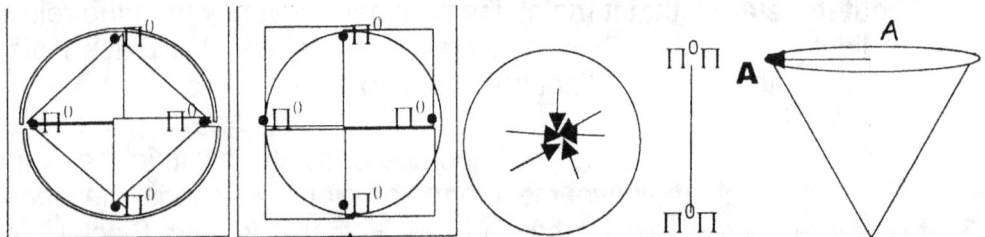

At that point space developed to a sufficient degree as to allow material, space and time form units being confined structurally while standing also individually apart. It was the pre runner to what then later became galactica filled with stars that was filled with atoms. But this was the prelude to all of that.

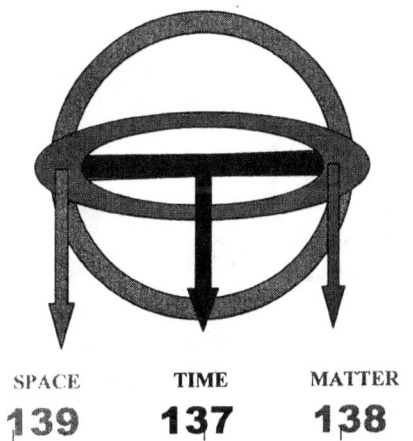

SPACE	TIME	MATTER
139	**137**	**138**

The period I refer in the space above was when space parted with time. It was when Π^0 moved to Π or the spot became the dot. It must no be confused when matter entered the equation. In a while I shall explain what happened at a proton displacement value of 139, 138 and 137 when space filled with the heat that we now think of as material. What I refer to above was in process where cold for the first time parted from heart before any time delays became a reality in space-time.

The motion is still within singularity because moving involves forming a relevancy between heat and cold and between infinity and eternity, between space and time and most of all producing what will in the far future develop into a Universe that can even be a host for life albeit on a very small spot for a very short while in relation to the vastness space has and the duration cosmic time has. This where time started and time remains at this edge of forming space by motion from singularity that cannot move because it has no space.

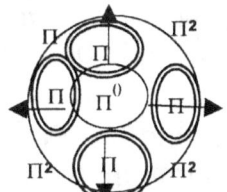

By moving from 1^0 to 1^1 and from $1^0\Pi^0$ to $1^1\Pi$ requires space. Yet, when form came into form such moving did not leave the realm or the domain of singularity. That is still with us since the principle has nowhere to go but to remain in the universe. The motion brought about Π as the motion brought about Π^2 using the same motion. It is the motion that moved 1^0 from 1^1 or $1^0\Pi^0$ to $1^1\Pi$ that became time in the square and the motion including time became space $1^0\Pi^1\Pi^2 = \Pi^3$

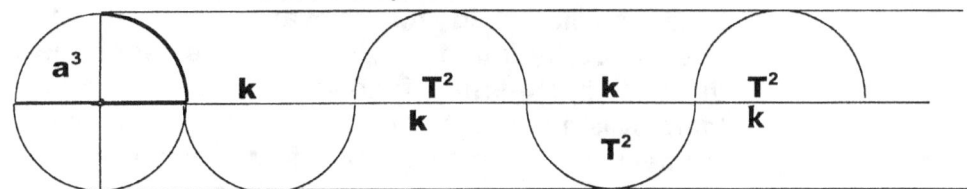

The graft is a basis on which the entire Universe was formed when only form was available. In that scenario the atom was set without using space...yet when the atom developed before the Big Bang.

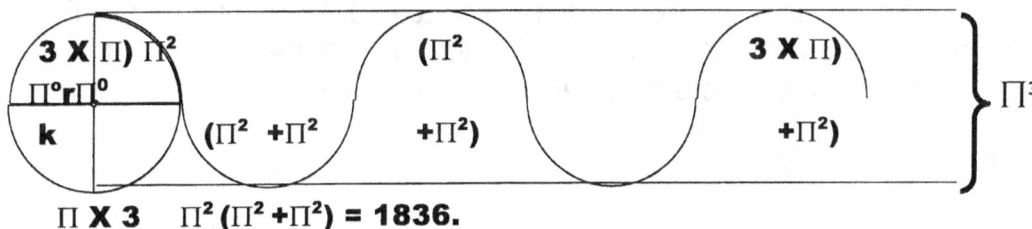

$$\Pi \textbf{ X 3} \quad \Pi^2 (\Pi^2 + \Pi^2) = 1836.$$

Time started by placing the double proton in space and in matter, where space and matter will always be in a sphere. The sphere always forms 7° from one point to another running outwards.

$$\$T = 7(\Pi^2 + \Pi^2) = 138$$

Explaining is as follows:

T is space-time (\$) in the time sector (T)
7   This indicates that the Coanda principle placed motion to space.
$(\Pi^2 + \Pi^2)$ Refer to the proton committing the Coanda effect.

At the same time space formed as a consequence to the Roche limit and so did matter. Forming the Aanplasings -Atomic-Epitome (the point where matter breaks in singularity). This was where the three-dimension concept was introduced but not quite accepted in practise because form still rules.

Being part of the 3D we have the inclination to think of something and then we also includse in that something the space we experience. Thinking of the spinning top we will think of the edge (A) forming the end of the line, the position of seven. This cconcept we have is part of the 3D Universe. To understand cosmoogy we have to rteturn to the biginning of cosmology. The relations then was when Π^0 formed Π.

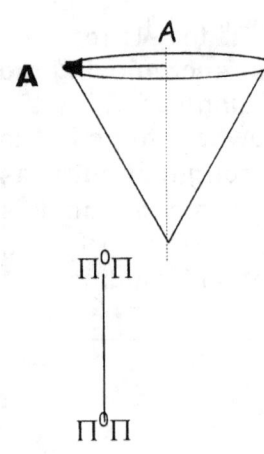

The moving of Π^0 to Π involved relegation and not motion as we consider motion. It was Π^0 getting a side and that is all. There was no true side but only a form that came into place. Singularity (A) received singularity (A) and no more of anything but the shift to comply with having a relevancy forming in relation to singularity. The dots had no sides, had no length or diameter. There was not measurable space or measurable time involved. The time could have been a micro, micro second as much a trillion millennium because time had no relevance. It was eternity interrupted by infinity, as it still is the case, however the line that eternity followed was no line because there was no space to hold the line. The line was momentarily interrupted by infinity, however with no one there, there was no one to notice. The lines were not lines but relations to sides being formed.

There was then an outer line forming time in space 10/7. Then there was the inner line forming space-time being 7 /10. The there was material filling space at $(7)(\Pi^2+\Pi^2)$ forming the sphere as it was filling the sphere.

$$\$T = 10/7 \, (\Pi^2)(\Pi^2+\Pi^2)$$

This became the atom $(\Pi^2+\Pi^2)(\Pi^2\Pi)(\Pi^0+\Pi^0+\Pi^0) = 1836$ and the atom formed stars that still act in accordance with and to the atomic relevancy

Outer space substantiate the atom as $(\Pi^2+\Pi^2)(\Pi^2\Pi)$ 3

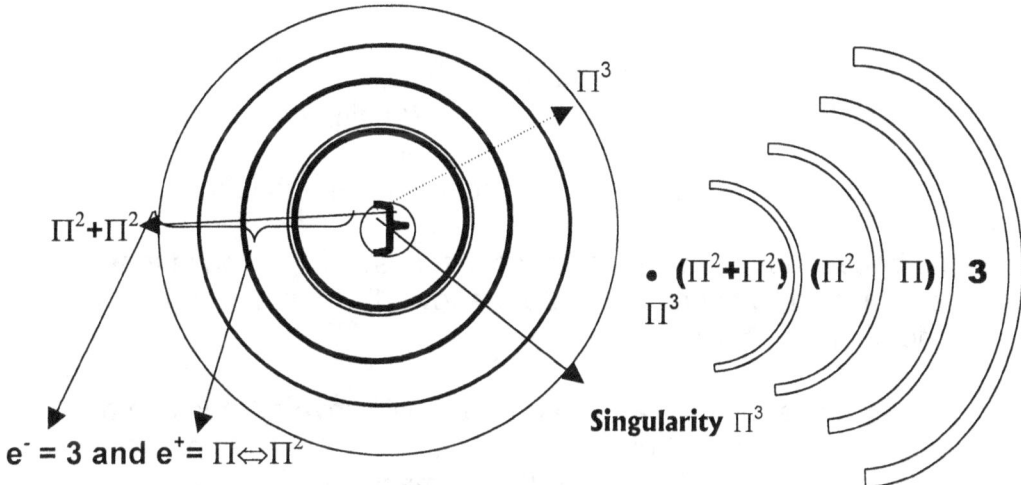

$e^- = 3$ and $e^+ = \Pi \Leftrightarrow \Pi^2$

Every layer in the star represents one factor in the atom since the star is just another cosmic atom securing strings of atoms that as a unit aims for one goal and that is to secure one singularity within the star.

Explaining as follows:

10/7   The limit between what is part of the atom and what is excluded.
7/10   Forming the matter factor.
$(\Pi^2/2)$ Indicates a deliberate inclusion of an atmosphere or a liquid or a neutron
$(\Pi^2 + \Pi^2)$ Shows that the proton still had total control

$\$T = 7/10(\Pi^2\ (\Pi^2 + \Pi^2) = 138$

Explaining as follows:

7/10   Heat flowed to material supporting gravity within the centre of the sphere.

$(\Pi^2 + \Pi^2)$ the double proton

(Π^2). The boundaries were set by the motion that the neutron provided.
The atom was born

$\$T = 7(\ \Pi^2 + \Pi^2\ ) =$ 138 The circle the atom has.
•$\$T = 7/10\ (\Pi^2(\Pi^2 + \Pi^2)) = 136$
Motion towards the inside of the atom;

$\$T = 10/7\ (\Pi/2)^2(\ \Pi^2 + \Pi^2\ ) = 139$ the relevancy of space carrying time to allow material space to apply motion within.

The atom formed a circle and that placed the Coanda effect in control of the Universe. For the first time there was matter in relation to time 10/7 in space in relation to space-time 7/10

This is what the Universe consisted of, everything that is today, was then, in a dimension that only holds "gravity" or the " gravity - motion". The Coanda effect took the Universe into the three dimensions.

$$\$T = 7/10 \ (\Pi^6) / (6 \times 10) = 112$$

7/10   is the matter has the dominant value.
Π^6 matter has the six sides it holds in the fourth dimension.
6   are the six sides to space occupying matter.
10   are the value or dimension in which space holds a ratio to time.

The cosmos began, not to a specific space, because all the space that was there, initially is still there at present. Any atom above 112 cannot apply to the fourth dimension not then and not now.

A proton with a "mass" of say 12g/mol on Earth will have a "mass" in accordance to Earth standards of 25g/mol on Jupiter and it will hold a comparable "mass" of 100g/mol within the sun. This "growth" in mass of any molecule within the structure's potential occupation of space-time increases. With this in mind, one cannot merely bring in such a relation to the "mass" of the proton in the beginning or within a star, or as it is on Earth. As the atom's spin increases or decreases in the relation to $\Pi^2 + \Pi^2 \rightarrow \Pi^2 \Pi \rightarrow 3$, and with it, the "mass" will subsequently alter.

In explaining all of this, it is quite impossible for me to give it a value in mass, or time as both these factors alters in space-time occupied.

I INCLUDE THE TECHNICAL DETAIL TO SILENCE THE "SUPER-EDUCATED-KNOW-ALL" THAT IS FLOATING ON THE "CUTTING EDGE" OF SCIENCE. FOR THE AVERAGE PERSON THAT DOES NOT HAVE ANY CLAIM ON THE IMPORTANCE OF A TITLE IN BEING PART OF THE ESTABLISHED "SUPER-EDUCATED-IN-XEPTED-SCIENCE-MOCK", FEEL FREE TO READ THE TECHNICAL EXPLANATION, OR IGNORE IT, IT DOES NOT CHANGE, ADD, OR DISCARD ANY LATER EXPLANATIONS.

Whatever one believes, one has to be honest by admitting that time had to start somewhere. It proves only shortsightedness on the part of the Newtonians, to conclude that "gravity" started at 10^{-43} sec after the "Big Bang". This only concludes that the start was with the "Big Bang" and little else. Nothing is said about what caused the "Big Bang". Beside the point, but still very valid is the fact that I have no words in expressing my resentment with the term used as the "Big Bang" being the start of the Universe. This name only explains how little science understands about the cosmos.

NOTHING WAS BIG BACK THEN AS NOTHING WAS SUDDEN, OR QUICK OR BANG.

No person ever came up with a logic and scientific explanation to what brought about the process of heat expansion. What is irrefutably true however is that it came on route from eternity or timelessness or whatever one wishes to call it.

At this point, I have to explain the mistake we go about thinking about science and time. At first I was arrogant enough to think I was the first to understand the way time works. After all, it took me some time to figure out how the handle fits the fork. Then to my shock, I found that H.P. Wells already concluded my way of thinking about a century before I have. Well that proved so much for my personal brilliance and modesty once again, returned to me.

When witnessing an event we regard as an explosion we surmise that what happens on the inside of such an explosion is extremely fast, but to the contrary, it is very slow. In the explosion, the duration of time extends, becoming longer.

To explain this we take two persons, one watching the other runs a mile. We place both persons initially in the same duration where both will endure four minutes of time lapse. The spectator will see in real time how the competitor takes four minutes to complete the mile.

Then in the next scene, we increase the duration of the competitor to 1 : 60 and the spectator's time remains the same. To the spectator the athlete will be covering the distance sixty times faster, and to the athlete the spectator will be cheering 60 times slower. The spectator would not believe his eyes because of the athlete's abilities in running that fast while the athlete will think the spectator is in frozen state of admiration.

In the third scene, we enhance the duration in the athlete's time zone by another 1 : 60. This will bring about that the athlete, in the view of the spectator, will be running the mile in less than 7 hundreds of a second and the athlete will be watching the spectator trying to wave while the action of the spectator will last 240 hours. In the eyes of the one person, the other's time span will be either an explosion or, everlasting, depending on the person's point of view from the space in time that he holds.

To each one, the spectator and the competitor, a time lapse or time duration of 240 minutes occurred, although the actions in both sectors would have seemed to alter severely.

Any confusion coming about from the explanation above, I wish to remind that it is a common and well-accepted fact that time slows down as "gravity" increases.

Behind all of this explanation is one obvious rule. When the one subatomic particle positions in such a way as to displace space-time in the form of heat breaking down, the value of space, in order to meet the requirement of time, is once again all a relation between space and time. The less space heat has, the less the value of space becomes, because space has no value. Without heat in space and without matter there is no such a thing as space. Therefore, space does not exist, but for matter and heat valuing space to form time.

To understand this one must firstly understand the principle behind the theory I introduce.

At the most inner point one find time or if we can supply it with a completely fictional name: "The gravity - motion". The gravity - motion carries the value of Π^3. This value determines time in eternity a position matter has no space, but is occupied in singularity.

Taking the neutron position to that of the proton we find the value created when the three dimensions (six sides) came about $\Pi^2 + \Pi^2 + \Pi^2$, which carried to the fourth dimension in cosmic or geodesic space-time becomes Π^6. When relating Π^6 to singularity it becomes Π^1 (space) x Π^1 (time) x Π^1 (matter). I do realize this explanation does not suit normal mathematical principles but we are working in dimensions and Π^1 (time) in a straight line is 180° and Π^1 (matter), which is half a circle is 180° and Π^1 (space) which is a triangle is 180°. As each Π^1 represents one dimension establishing another dimension and providing that dimension's existence.

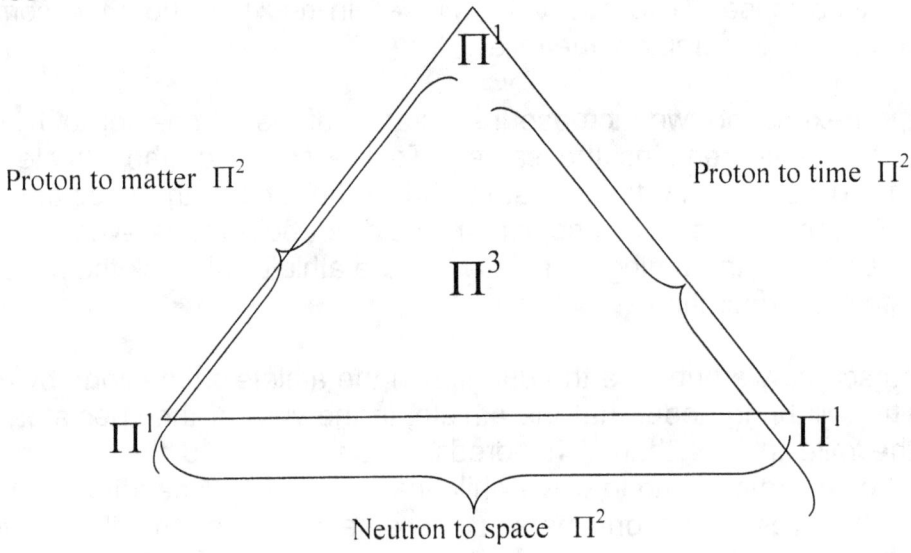

Proton to matter Π^2 Proton to time Π^2

Π^3

Π^1 Π^1

Neutron to space Π^2

FROM THIS SPACE HEAT AND MATTER DEVELOPED

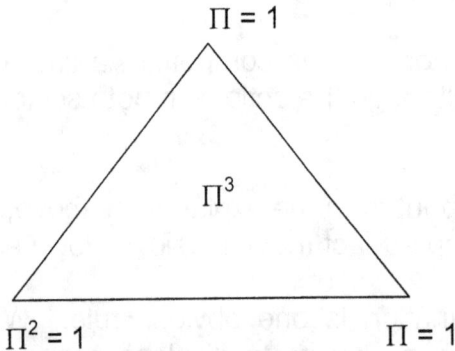

$\Pi = 1$

Π^3

$\Pi^2 = 1$ $\Pi = 1$

In relation to the " gravity - motion" space through a straight line will be Π and through the half circle matter with heat positioning space ($\Pi = 1$) to 3 sides 3. Relating the gravity - motion Π^3 three will always be a Π^2 and a Π combining 3. The half circle will be Π^2 and the straight line Π.

Behind all of this explanation is one obvious rule, the one subatomic particle positions in such a way as to displace space-time in the form of heat breaking down the value of space in order to meet the requirement of time. It is again all a relation between space and time and the less space has heat, the less the value of space becomes, because space has no value. Without heat in space and without matter there is no such a thing as space, therefore space does not exist but for matter and heat valuing space to form time.

It would be far too complicated to explain why space-time and water share so many characteristics but they do. I have, to some extent, tried to explain it but I am aware that the explanation falls short of satisfying. I will repeat it once more, well aware that it cannot bring acceptance. It is heat that produce gas or liquid or solid. The period before light, everything about the cosmos was a soup cocktail, heat was liquid and space in singularity was liquid flowing heat that appeared like water. In The Thesis I spent many pages in explaining this fact, but I do not wish to overcomplicate this book as I wish to bring across the scientific proof about the seven days of creation from a realistic scientific stance, for all persons to understand.

It seems very ironic that science with all its bravado, money, wisdom and splendour, can only begin at the point where light came to the Universe, while the Bible explains the creation in detail, long before the "Big Bang". The reason why there was no light before the "Big Bang" was that the spinning matter exceeded the speed of light, being $3\Pi^2$. The Authentic Author of Genesis refers to this as a mighty wind and this leaves a question. What better name can one give to this occurrence?!

With time in eternity, space in zero and matter being time and space, what would ever bring about that this situation changed? No Super-Educated-Wonder ever came forward to explain this. Why did the "Big Bang" start and what brought the "Big Bang" about? Only the Bible produces any logic to this question. Time, matter and space froze in one, there was no reason in nature for things to change, since this situation lasted eternally. Nature with all nature's laws did not apply, therefore one cannot say that nature started it. Nature was frozen. Nature was not even solid, it was in a state beyond being solid. Nature was nowhere!!

The spin in the Universe slowed down, up to a point where the spin was equal to that of the speed of light. At the point where the Universe spin equalled that of the speed of light, the Universe was still in total darkness. The light (photon) was there, but did not yet produce light. Light only came about as the spin reduced to below the speed of light, and only then light became obvious. The Universe grew away from darkness as this event lasted many eternities, during the period where the light separated from darkness.

How did this "Big Bang" take place? The best way to examine the reason is to see why anything in the Universe expands. To get anything to expand one has to heat it. All matter expands when overheating. Science may come up with whatever brilliant theory, the fact of the matter is that when matter overheats it expands. The bigger the overheating, the bigger will the expansion be, it is as simple as that.

This means whatever leads to the forming of the "Big Bang", whatever preceded it, it had to come about from matter that overheated. With the event of the nuclear age, the proof came about that matter is heat in some frozen form. Unleashing heat from its frozen form brought about a jolt of heat, never yet experienced by man. By breaking matter from the frozen state, of which it is in, within the atom, heat produces light and heat. Where this process clearly shows how new space-time forms is where the releasing of heat caused winds that stun man's logic.

The nuclear explosion shows quite clearly what the "Big Bang" was, with the nuclear explosion being a very minute form. Yes, we have all heard the rubbish about matter and anti-matter. What can be anti-matter, since matter is heat, defined to a certain space occupied for that time. With matter being frozen heat, what would form anti-matter. Anti-matter means the opposite to matter, and if matter is frozen heat, anti-matter must then be overheating heat. This in itself is quite ridiculous. Anti-matter can only be matter with an opposite spin to that we think of as matter.

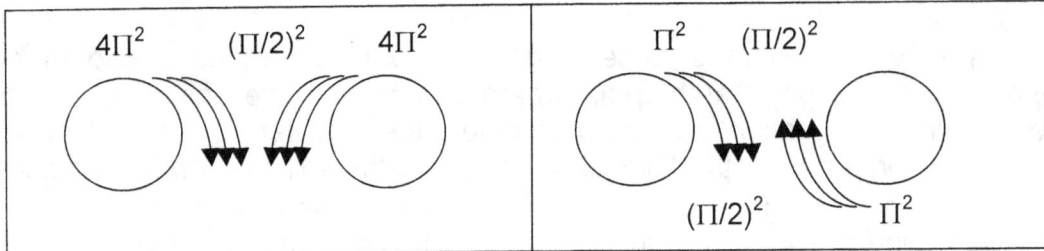

In the second sketch Π^2 $(\Pi/2)$ / $(\Pi/2)^2$ Π^2 the Roche limit cancel each other and with that the space of the neutron effectively disappear. The two protons touch destroying each other and the neutron as well as the proton demolish and became heat 3^3.

The process where matter then touches matter, it will bring about a reduction in the feeding process of heat,. where all matter in that space will overheat and expand, producing unfrozen heat. This means there was heat occupying space, and matter with both in relation to time.

The proton with a positive space-time displacement less than 136 placed its displacing properties in negative space-time displacement. In short, to substitute for mass shortcoming of less than 136 grams/molecule and still finding sufficient cooling properties for the proton to survive the overheating deficiency, it has to spin more rapidly, therefore by spinning it makes contact with more heat than it would otherwise do. One may consider this as "breathing". If the proton does not find adequate supply of heat by being motionless, the atom has to substitute the movement through motion. Forming an object that has an increase in heat supply through work commonly uses this fact. In nature, just the opposite is true because of the motion by movement, as one find in the case of wind.

There is no "force" in the cosmic flow of time because everything is a "force" in one way or another. Everything is in a 90^0 angle with time therefore the cosmos is restraining time while it is retaining space it is moving in relation with time but also opposing time all the way. The proton takes heat from space in an effort to maintain temperature and stability. This flow of heat brings about the reduction of space by increasing the heat in that space. The flow of the heat, through the electron by means of the neutron to the proton is time. The amount of heat taken by a proton is a constant throughout the Universe but relative to the space reducing effort of all the protons influencing that specific space. That is "gravity" (a term I denounce and reject). "Gravity" is nothing else but additional application of time. The higher the gravity is, the slower the time will become by prolonging the duration of time,

By taking this statement and introducing that to a Galactica, the shining luminous middle part, holds time to eternity through movement. The atoms in the middle admits light because it holds time duration to the speed of light, the longest duration that time can be and still remain in the fourth dimension.

The centre of all cosmic structures determines the time that applies to the structure itself. As the cluster of protons supply the density that influence the space of occupation, the density is a collective reducing of space with the increase of heat. This can apply through an object relating to space through movement by the object and by the object reduction of space through the density of the accumulative effort of the cluster of protons we named elements.

By referring to "gravity" only one aspect of the space-time relation of any elements apply. There is no mention of the second and crucial part of "Gravity" where the motion of the object brings about the space-time relation, or if you wish, providing the cooling aspect. At first, the proton cluster's total positive space-time displacement has an insufficient "gravity" to secure a stable cooling effort. This spinning of the element clusters is inherent from an event, even predating the "Big Bang".

The spinning motion of the element clusters (or proto stars or if one wishes to use the name of future stars, it will be just as applicable) hold their relation to heat secures by maintaining motion. As the time value in the clusters space occupation (mass) secures an era related value, the structures that were spinning, reposition in such a fashion as to apply a new linear displacing value. At first the motion is such that the linear position is negligible, but as the mass grows, the linear distance grows accordingly, placing the revolving structures further apart and at the same time, "pushing" the rotation of the objects in a wider revolving orbit.

By widening the rotation circle, the objects rotate at a "lesser" pace and this pace coincide with the space-time occupation ("mass") of the totality in the effort of all the protons put together. In this one will not find a "force" but it will be a complete balance between matter, space and time. By securing an ever-increasing space-time occupation (mass) the future star will reduce its negative space-time displacement (motion) and increase its positive space-time displacement (gravity). The higher the positive space-time displacement (Gravity) becomes, the lesser the negative space-time displacement (motion) will be. At present only stars holding an iron$_{56}$ inner core can maintain a star status, and any object with a lesser element in the inner core will not bring about fusion, or in fact, any form of luminosity.

For instance by the time the "Big Bang" arrived, only elements with a "mass" of 112 had the ability to release from the Galactica luminous core and during the Era of the Quarks, the releasing mass of the time determining elements carried a combined proton-cluster "mass" of 88. As the single proton's time holding value increased (molecular mass) the time grew less and the Universe "grew bigger".

	Sun's age		Interval		Time in relation To the past	
First	1,5	$\times 10^9$ years	1,5	$\times 10^9$	3,5	$\times 10^9$
Second	2,5	$\times 10^9$ years	1	$\times 10^9$	2,5	$\times 10^9$
Third	3,5	$\times 10^9$ years	1	$\times 10^9$	1,5	$\times 10^9$
Fourth	4	$\times 10^9$ years	0,5	$\times 10^9$	1	$\times 10^9$
Fifth	4,25	$\times 10^9$ years	0,25	$\times 10^9$	0,75	$\times 10^9$
Sixth	4,3	$\times 10^9$ years	0,05	$\times 10^9$	0,05	$\times 10^9$
Seventh	4,45	$\times 10^9$ years	0,075	$\times 10^9$	42 500 000	
Eighth	4,525	$\times 10^9$ years	0,0025	$\times 10^9$	40 000 000	
Ninth	4,5265	$\times 10^9$ years	0,0015	$\times 10^9$	38 500 000	
Tenth	4,526575	$\times 10^9$ years	750 000		37 750 000	

Up to this point all arguments came about from the theory about the "Big Bang". In The Thesis I show mathematically that the Universe will last seven cosmic days. This however is not the seven solar days and under no condition may one confuse the two.

We are in the fourth cosmic day calculated from the Big Bang as if the Big Bang was the first day. To understand the process of the cosmic days, please study the cosmic almanac as seen on the last page of this letter. While I am saying this, this book is about proving with undeniable facts. That I leave to The Seven Days Of Creation ISBN 0 – 9584410-4-9. In this letter I only show it is possible to prove what I say I proved. The Bible speaks of seven days of creation; therefore we must look for the seven days the Earth formed. According to the cosmic calendar, we presently find ourselves in the fourth day, with three more days to follow. Why do I refer to these periods as days? Well the term "day" is as manmade as clothes, buildings, trains etc.

In this is another point that proves the technique science applies at present, does not nearly give a near value to the time duration of development on Earth. It should be out by as much as a few billion years for all we know. To indicate the meaning of what I am trying to bring across I shall illustrate a time scale in which the development might have taken place. I do most strongly disagree with the age the Brainy Bunch hands out to the solar system but since I have no better time to give I shall use the Xepted table ONLY AS AN INDICATER SERVING TO PLACE RELEVANCIES.

It is not an accurate and tested scale in time but merely as a presumption in order to indicate how the frequency will relate to the time duration. From where we stand, we may have a perception that the frequency is getting shorter, but as seen from within the sun, the time duration would be precisely the same value each time. This is the process in which time is concentrated in the space confinement and the relative space-time is amplified to extent the duration. In time, this variation is perceived as flair and later as a pulsating readjustment. I hope it now will be apparent just how small and under developed our Sun really is when compared to other structures.

This comes about because the relative size of a star is based on its space volume that contains matter and the incorrect way in which the density of stars are calculated. At present the frequency could have come down to as little as 15 000 years, maybe slightly more, but who knows. However, it is not the frequency that is the problem, but the way in which the frequency is measured that is of concern. At present, we relate to the duration of time laps relevant to the magnitude in which the Sun presently is. This might be a problem to all of mankind and civilization. There exist neither method nor means in which one can determine at what stage of progress the Sun is in at this moment. All that is extremely clear is that at one stage, the Sun becomes a raging bull, and a sleeping bear follows this. In between these two possibilities of time duration, time can become double the value it holds now, which then is followed by a period where time might have half the duration, we experience at present. The first thing that springs to mind, is that we find ourselves in the middle, which averages out the extremes. That is not the case.

Let us start by taking the size Jupiter is today. We know seven events happened and each event had influence on the Jupiter distance.

The relevance of the actual distance is however, 16,5; 15,4; 10,6 and 8 respectively in relation to the others of 2Π.

With this knowledge secure, we have to seek the positions evidence as how the structures came to be in that place as the inner planets do not confirm their position in accordance with the rest of the solar objects. When we give the distance that should apply if the inner planets were also just orbiting objects it would then be

The official average distances from the centre of the Sun the average distance of rotation that each inner planet completes in relation tot the governing singularity.

At first I thought the way I presented my first impression of the solar development was correct in as much as the way I first introduced the image. Back then I was still very much under the influence of the Newtonian conceptions of a runaway star, and other misconceptions I later found to be alien to cosmology. There can be no runaway star precisely for the reasons I explain the constitution of Galactica. A star with an individual developed space in the time of this era the iron$_{56}$ era, will then establish a circular "gravity" that is able to withstand the influence of the linear gravity. The higher the circular "gravity" becomes, the more static will the linear "gravity" be. In the case of a Proton star (Black Hole) the linear component lies with the fact that we can actually see matter performing its linear component by not curling as lesser stars do, but placing the circle and linear components all on the matter as the Proton star pulls matter, space and time into its pace-time occupation. A Proton Star is unmovable, static, and stationary and every other name you wish to connect to its immovability. It can no longer go anywhere. The stars within the sphere where doctor Hawkins identified a Black Hole to be, within the very centre of a galactic, holds all occupation relevance to the spin or linear component of space-time occupation and only a very minute part to the circular "gravity" component.

I bring evidence to prove my personal theory development and showing honest misconceptions on my part. In that view, I wish not to remove the first suggested solar formation but to replace it partly with facts that I became aware of, as my personal insight grew.

In the case of the extended duration, we might live to a relative ripe old age of 120, but in geodesic terms this might only be 20 years or so, because our hearts are programmed to last that many beats and then dies on us. To what we then perceive as one hour, would in fact be 10 of the local minutes. So, our biological clock would be completely mesmerized because the time ratio has gone out by 6: 1. No mention is made of the fact that we shall have to endure the extreme temperatures. Under such conditions, there is not even a thought of ice caps on the poles, and the tropics suffer from continuous rain all year around. In those wet conditions, nobody would be able to survive, because of the heat and the variation in river water levels, the ongoing mudslides and the exploitation of bacteria and ...... of these conditions that they would inevitably find most favourable. Still we can populate the icy parts we cannot use at present. The radiation of such an event would leave most white people to die from various forms of cancer. The Negro is therefore least affected; somewhere in their genes, they have already survived a number of these events. Nevertheless, it is not only the violence of such an eruption causing uttermost expansion through such eruption, but it is the manner that life has to learn to cope with the extraordinary changes that occurs and the way life has to find a way to protect the body from such high gravity or heat. The time in which the eruption comes about that is a factor. It is not only heat as we would think of a hot day but it is more radiation like heat with higher electricity and magnetic sun waves that penetrate the body and accelerates growth of sells. To the sun, it might start as a smooth and even process that then proceeds in certain time duration, but to us on the outside, the whole thing is quick and after all, to us as humans our perspective is the most important.

There is another way of looking at the effect the Titius Bode law applies to cosmic atoms and this is very important within stars. No person can deny the fact that the Earth is a sphere, excluding outer space, where our need to apply entry into the sphere of inclusion ($\Pi^2\Pi$), and the law to abide by is the four cosmic pillars. You have to abide by the Roche limit where rules allow you entry, or destruction. No Newtonian can deny that.

When you cannot deny the fact that the Earth is excluding space as it is including time the rest is beyond denying also. One has to seek the evidence where the evidence is, where one can locate such evidence and above all, read the evidence correctly. The evidence proves the existence of a binary before the Earth came to be. The four inner planets are left over parts, a reminder of a star that uses to be at the other side of the atom. While the one side of the atoms in a star relate to the square of space 10 / 7, the other part of the atom in the star relate to the matter to matter (neutron $(\Pi^2\Pi)$ holding matter to space while space becomes time $(\Pi^2+\Pi^2)$.

The Sun was in a binary with another cosmic structure that has no longer have a full place in the solar system as the solar system stands today. The second object had a good measure of the suns' potential, but not adequate to survive. If the Sun was the size of what Jupiter at present is, the second binary was then about the size the Earth is today.

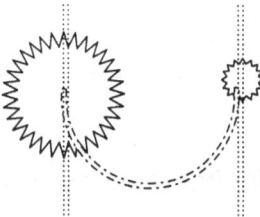

Both had individual singularity Π placing a value of $2 \times \Pi$ in space, as well as a common singularity $\Pi^2 + \Pi^2$.

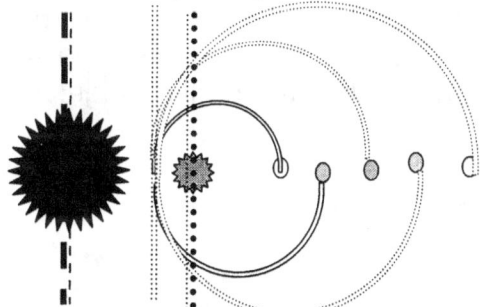

From this binary extended a singularity connecting five rock ice cubes, each holding a point of singularity, with the electron position of the binary where the binary holds the mutual point of singularity. This had nothing to do with 5 or 4.5 X 10^9 years in time laps.

What I am about to do, is very unscientific and the next few pages must be regarded as pure speculation, brought into the book for one purpose and that is to amuse. That is all value that the next calculations have. I dispute the fact that any calculations can ever determine the precise size the structures had because the structures at present hold the very same size that it held when the solar system formed. However, life is not only about proof and fighting dispute of proof, there has to be some entertainment in the book, merely then to satisfy our need to gossip. It is far better to gossip about the planets than it is to gossip about one another, because I do not think it will hurt the feelings of the solar objects at all. It is utter speculation and a needless process and I do not wish to encourage such wild guesswork in any way what so ever.

From this binary extended a singularity connecting five rock ice cubes, each holding a point of singularity, with wit the electron position of the binary where the binary holds the mutual point of singularity. This had nothing to do with 5 or 4.5 X 10^9 years in time laps.

What I am about to do is very unscientific and the next few pages must be regarded as pure speculation, brought into the book for one purpose and that is to amuse. That is all value that the next calculations have. I dispute the fact that any calculations can ever determine the precise size the structures had because the structures at present hold the very same size that it held when the solar system formed. However, life is not only about proof and fighting dispute of proof, there has to be some entertainment in the book, merely then to satisfy our need to gossip. It is far better to gossip about the planets than it is to gossip about one another, because I do not think it will hurt the feelings of the solar objects at all. It is utter speculation and a needless process and I do not wish to encourage such wild guesswork in any way what so ever.

For your entertainment alone, I shall go about trying to determine the size of the Sun back when…as the size of the Sun and other structure was when the dual came to its final resolve.

Let us start by taking the size Jupiter is today. We know seven events happened and each event had influence on the Jupiter distance.

With this knowledge secure, we have to seek the positions evidence as how the structures came to be in that place as the inner planets do not confirm their position in accordance with the rest of the solar objects. When we give the distance that should apply if the inner planets were also just orbiting objects it would then be

1	Mercury	$58 \div 2\Pi = 9,23 \times 10^6$km
2	Venus	$108 \div 2\Pi = 17,188 \times 10^6$km
3	Earth	$149 \div 2\Pi = 23,714 \times 10^6$km
5	Mars	$227 \div 2\Pi = 36,12 \times 10^6$km

The relevance of the actual distance is however, 16,5 ; 15,4 ; 10,6 and 8 respectively in relation to the others of 2Π.

In the case of Mercury, Mercury is 5Π further than the $(\Pi/2)$ (Π^2) of the others are and Venus is $(\Pi^2/2)$ (Π) ; Earth is Π (Π) and Mars is almost $(\Pi/2)^2$ Π away from the sun. I admit that the distances do not apply to the millimetre but that will become apparent soon. The importance here is the relevancies Mercury is $(\Pi/2)$ (Π^2). That is the place where the binary minor should be in if it was still there.

Venus is $\Pi^2/2$ (Π), which is in the space-time that binary minor held at a position it would hold its value to Π.

Earth would be in a position of one Π more than where binary minor would relate Π. This then is $\Pi \times \Pi = \Pi^2$. Mars would be $(\Pi/2)(\Pi)$ (half a Π) even further away than the Earth's position of Π (Π). We shall get to Ceres (a fragment of what was another planet in due time. Therefore let us establish the relative positions to Mercury, the sole holder of the star binary minor in relation to the other fragments.

Mercury = 0.
Venus = $(\Pi^2/2)\,\Pi$. The edge of the entrance field that Binary Minor had.
Earth (15,4 + 10,6) = 26. That has a relevancy of about $(\Pi/2)\,(\Pi^2)$ = 24,35. This is where we must not forget that the Earth too is a binary and the moon played its part in the drift relating to a position of 26 instead of 24,35 or $(\Pi/2)\,(\Pi^2)$.

Mars holds a relative position of 10 (Π) and we know that 10Π are the relevant space position to $\Pi^2\Pi$. This indicates that anything outside 10Π will be far outside $\Pi^2\Pi$ and this places the object in that space, without space. Any object without space will be directly into time and this will mean total destruction of that object. It will have the very same consequences as having a cosmic body holding an iron core in the previous era, or having a cosmic body with a silicon core in this era. It will and must disintegrate, there is no question about that. The space-time applied to such a core will be double than what is to the other five inner planets. It will destruct, in the same manner as did the Shoemaker Levy 9 comet with the one exception, it held a relative position where the Sun could not get hold of the fragments as Jupiter got a grip on the Shoemaker-Levy 9 fragments.

At first I thought the way I presented my first impression of the solar development was correct in as much as the way I first introduced the image. Back then I was still very much under the influence of the Newtonian conceptions of a runaway star, and other misconceptions I later found to be alien to cosmology. There can be no runaway star precisely for the reasons I explain the constitution of Galactica. A star with an individual developed space in the time of this era the iron 56 era, will then establish a circular "gravity" that is able to withstand the influence of the linear gravity. The higher the circular "gravity" becomes, the more static will the linear "gravity" be. In the case of a Proton star (Black Hole) the linear component lies with the fact that we can actually see matter performing its linear component by not curling as lesser stars do, but placing the circle and linear components all on the matter as the Proton star pulls matter, space and time into its pace-time occupation. A Proton Star is unmovable, static, and stationary and every other name you wish to connect to its immovability. It can no longer go anywhere. The stars within the sphere where doctor Hawkins identified a Black Hole to be, within the very centre of a galactic, holds all occupation relevance to the spin or linear component of space-time occupation and only a very minute part to the circular "gravity" component.

I bring evidence to prove my personal theory development and showing honest misconceptions on my part. In that view, I wish not to remove the first suggested solar formation but to replace it partly with facts that I became aware of, as my personal insight grew.

Let us establish a line of evidence and fill the puzzle.

1.   There was another star with the Sun where the Sun was Binary Major and Star Unknown was Binary Minor.

2.   The Binary system catapulted the solar system out of its frozen eternity, way ahead of its time of development, bringing along the rest of the micro stars. The outer "planets" are not planets, they are micro stars in development.

3.   The Binary system formed part of a Lagrangian system holding the Binary in the centre and with Jupiter as the first orbiting satellite.

4.   The position Jupiter held for most of the developing period made Jupiter the second main benefactor of the dual, with the Sun the major winner and Binary Minor the major loser. This will also explain why Jupiter has such an advance in space-time occupation when compared to the other micro stars.

5.   The position where the (six) inner planets find themselves, were a void, AS THE BIBLE CLAIMS.

6.   The relative positions were as follows:

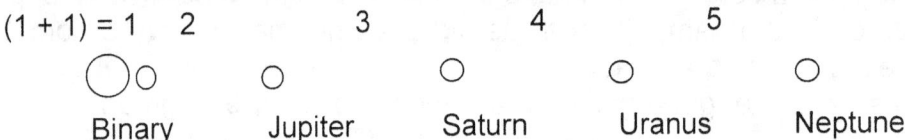

$(1 + 1) = 1$ 2   3   4   5

Binary   Jupiter   Saturn   Uranus   Neptune

Then Unknown Star capitulated as it could no longer serve the dual it fought.
It fragmented into possibly 9 major parts and many minor parts, (the comets.)

7.   As the core fragmented the brittle parts dislodged in a position each to a relative neutron position in the space-time binary minor held relating to its point of $(\Pi^2+\Pi^2)$ $(\Pi^2\Pi)$ 10Π. Whichever way we Earthlings will look at our position, from whatever angle and by whichever calculation we devise, our relevancy will be 1, will be 10, will be Π^2. Should any person ever do a calculation and find his answer does not bring this fact to bear, he must go back and fix his mistake. There will be a mistake on his part.

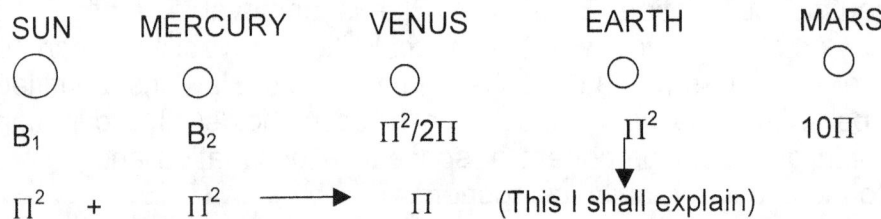

SUN   MERCURY   VENUS   EARTH   MARS

B_1 B_2 $\Pi^2/2\Pi$ Π^2 10Π

Π^2 + Π^2 $\longrightarrow$ Π (This I shall explain)

Binary stars, spinning to self-destruction will produce significant heat. Heat create space, space forms winds.   That is facts that the Bible presents and is indisputable.   Where the Earth was, was still a void, containing a sphere of circular displacement and this will reduce linear displacement to zero. Linear displacement is space and circular displacement is containing heat for matter survival.

Binary Star Minor overheated. That is why the core brittle and fragmented. This action will release tremendous contained heat, the heat will produce magma flowing in space like water in space and this eruption of heat space that created winds. Once again the recollection fits the scenario. Releasing the heat and producing space will establish space-time and fill the void where the Earth should fit. This is fact and if anybody even tries to dismiss this will be because of abstinence on his or her part. I did not prove the Bible correct. The Bible told the truth and in such correct detail, it is beyond human comprehension, but sublimation on the part of Newtonians and science before them, disallowed their ability seeing it.

I found no one could look past me and see my formula $R^3/T^2 = 1$ and $\$T = (\Pi^2 x\Pi^2)$ $(\Pi^2\Pi)$ $3 = 1836$ which is the relevance of the cosmos. By not finding a person that could see past me, I knew that person will not be able to look beyond "a burning Sun and see the frozen state in which the Sun is. Without noticing such crucial evidence, the rest goes lost. That person that sees me and not my formula will never see the cosmos for what it is.

Slightly of the mark but duly valid I wish to make a brief remark on the Sun/ moon binary. As the moon is also in a binary extended position with the Earth I wish to take this quick opportunity to show that the moon was never part of the Earths proton - proton value $(\Pi^2 + \Pi^2)$ value but is in a neutron to space position $\Pi^2\Pi$. This can only apply when the one object occupying less space-time has a proton value $(\Pi^2 + \Pi^2)$ that is less than the superior object's position on $\Pi^2\Pi$. In other words, when the total core value of the lesser structure is in any case less than the neutron value that the larger object relates to, concerning the smaller object. This means the one is totally dominating the other in all aspects.

Some quarters of the Newtonian High Priest in High ranking made claims that the moon once formed part of the Earth. In the following elaboration I shall prove why I dismiss this claim as utter nonsense.

From these facts about a binary, one can then clearly see that having two structures in a position overshooting the Binary scenario, is very much fantasy. It is just not possible because the valid space-time will exceed 112, and the structure will not have the ability to hold position in the universe that is limited to 112.

The proton value of the Earth is $(\Pi^2 + \Pi^2)$ and it will hold the second object (the moon) at $\Pi^2 /2$. This is because the second object is in the "gravity" application of the larger object (the Earth) and the "gravity" factor of the Earth takes on a linear value, half that of the gravity factor of the Earth. The Earth will not allow any linear action to exceed 10Π and at $(\Pi^2 + \Pi^2)$ $(\Pi^2/2)$ it exceeds that value.

As the two core has a dual space-time occupational value of $(\Pi^2 + \Pi^2)$ $(\Pi^2/2) = 97$, and the core value of the Earth is at $7/10$ (4 $(\Pi^2 + \Pi^2)$ the combined value will even exceed the critical space factor of 3 $(\Pi^2 + \Pi^2)$ applying to stars holding space, therefore the space separating the two objects will vanish into singularity. The reason why the Roche principle maintains core separation is that the core combinational value , seen from one or the other objective, is $(\Pi^2 + \Pi^2)$ $(\Pi/2)^2 = 48$. The individual space-time factor of each core is $7/10$ (4 $(\Pi^2 + \Pi^2)$) = 55, therefore the space holds less heat and therefore more space.

Where two structures go into a Roche Lobe and the one structure forms a proton value of Π^2 , but the comparable space-time occupation is less that Π the Shoemaker Levy 9 structural fragmenting will take place.

As larger structures will have no occupational space loss due to overheating, but the one holding a Π value has great concerns.

From the superior object the occupational distress will be $(\Pi^2 + 2\Pi)(\Pi.2)^2 = 39,8$. The geodesic space value as a factor is $\Pi^3 = 31$, therefore it will bring about a "gravitational pull" revaluing the relation to $(\Pi^2 + \Pi)(\Pi^2/2) = 64$. Being at 64 it means the smaller object holds a position of space reduction as the space value is twice that of the geodesic value. The conclusion is that it will fall under the invert square law of spheres. By looking at what happened to the comet, one can see that such estimation will be correct. When taking that formula and applying it to the position that the smaller objects holds, one cannot surmise immediately that it will be part of the atmosphere, therefore the 7 in the formula in atmospheric heat income will change from $7(3(\Pi^2)(\Pi^2/2)$ to $(4\Pi^2 + \Pi^2)\Pi^2/2 = 121,36$ because the second object holds a far superior occupational position in its application of "gravity". With a relative value of 121, overshooting the highest atomic occupational possibility of $7/10 \Pi^6/60 = 112$, the atomic structure that the smaller object holds will diminish to heat and photons. It will break up, turn to heat, photons, and dissolve, which is precisely what happened to Shoemaker Levy 9. One can witness the structure demolishing in heat, light and fragments.

With an object larger than that of Shoemaker Levy's relevancy to that of Jupiter, the same laws apply but the values derived from it bring about a different end result. The only change will be in the position of the relevancy where the one object being the superior will again apply the same formula in establishing its position. $(\Pi^2 + 2\Pi)(\Pi/2)^2 = 39,8$. With this value being the same as $2(\Pi^2 + \Pi^2) = 39,47$, it will hold the structure in a cosmic orbit, not being able to reduce the space-time separating the two, and applying the gravitational equilibrium of $2(\Pi^2 + \Pi^2)(\Pi/2)^2(2\Pi^2 + \Pi^2)(\Pi./2)^2 = 73$ and with the space-time occupation not only exceeding $3(\Pi^2 + \Pi^2)$ where space destructs but going another half a Roche factor down $(\Pi/2)^2/2$ above and beyond the space demolishing value of 58, it means there must be a total structure space decrease of some sort. It will not be a structural break up and fragmenting as in the case of Shoemaker Levy, but still a space-time occupational re-adjusting. This one can witness by studying the evidence Hubble's photos brought back. As indicated the superiority of the proton rules, not only the atom, but also the universe. The volumetric size matter holds, is in precise ratio to the space value of the protons. Apparently all protons hold the same space-time value ("mass") with only the space that changes holding the protons. This factor indicates the density of the star and it is a far greater asset to space – time occupation than merely mass. In this aspect of the proton is the universal equilibrium that produces universal time as matter takes heat in unoccupied space-time directing it to densified space-time through occupied space-time and then finally to time. The progress in the proton is the demise to space. As space is in singularity, space cannot demise. If space demises the singularity within the proton, which controls the space-time occupation has to grow.   When the space-time occupied grow, it will control the space-time unoccupied.

The simplicity in proving this is laughably stupid.   Photons travel through unoccupied space-time, and if the amount of protons can influence the travelling light, the protons in that particular space during that particular time, also influence the unoccupied space-time. There is more heat around the Earth than in outer space.   The protons therefore that controls and maintains the Earth's "gravity" also has to draw the accumulation of heat to the Earth.

Saturn and Uranus which is much further from the sun, is immense hotter on the surface than in the case of the Earth.   That fact has to be a sure indicator that the application of "gravity" has to have something to do with the attraction of heat.   If heat will only flow from hotter regions to colder regions it indicates that the proton has to be a lot colder than even outer space. By moving particles through spin brings about cooling, therefore the proton has to spin much faster than the speed of light to be able to draw photons from the unoccupied geodesic space-time (outer space) to the proton.

If the proton draws heat it can only be to cool the proton and therefore the proton has to accumulate heat.   Through this then the single proton grows in "mass" or densified space-time. This brings about that all matter becomes larger through the development of time.

The "mass" will deform, possibly brake up, as the space within Jupiter will revise the time.   The space, which the wood occupies, will reduce to the extent that the structure may brake into a liquid and even a gas.   Through this the "mass" will not reduce, it will increase as the heat component increases.   As the heat component increases, the matter will grow faster than it would in outer space.

The formula science uses in determining time is $t = 1 - \sqrt{C^2 - V^2}$ in as much as the photon's speed (square $\sqrt{}$) minus the speed of light (C) square minus one representing time will produce time.   This formula does not allow for any change in time. With this view, science is also in solidarity about the fact that everyone in science accepts that "gravity" influences time.   This fact was tested in launching the most accurate chronometers man can devise and found positive results.   Yet, not one formula complies in any way of this change to influence the universe.

I indicated the influence density has on the "gravity" by showing the relative difference planets hold.   Presumably this influence of density will multiply by billions of times in one Black Hole, or as I wish to call it, a Proton Star. If "gravity" influences time duration to retard in a minute environment such as the Earth, how much more does time retard within a Proton Star?   Time would literally to all human measures stand still.   It will become eternity because that is what time in a still standing mode is to us humans.

A Proton Star is just the first star with the uttermost fragment in space (almost to the point of singularity) of the universe as it came out of eternity, equal to the "gravity" endured by matter back then, during the "Big Bang". Even if one use the Newtonian formula the measurements must be beyond calculation, bringing the time duration that applied during the "Big Bang" also to eternity. To us non-Newtonians this conclusion is obvious, but to the Newtonians it is far too simplistic. Not surprisingly, the logic behind the argument and facts are far too simple for our Super-Intelligent-Super-Educated-Wise-of-the-Wise. Being as super intelligent as they are, the cosmos has to test their own brilliance by introducing problems only those with their super intellect can see, understand and solve.

The matter of the fact is that when time slows down to a minute pace, it will seem everlasting to us. This fact is beyond any argument. Another fact is that heat does not bring about fusion, but it does bring about change in the application of the duration of time, affecting the space in which the time is.

This rather lengthy elaboration of repeating facts already explained in detail is to bring across how little science can piece together the most obvious and logic of facts, which they supposedly are the masters of. Life is fare more complex than anything in the universe and because we are part of live, we can only view life as life reflects the history of time. We humans are part of life, yet with all the research, no one ever came up with a definition about life.

Life is an energy with the ability to manipulate space-time occupied and unoccupied. To change the body, which holds life, is only part of the manipulation of space-time occupied to the benefit of live occupying that space-time.

Because of the atmospheric and surface heat they believe the water formed vapour and the vapour vaporized and disappeared into the vastness of outer space. It is this part of the theory that makes the theory completely unnatural and bogus. What the scientist wishes to imply, is that the sun's solar winds will be stronger than the "gravitational pull" of the planet. The minute the vapour becomes a solid, which water is when frozen, it will be heavier than air and it will fall back to the planet surface. Even when evaporating again before the water reaches the planet surface, it will evaporate, but again it will form water and ice, and this process will continue indefinitely.

As for comets with boiling water forming the tails as the Sun "heats the surface of the comet". I am not willing to waste any space or time in this book by dealing with such illogical nonsense. I do explain the misconception about comets and their tails rather extensively in "Matter's Time in Space".

The Earth has an abundance of water. The question arises: Where does it come from? The answer is in the closeness the moon has with the Earth. The moon is not a moon to the Earth, but much more a sister planet. When studying the effect of the Roche Lobe and interpret this to the relation the moon and the Earth once must have had, many unexplained questions find answers.

Examining all the facts about the dual planet system, it seems one is blessed with all the cosmos can offer, and the other one is dead and docile. The sister planets are in the most extreme of positions of all planets in the solar system. Science has developed the knack to apply circumstances they find today and interpret it as if it has been there since time began.

Let us reflect what happens in the Roche Lobe and apply this to the sister planet system. Even if the distance between the Earth and the moon does not fully comply with the necessary Roche distance today, one has to bring into the equilibrium the fact of solar development, which would be in the category of the Hubble Constant. There is differentially growth of the Titius Bode application to consider.

As every one knows, water form where one oxygen particle forms a compound with two hydrogen atoms. When any two structures go into a Roche Lobe they cut the circular motion (R^2/T) off from the geodesic space-time.

Through this action one find a secluded system, cutting off all influences from the outside.

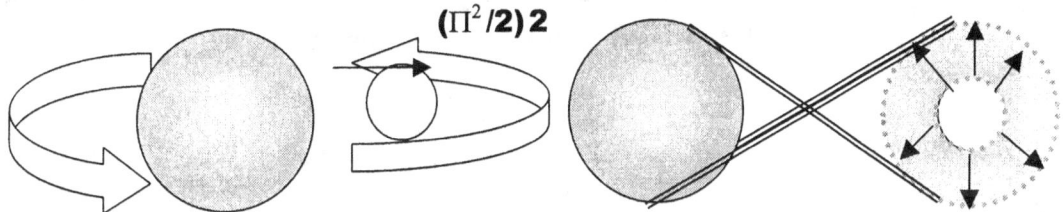

$$(\Pi^2/2)\,2$$

On every cosmic system "gravity" will always be Π^2, but the value of Π^2 will be different. As both systems share a common linear "gravity" (R/T) of $\Pi/2$ at point L, both structures will have the same atmosphere. With both structures forming the atmospheric value of $\Pi^2 \, \Pi$, this will allow the perfect condition to form compounds. In view of the fact that the Earth will have a dominant Π^2 value, it will take up all the progress that the double structures can produce through the Roche Lobe.

The void to which the Bible refers, is the effect of the Roche development between the two systems, as an atmosphere formed in the Earth section, destroying any possible chance of atmospheric development on the moon.

However, another major factor of development is that the core of the Earth will benefit largely from the Roche system, as the Earth will be the major benefactor of the heat increase, deriving from the large spin the circulate motion of both structures increase. The Earth therefore has a double development in progress denying the moon its fair share in normal development.

On the surface of the Earth a great amount of water developed, cooling the Earth's atmosphere drastically. The cooling will accompany a huge vapour of water, as the water formed clouds. With the atmospheric temperature being this high, the clouds will evidently be extremely high, forming a massive and thick cloud layer. There will be little chance of rain, because the water will form back to vapour as the water rains down on the surface, never reaching the surface.

Who can, without the support of a fully developed technical language, describe the Roche-developing factor between the Earth and the moon, in a better way than did the Authentic Author of Genesis?

After separating the waters from the sky, came the third solar day. The Bible verse is as follows: "Let the waters under the heaven be gathered into one place. so that dry land can appear."

One has to remember that the Roche principle was still in full effect, much more than as it is presently in effect. During the Solar Day, it is only "non-conviction" and during the "Solar Night" it becomes conviction (if you will).

To understand this commissioning is quite simple. Pour a steaming, boiling hot cup of coffee to a specific point. Then let the cup of coffee cool to room temperature and you will find that the mass of the coffee reduced by a millimetre or slightly more. This is not due to loss of matter through vapour, but through the loss of heat. The mass of any heated substance swells. The very it apply when the Earth and the moon surface heats extensively. To all natural principles, we are still in the Roche-Lobe but as the source of the heat is external, and not internal as in the case of stars, the Roche Lobe comes into effect with a solar expansion.

In the next phase one can clearly see that the "day" the Bible refers to consists of many seasons, with many growth periods.

It will be of little use to remind the readers that man has a written memory of a few thousand years, where even with such little information, the biggest amount of the evidence is lost through time.

In the Roche-Lobe the following principle becomes a major factor:

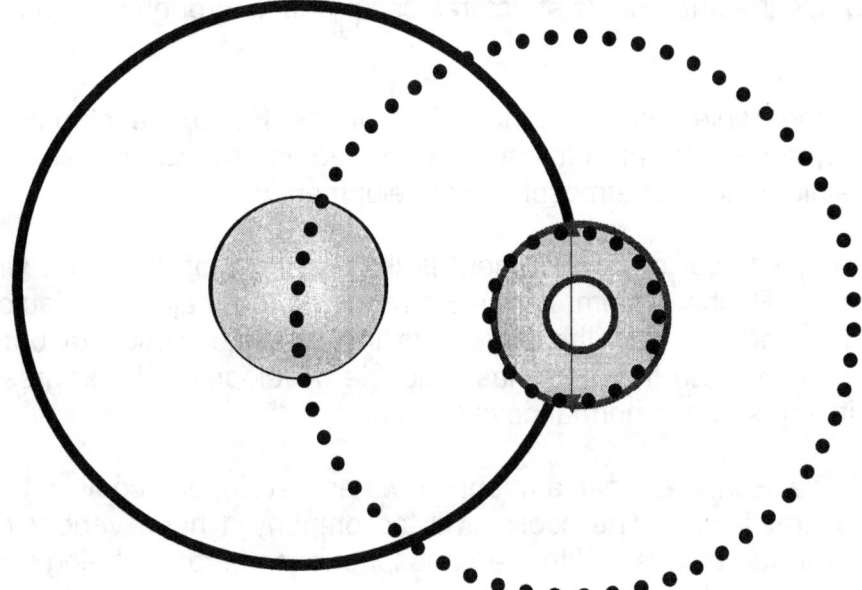

With the Sun blowing heat in thick clouds over the planets, the heat that the Earth and the moon detains through "gravity" is many times more than it is at present. Because the iron$_{56}$ core will reduce most space, densifying heat, it will also retain most heat. As it cannot accommodate all the heat retained, it will relocate heat through space forming, to the surface. In relevancy to the iron-core the silicon is space, therefore rejecting surplus heat will bring about introducing excess space amongst the silicon. This is the same that apply when baking bread and the bread "rises" in the oven. The silicon layer "rose" as the heat, coming from above, as much as from below baked the silicon to rise. From the top, the water still formed vapour, taking all the heat that the moon and the Earth holds as a compliment and then with the core of the Earth being the dominant, directs almost all heat to its core, because the Earth's iron core brings about the most space depletion ("gravity"). Through this process where the combined effort of the moon and Earth removes heat by space depletion a large area becomes effected on the space end ... but, on the inside, at the time end, the iron core of the Earth is the almost sole recipient of heat, leaving the silicon of the Earth as the sole benefactor of space-incorporation (bread rising). Said in another way, the moon helped in doing all the work, but the profits of the work went entirely the Earth's way. That even includes the vapour where oxygen and hydrogen combined through the excessive heat to form water. The vapour from the charging of hydrogen and oxygen, discharged again as the heat moved by lightning to the Earth. From this water formed in abundance as the moon and the Earth both collected, both stirred oxygen and hydrogen into a mixture, but only the Earth collected the end product.

This process became as seasonal as winter and summer now are, as seasonal as rain spells and drought spells are or as ice age and heat spells are. Who would know the intervals, and the intervals are not important, because time back then is not time at present. The important issue is the evidence left in the Earth.

As the cumulative positive space-time displacement rises above the value of the other surrounding protons in surrounding atoms, the spin will exceed the average inner-Core-value of the other protons, thus "freezing" the nucleus of the atom in fusion in the time zone of the major element.

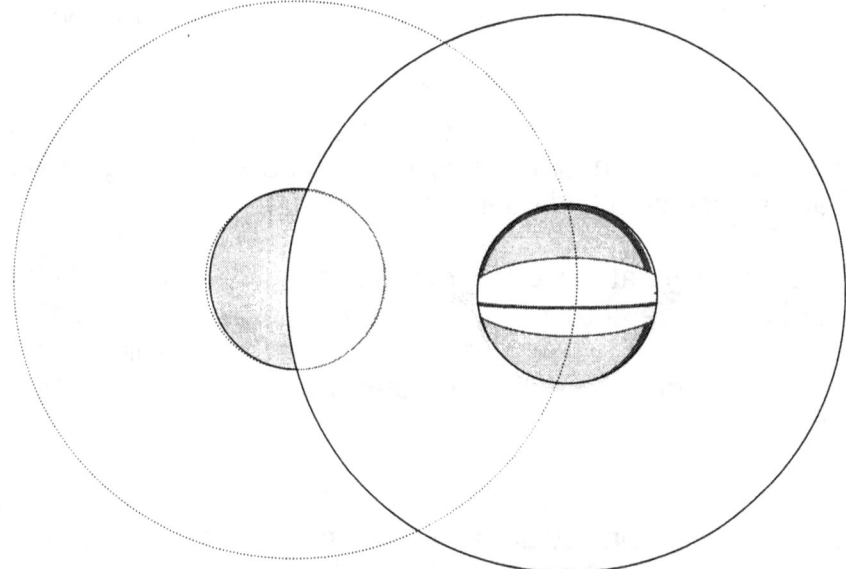

As the time duration extends further with the growth of the proton, that specific atom will develop a time duration much higher than the surrounding atoms. An accumulation of heat above and beyond the actual time duration develops, placing the space-time of that atom much higher than that of the actual time value within that layer. The "mass" increases and this causes the **aanplasing** to grow "pushing" the atom in the direction of the centre of the star where the time duration matches the value of the growing atom.

The ratio imbalance that occurs within the atom can be displayed as follows

$10 (\Pi^2+\Pi^2) \rightarrow 10 (\Pi^2+\Pi) \rightarrow 10 (3)$
$10 (\Pi^2+\Pi^2) \rightarrow 12 (\Pi^2+\Pi) \rightarrow 10 (3)$ **as the time value grows**
$11 (\Pi^2+\Pi^2) \rightarrow 11 (\Pi^2+\Pi) \rightarrow 13 (3)$ **as fusion occurs releasing heat and light.**

Because of the growing demand for heat caused by the increase in space-time density and therefore bringing about more use within the proton, the neutron, which is the balancing factor of the atom, has to readjust and apply more heat in order to keep the neutron form overheating. A point arrives where the accumulation effort of the neutron can no longer sustain an effort in supplying the accumulative call for heat by the proton.

In order to apply a balance in space-time density (mass) the neutron captures a proton that is in verplasing and this proton freezes into the density of the overheating atom. With the growth in the density of the proton, the neutron's accumulative cooling effort does no longer need to sustain such an effort, therefore rebalancing the heat application. The heat excess and rebalancing releases a great amount of heat built-up in the unoccupied sector. As this heat release comes about from the neutron reaction, it comes out as radiation and photons. In the lesser dense (top layers) area the release of heat overshadows the requirement of the proton, therefore a lot of light and heat discards back to geodesic space-time. However, in the inner star structure this readjustment will be at an equal balance and other atoms within the space confinement will apply the heat to suit their need. This produces much less of heat to the outer regions of the star.

Should a value of $7/10(\Pi/2)$ represent the hydrogen atom and $(2\Pi^2/4) \times (\Pi^2/2)$ that of carbon. By inflating the carbon atom's unoccupied space-time slightly one can see that it would accommodate a hydrogen atom.

In this, the Earth iron core grows at twice the ratio of the silicon layer. As the Earth grows, the Earth has to rise above the water at certain points. Therefore the Bible once again is correct by declaring that the water mass, which at first covered the complete surface of the Earth, separated from the water by rising above the surface of the Earth.

In the beginning of the part, I proved the ratio that applies to the atom ratio being $(\Pi^2 + \Pi^2) \times \Pi^2 \times \Pi \times 3$. This particular ratio not only applies to the atom determining the "mass" of the proton (Mp) in relation to the "mass" of the electron (Me). This ration extends far wider than only the atom, as it is an indicator to the revaluation of time duration application and plays the major role in determining the "sound barrier".

The "sound barrier" is a sure indicator as to how heat relations affect the atoms. By intensifying the heat ratio between atoms, the time to space of the atom changes completely to a point where the time overshadows the transfer of sound. It is all a dependence to heat forming time or on the other side space. The less heat there is between the atoms in the unoccupied space, the less the time will be affecting the atom, and the reverse is also true.

It is not only the heat that **one finds between the atoms** that influences the proton – electron ratio and this is the major part of the huge misconception in the view "gravity" applies, because "gravity" not only relies on the "mass" of the protons, which makes up the number of the protons, and it is not mainly the density in which the number of protons are, but it is just as much dependent on the "speed" that the protons travel in relation to other protons.

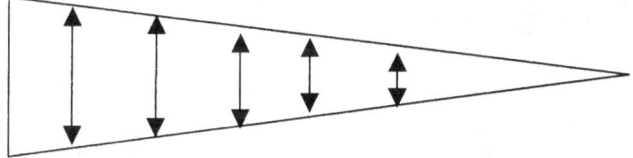

As the electron "travels" through space the time relation increases. The proof in this is that when a wind blows over an open fire, the heat in the fire increases. Oxygen as such, does and cannot burn. It is the special relation that oxygen holds with heat that increases the heat by speeding the flow of oxygen and nitrogen through the fire that increases the amount of heat.

By reducing the flow of air (oxygen and nitrogen) the fire smothers and the heat contained or regulated in the process. This does not change the transformation from wood to ashes it retards and controls the process.

While the one proton connects to space in singularity (Π^2 going to singularity) and connects space Π (in singularity) the other proton brings time Π^2 directly to space Π, re-uniting space-time as a unit to singularity (Π^3). We may call this re-unification unifying the "Graviton" in order to identify the one proton unifying time with space, while still in contact with the other proton holding (Π^2) time to (Π) space in as much as being occupied by matter (Π^2) and unoccupied heat Π forming 3.

This places Kepler's formula at a relevance s $a^3/T^2 = k$ where k^1 will hold a relation to heat with time in eternity, while AT THE SAME MOMENT infinity also applies, giving k^0 the value of singularity. This is all to do with the Universe remaining in contact with singularity while keeping the Universe in matter and heat, in the dimensions of space and time. This explains then the absolute value of time as a square, with the square having both a circular value of R^2/T multiplied by the linear value of R/T producing the link to singularity in time and singularity in space. That is what is the significance about the formula $R^3/T^2 = 1$ where R^2/T x $R/T = 1$ represents space-time in singularity as well as space-time in densified, occupied and unoccupied format. That means the everything of the whole lot, or as we say in Afrikaans, the "Heelal" meaning Universe. It is a replica of Kepler's formula $a^3 = T^2 k$ but it brings space in line with both aspects of time being $R^3 = T^2$ if $k = 1$ where T^2 then become R^2 / T and k becomes R / T. It put space in relation to both aspects of time because the moving of space is both in a circle and also in a straight line at the same time. The one cannot be without the other.

When saying this, I wish to include the following explanation for those that may have an interest in the technical aspect.

Energy is a term for a power or a force, an effort to get work done. Energy relies on movement. The influence one object holds relating its position in accordance to another position.

When an object remains in one position relating to the rest of the surrounding objects, the time it remains in that position is unchangeable. Therefore, for that duration it will remain eternal. To reflect this relation to a formula will be $R^3 = 0$ and $T^2 = \Omega$. Once the movement starts the position changes, therefore the space relation changes. The movement relates to time because any movement takes time, even with an uncontrolled explosion. Changing space always takes time.

Science agrees on one aspect of time; they have no idea what time is and when anything has the value of being unexplainable to science, it does not exist. Should one not believe my suggestion that science do not accept time as a part of the law of science then think of the money science spend on a ridiculous conception in as much as "time-travel". Even discussing such a conception is time wasting and science should discard it as time wasted. That topic is beneath my dignity as much as it is below my mentality.

Applying energy ALWAYS entertain a synopsis of space converting to heat through a period measured in time. Should one not believe me, think of a lamp, a heater, a stove, they all work on heat where heat is applied by some or other man made device converting heat from space surrounding a generator to heat travelling through some element and when oversupplying the element with heat in a controlled manner. It will readmit the heat to space in the space life wishes to apply the heat. By oversupplying the conductor with electric current, the conductor, resistor or whatever will burn. Anything can only burn with excessive heat applied at one place. When an electric device burns, it is just that. The heat we concentrate by spinning iron through a concentrated excited space filled with polarized (excessive spinning) heat, and by spinning excessive, that heat distinguishes itself from the rest by maintaining a higher spin than the rest of the heat and therefore create an individual time to space than the rest of the environment. This allows the iron, also spinning therefore applying an individual time in space, to place the heat in a separate time than the surrounding atmosphere. It will take heat directly to the Earth in a time- frame where the space is much less than the environment.

The flow of electricity is not a force, it is energy where heat receives a separate value of time, distinguishing it from the rest and as with all energy, it will bring about a reduction to space. In short, electricity is heat flowing. Electricity is space converting to heat and the proof is by investigating electrical human interactions. When human flesh makes contact with a high flow of current, the flesh shows sings of burning. In all events of applying energy, it comes down to the conversion of space to heat through time. It is in all cases, the heat (of the sun) that supplies the heat (of the Earth). All considered; energy is all about interchanging and converting heat to space and space to heat and relating this action to time. When converting elements to stored heat, which fossil fuel is, is in fact transferring heat from the Sun to chemical bonding. This is transferring heat to space in a natural surrounding. Forming that compound to oil, coal and gas is storing the transferred energy in time.

Unleashing the energy for use to extend the influence life holds on our region of the universe, is again all about the transfer from heat back to space, and using the conversion to the "benefit" of mankind.

When dissecting the "Big-Bang-Theory" it is all about converting heat to space. It was rather exceptionally hot at the start in a surrounding which seem to u as considerably small. This however is only a perception we form from our perspective in retrospect. In truth, the universe was just as large back then as it is small today. Nothing grew and nothing shrunk, because nothing goes wasted.

Reflecting on what King Solomon said, three thousand years ago: "There is a time for everything." In this sentence, the primary word is time.
This picture applies. Whether our Newtonians want to accept it or wishes to understand it has very, very little to do with what reality is in science. This is the evidence and any child will see it.

There was time in singularity, as there is still time in singularity and everything will end in time in singularity. However singularity means just what it says, there is no movement of time, no movement in space, there is no space. Every aspect was frozen at zero to everything, what ever you may think of, it was frozen solid. Then came the Creator's command and everything responded immediately. That immediately is not our immediately. Ask any Theologian proclaiming knowledge about the Bible and he will tell you that the Creator is, according to The Creator Himself, FROM ETERNITY TO ETERNITY THROUGHOUT ALL ETERNITY. By responding does not mean everything started running frantically out a blistering pace. It means eternity ended. That does not mean time as we see it at present started, but at a pace ten times faster. It is a pity that the Theologians never read the Bible, because it the Bible documents it all. If they read the Bible as they claim, why would they then insist upon an Earth being seven thousand years old? The Bible states that to our Creator a thousand years (another way people used to express a time back when the recording of time lead to misconceptions), has the same value as one nights work. THIS IS ALL ABOUT EXPLAINING THAT THE DURATION OF TIME IS PURELY A MAN MADE CONCEPT.

Time started in infinity or in eternity, whatever you wish to say, as long as you say time did not move at all. Then the command came and time overheated for the first Π^2 in time. That brought space into play.

When a bowl of soup is boiling, have you seen the bubbles of air rising from the soup? Has any Newtonian ever took the time to explain that process in detail? I think not, because such explanations would be far too "everyday-like" to bother their mighty brains. Black Holes and finding the mass of the neutron, and such mighty brainpower cannot bother with small events.

Well, that boiling soup tells the complete story about the creation. Poets and painters and writers always wishes to say how "they created their creation". That is rubbish; they created nothing. They brought nothing new to the cosmos, they only rearranged what was a small part of the cosmos into a new order, that one can detect a distinction from. Creating is producing what never was before.

When looking at the boiling soup, one sees bubbles rising from the soup at the top. In the soup's brew, there are only liquids and solids before the heat came. No one placed air in before the event or during the event and any time. Yet from the brew of liquid and solid rises gas, or if you wish space. That space was not there previously. That SPACE WAS CREATED. That space is energy en energy is the interaction between heat and space. As space becomes a part of the soup, a part not there before, with no room to be, it moves out. We refer to that process as boiling. That space creation is applying heat to time, and time in singularity will respond as space in singularity. The space created will vanish just as it came, back to singularity. By applying heat to time, brings forth space, and from the three components, only the heat factor is not in singularity. It removes space in singularity from time in singularity to establish room (space) for heat (time).

That is how creation started. Time in singularity overheated and the product of that was space.

Because time and space both, is part, of the same thing time became space and space became time. I prove that the repeating of this process happened about seven times, in seven different ways and explaining the other five ways is rather complicated and tedious. Therefore, I shall only give two explanations in this book. One is as I explained Π^3 (singularity time) parting with Π (singularity space) leaving "gravity" Π^2. What happens to space happens to time. Space holds three parts with six sides, of which only three sides directing towards any object at any time. Therefore 6/2 (half of the six sides are valid at very instant, only 3 has an effect.). Time in singularity holds a line directing to time. A straight line is 180°. Matter Π^2 holds Π^3 from Π, being valid in forming $\Pi \times \Pi = \Pi^2$.

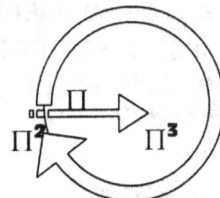

The Π^3 is matter in singularity
The Π^2 are motion or heat.
The Π is space in singularity.

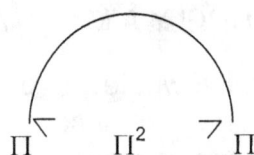

The half circle is 180° placing matter in a circle but because space only applies, to one half, 6/2 and matter holds space to value 6/2=3 only half the circle comes into effect. Half a circle is 180°. Because space has three parts in effect, it also becomes a triangle.
That means

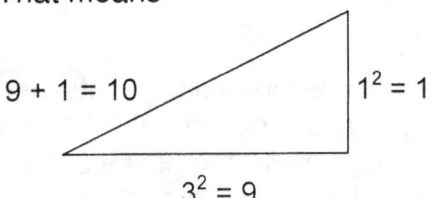

$9 + 1 = 10$ $1^2 = 1$

$3^2 = 9$

Where space holds three and time is one, the heat within that space becomes another dimension, the fourth dimension holding space-time $(3^2+1^2) = 10$. That changes the matter inside space in singularity at ten and "gravity" at Π^2. This is why "gravity" Π^2 is space (10) losing one dimension (Π^2). "Gravity" is all about space (occupying matter and heat) losing one dimension back on a long journey to singularity.

As time is in singularity, and space is in singularity and both are the same thing $(\Pi^3 \rightarrow \Pi^2 \rightarrow \Pi)$ the 10 of matter (heat) that affects space (10 Π) will also affect time (Π^3) and therefore time carrying heat will become 10 (Π^3) with space 10 Π. Anyone with a simple calculator can divide 10 Π^3 by 10 Π and see where Π^2 fits in.

The reason why time in singularity is Π^3, lies in the two components of space-time occupation or "gravity" manifested in the Roche limit. All object spin and spinning is a circle Π^2 while all objects are moving in a direction $\Pi^2/2$. Again only, half of Π has any dimensional validity at any given time, therefore the dimension surrounding an object is Π. Because the "gravity" of the surface of the cosmic object extends from Π^2 to Π but only half of the circle of Π (180°) can apply to time (Π^2) being in a straight line $\Pi^2 \rightarrow \Pi$, "gravity" will form at that point of $(\Pi/2)^2$ giving the Π in space the "gravity" to hold.

$\Pi^6 (\Pi^2 + \Pi^2) / (6 \times 10)$

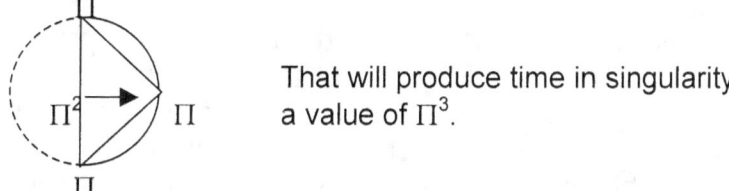

That places Π in a total of Π^6 with 6 sides in space (10) affecting the proton $(\Pi^2 + \Pi^2)$

That is why space will forever comply with $\Pi^6 (\Pi^2 + \Pi^2) / 60 = 112$, and time forming the line (180°) between the half circle (Π to $\Pi = \Pi^2$) at a 180° will form the triangle of space in half (180°). The matter component of the Titius Bode law effectively apply to the value of space, therefore 7/10 comes into the calculation. That places any atom with an existence in space at a premium of 7/10 (Π^6) (6/10).

The reason why plutonium at $5(\Pi^2+\Pi^2)(\Pi/2)^2(3/5)=244$ is at the element limit is obvious; when dissecting the relevancy in detail.

That will produce time in singularity a value of Π^3.

Explaining the other five stages of gravity (Π^2) development is extremely complicated and for that there is no room in a commercial book such as this. My motto in this book is "Keep it simple"

$$\Pi^2$$

$$\Pi^2 \qquad \Pi^2/2$$

$$(\Pi^2+\Pi^2)\,(\Pi^2/2)+(\Pi^2+\Pi^2)\,(\Pi^2/2)$$

$$= 2(\,\Pi^2 + \Pi^2\,) \times \Pi^2 / 2 = 96$$

= 192. This takes the relevancy back to before the "Big Bang" in fact to before the advent of the neutron.

$\Pi^0 \Rightarrow \Pi$

With time in singularity, time was eternal. Time is the spin rate of heat in space. That means the way the movement changes where matter and heat relate to other matter and heat in space. With all movement relating to a circle (Π^2) going somewhere (Π) in space 3. The Π will form the radius to the circle (Π^2). Any novice can see that the longer Π becomes, the wider Π will be and therefore the longer change in the repositioning of matter will be.

It took some while to realize that the Sun is one huge gigantic, awesome (there is no word to fully convey the thought!) electrical short circuit that is why all stars in this era, (a star with an individual developed controlling the flow of heat to singularity in a controlled manner), without creating space as space converts to heat space in the time of this era (the iron 56 era), will establish a circular "gravity" that is able to withstand the influence of the linear gravity. The higher the circular "gravity" becomes, the more static will the linear "gravity" be. The electricity we generate on Earth is not even a thought comparing it to what is applying in the sun. However keep in mind that, we must not over estimate the sun's ability to that of the Earth and think it to be big. In the case of a Proton star (Black Hole) the linear component lies with the fact that we can actually see matter performing its linear component by not curling as lesser stars do, but placing the circle and linear components all on the matter as the Proton star pulls matter, space and time into its pace-time occupation. In the Sun the Sun is concentrating heat by a huge generator it holds in the inner-Core-value and that is what "gravity" is.

Electricity is all about a process condensing heat and "gravity" is a process concentrating heat. There is an enormous difference but takes some explaining and that is not applicable to this book. A Proton Star is unmovable, static, and stationary and every other name you wish to connect to its immovability. It can no longer go anywhere. The stars within the sphere where doctor Hawkins identified a Black Hole to be, within the very centre of a galactic, holds all occupation relevance to the spin or linear component of space-time occupation and only a very minute part to the circular "gravity" component.

(1) A binary formed from the pre Big Bang frozen blanket.

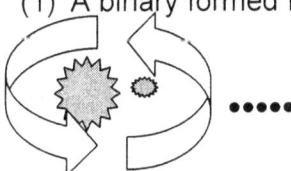

(1) Outside the spinning dual were 5 dots minding their own frozen space and concerning their own singularity, which the binary confined as space under influence.

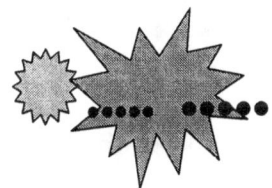

(2) Then the one factor of the binary dismantled by allowing the singularity to brake down in heat and fragments holding singularity.

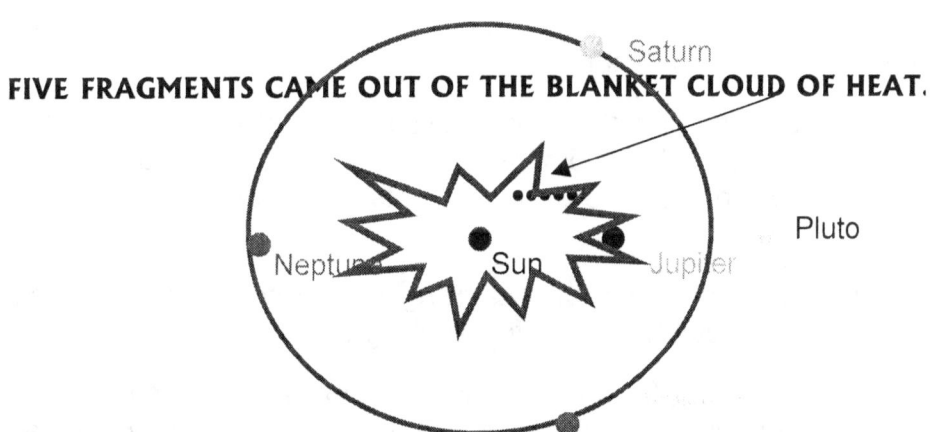

FIVE FRAGMENTS CAME OUT OF THE BLANKET CLOUD OF HEAT.

This debris can be witnessed as meteor craters on the solid planets. The bigger structures were the first ones to become cosmic missiles and disappear from sight, because there were less of them to go around in the first place. Afterwards only smaller particles remained and they too became ill fated. However, none of these fragments became part of the planetary system. As the inner structure changed its orbital course in both the Binary minor and the Unknown structure, its gasses became reduced in size in other words compressed, while the inner heat in the gas rose to extreme values, creating an enormous wealth in magnetic space time (how and why this happened, will be explained). However, the inner core could not maintain the reduced value of the stars, as they were firstly far too small, and secondly the core was in process of becoming fragmented. Therefore, it did what all unsuccessful stars do. The magnetic space-time at first were deprived of its negative space-time displacement because the increase in the stars overall growth in negative space-time acceleration. So all the matter was compressed up to a point where the acceleration of the stars became a value of evenly linear motion. As the momentum did not accelerated any longer, the compressed matter began expanding again. Then the magnetic space-times own negative displacement took over and the gasses of which the two stars comprised of, became known as Oords Cloud by humans after a duration of $4,5 \times 10^9$ years.

Then soon afterwards another two tragedies followed and only because of these two tragedies, life became possible on Earth.

THE HEAT OF BINARY MISSING CONFINED FIVE FRAGMENTS WHILE FROZEN SPACE-TIME CONFINED FIVE UNDEVELPOED MICRO STARS WHILE THE SUNAS THE PRIZE WINNER CONFINED THE LOT

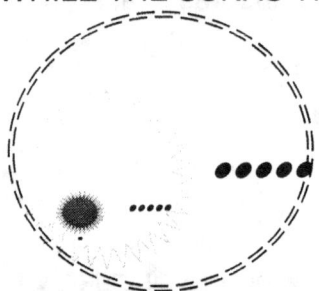

Let us establish a line of evidence and fill the puzzle.

1. There was another star with the Sun where the Sun was Binary Major and Star Unknown was Binary Minor.

2. The Binary system catapulted the solar system out of its frozen eternity, way ahead of its time of development, bringing along the rest of the micro stars. The outer "planets" are not planets, they are micro stars in development.

3. The Binary system formed part of a Lagrangian system holding the Binary in the centre and with Jupiter as the first orbiting satellite.

4. The position Jupiter held for most of the developing period made Jupiter the second main benefactor of the dual, with the Sun the major winner and Binary Minor the major loser. This will also explain why Jupiter has such an advance in space-time occupation when compared to the other micro stars.

5. The position where the (six) inner planets find themselves, were a void, AS THE BIBLE CLAIMS.

6. The relative positions were as follows:

$(1 + 1) = 1$ 2	3	4	5
◯o ○	○	○	○
Binary Jupiter	Saturn	Uranus	Neptune

7. Then Unknown Star capitulated as it could no longer serve the dual it fought. It fragmented into possibly 9 major parts and many minor parts, (the comets.)

8.   As the core fragmented the brittle parts dislodged in a position each to a relative neutron position in the space-time binary minor held relating to its point of $(\Pi^2+\Pi^2)$ $(\Pi^2\Pi)$ 10Π.

Whichever way we Earthlings will look at our position, from whatever angle and by whichever calculation we devise, our relevancy will be 1, will be 10, will be Π^2. Should any person ever do a calculation and find his answer does not bring this fact to bear, he must go back and fix his mistake. There will be a mistake on his part.

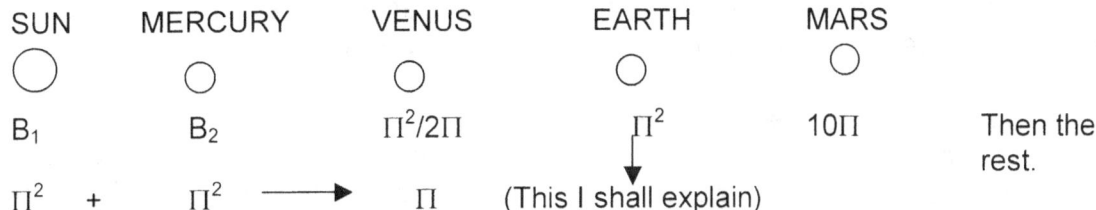

SUN	MERCURY	VENUS	EARTH	MARS	
B_1	B_2	$\Pi^2/2\Pi$	Π^2	10Π	Then the rest.

Π^2 + Π^2 $\longrightarrow$ Π (This I shall explain)

Our Super-Educated hope to establish Martian Colonists. They cannot even begin to comprehend their ordeal. Columbus only had to fight some sea dragons, a Wall of fire on the water and an ocean-sized waterfall on his way to India. Even if all that myth were true, and he encountered the hole lot in one day, his problems would seem minute when compared to what awaits our Martians. They have no clue of their folly. Mars is not another island in an ocean of water waiting anxiously to be discovered by a wind powered sea-faring vessel. Mars is an island, a piece of rock in the middle of a gas of heat, with total alien circumstances to life. Scientists can fool the public, they can fool themselves, they can fool with figures corrupting cosmic laws, but they cannot fool the facts. There is a reason why the Bible excludes the other solar structures and concentrate on the Earth. This is the same reason why any way we do calculating the solar system, the Earth will come to one according to human standards. Life from our perspective, from our perception, totally connects to the Earth. That connection is so deeply routed in man and other life, it is part of life. Newtonians may spread the rumour about life being ten a penny scattered to the social structure that man established on Earth through every social alliance or group, be it politics, law, medicine, science, theology or whatever denomination they wish to control, they will not create facts.

I admit, man derives all the above-mentioned lies mainly from corrupting the Bible. The Roman Catholic Church started the trend 1 500 years ago and still maintains it as best it can. You can prove whatever you wish to prove from facts you take from the Bible. The Bible is in support of whichever standing you may support. This is not because the Bible is incorrect, it comes from our insignificance to appreciate the full content of the whole Bible. The Bible promote only truth, man takes from that what man wants, divert the truth to suit his need by corrupting the lot.

I know presenting evidence as well as I do, will not change the course of science. I know too, that I am too small to pinpoint conclusively whether Authentic Author referred to the creation of the first cosmic period, or the first solar period. This is not because of inaccuracy on the part of the Bible. This inaccuracy comes from the human interpretation of Authentic Author on his vision, and then my insecurity to interpret his interpretations. It is human error bringing on misinformation. The Bible's recollection CAN and DOES apply to either period. Therefore, to respond in containing human misjudgement on my part, I shall re-apply the vision of Authentic Author in the context of the first solar day. I showed how it fits to the first cosmic day already.

Binary stars, spinning to self-destruction will produce significant heat. Heat create space, space forms winds. The Bible present facts that is indisputable. Where the Earth was, was still a void, containing a sphere of circular displacement and this will reduce linear displacement to zero. Linear displacement is space and circular displacement is containing heat for matter survival.

Binary Star Minor overheated. That is why the core brittle and fragmented. This action will release tremendous contained heat, the heat will produce magma flowing in space like water in space and this eruption of heat space that created winds. Once again the recollection fits the scenario. Releasing the heat and producing space will establish space-time and fill the void where the Earth should fit. This is fact and if anybody even tries to dismiss this will be because of abstinence on his or her part. I did not prove the Bible correct. The Bible told the truth and in such correct detail, it is beyond human comprehension, but sublimation on the part of Newtonians and science before them, disallowed their ability seeing it.

Let us envisage what factors applied to the third planet binary that set it apart from the other planets (not micro stars). First, the eruption occurred, and with the eruption came the release of massive quantities of hydrogen, oxygen, carbon and nitrogen. I respect to the quantities the Sun stockpiled for own "personal use it was not much. However, to what the other structures in micro planets had, it was exceeding their quantities by hundreds of times.

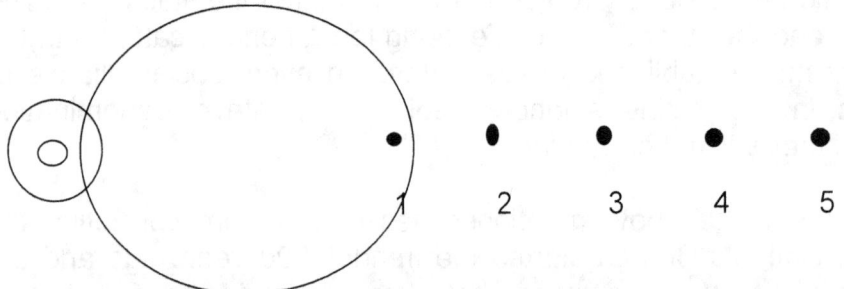

Jupiter (1) still enjoyed much of the vapour for the longest duration, and was a benefactor before the fragmenting as much as it was a benefactor after the event. As the clouds drifted from the position of point zero, the other micro stars also became receivers, in order of position. The mass they hold today, is still evidence of their position placing them as recipient 1 – 4.

1.   Jupiter   318 x
2.   Saturn   95 x
3.   Uranus   14.5 x
4.   Neptune   17 x

However, the Neptune micro star in all evidence has very particular characteristics that I cannot explain. I have a theory about Neptune, it is only a theory that I cannot substantiate with mathematical proof but nevertheless there is circumstantial evidence I shall present in order to prove that my theory is not merely wild guessing on my part. I shall return to this matter shortly.

The indication clearly points to the fact that the micro stars were in this cloud cover of star gasses, for a long period, and the period were substantially more as the micro star holds a position in relation to close proximity to the destructed binary. All the orbiting structures the micro stars hold, except (I suspect) Neptune, are fragments of Planet 5, the destructed one.

One must realize that Jupiter holding an inner core sizable enough to produce 2,34 times the "gravity" of the Earth were a result of benefiting from the Binary dual between the Sun and Unknown Star. Compare that figure to the rest, and it points to the fact that something major set Jupiter's development on a course of progress, where factors benefited Jupiter by far.

1.   Jupiter   2,34
2.   Saturn   0,93
3.   Uranus   0,79
4.   Neptune   1,2

Again we can see that Neptune had some other benefits to its progress, because, again Neptune diverts from the obvious sequence.

Once more, this phenomenon should not occur with Newton's presumptions about gravity. These bodies will collide and destruct, without a doubt. When the formula $F = \dfrac{M_1 M_2}{r^2} G$ apply, there should not be any force which is able to keep them apart. However, they do exist and what is more, they maintain a certain distance apart. Seen from this view, it is little wonder that the significance of this was lost in the notion that this is yet another "mystery" of the Universe. The scientists of the day (and the past) lost the importance, which this holds for us as Earthly dwellers.

As explained, there is no gravity, instead a balance exist in space-time, where a value of linear displacement relates to a value of circular displacement. $\dfrac{\frac{R}{T}}{\frac{R^2}{T}}$ (R/T // R²/T). Regarding this, the Roche lobe comes about because the Roche lobe forms a borderline between these two related values. Space-time lying within the Roche lobe stands at a value where the linear displacement is at a greater value than the circular displacement. In mathematical terms it expresses as follows in an equation:

$$\$ = R / T > R^2/T$$

Matter located on the border of the Roche lobe, will represent mathematically in the following equation:

$$\$ = R/T//R^2/T.$$

This is the position, which satellites have to comply with, in order to remain in orbit. Matter located on the outside of this border, which the Roche lobe holds, will mathematically represent as follows:

$$\$ = R/T \quad < \quad R^2/T$$

You, the reader, might react with surprise, because all structures with influences in this way, are far beyond having any influence on our life. This represents a great misunderstanding, as it has everything to do with life developing on Earth. In systems of unmatchable and unequal space-time values, the larger body will tend to dominate the smaller body in as far as high jacking the smaller bodies' space-time values are concerned. Please take note throughout all of this discussion, there is NO FORCE applied on any of these bodies, but only a balance, which maintains or goes array due to certain reasons. Therefore, NO STAR STRUCTURES (OR PLANETS) can ever collide, and a meteor colliding with a planet, is not a collision, but an imbalance of space-time manifestation.

To understand the meaning of this statement, I shall firstly explain why no star system can collide. When two objects i.e. double stars, come into a conflicting position as to sharing space-time, the two systems has a response in space-time values.

The academic world has treated me very poorly, because of my view on what they regard as Holy Scripture. I came across some brilliant scientists whom were able to form a conclusive opinion of my work after just reading the first four pages of a book containing almost two thousand pages of explanations and facts. By just reading two pages those highly informed professors decided that I am completely misinformed. The claim they made is that I am not familiar with Newton, and therefore do not "understand" Newton. They never even allow themselves the time to get to the part where I explain why I do not recognize Newton, let alone begin to introduce my opinion. Such intelligence, I must admit, is a true indication of just how their acquired brilliance can allow their decisions made in split seconds about a book that would take them at least one month to read extensively.

I know that if I went the conventional route by enlisting at a university and following such a course, I would have had to accept the institution's views on science or they would boot me out. I did not escape the booting process, as many academics discarded my work. I wish to put this to all of the religious scientists: Even if you will never admit it, you believe in Newton more than you believe in the teachings of the Bible, no matter how much you wish to deny it. The moment the Super-Educated faces up to any criticism about the work of Newton, they do not have the courage to even investigate the criticism. If one says that that person disagrees with Christ, everyone in hearing range is prepared to listen to such a view, although it must be considered by all believers as profanity against the almighty and this charge I direct directly to those so called believers that has Bible versus on their answering machines. Tell the physics department on any campus you disagree with Newton and they honestly treat you as a mad raving lunatic keeping busy with blasphemy!

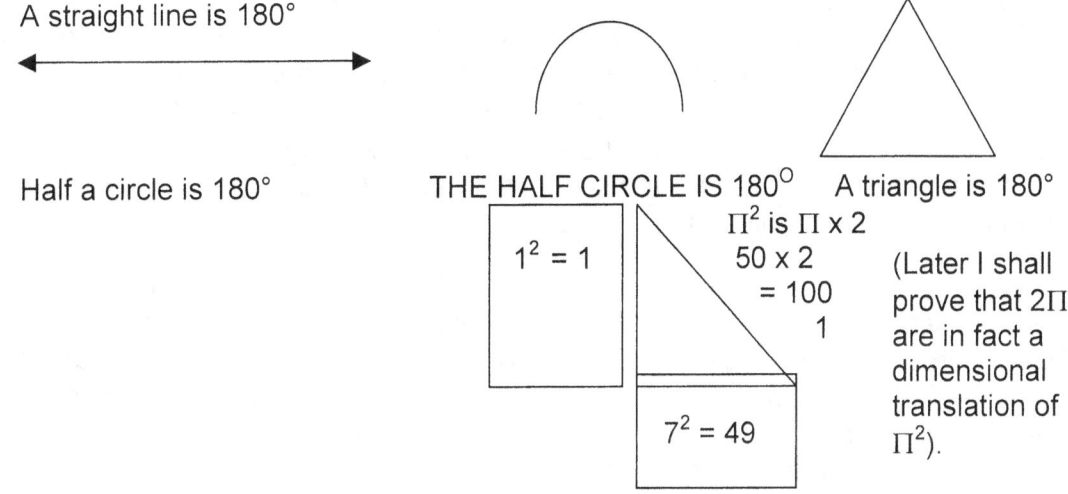

A straight line is 180°

Half a circle is 180°   THE HALF CIRCLE IS 180°   A triangle is 180°

Π^2 is $\Pi \times 2$

$1^2 = 1$ 50 x 2
                   = 100
                   1

$7^2 = 49$

(Later I shall prove that 2Π are in fact a dimensional translation of Π^2).

Π^2 because it holds a higher dimension is 49 + 1 (Pythagoras) = 50
and 50 x 2 ($\Pi^2 = 2\Pi$) = 100   (The two is a dimensional implication
$\sqrt{100} = 10$ carried by the value of 50)

Therefore space to time is 10 and matter to time is 7. Time to time is 1.
I base my facts that the moon never was or could be part of the Earth on the Roche limit that would never allow it. What is apparent though is that the Earth robbed the moon of all vital atmosphere and water.

This much I shall say: If not for the moon's position and size, life would not found such an acceptable environment in which to evolve.

The Titus-Bode Principle of heat growth and time growth works on a simple basis, the same basis as the speed of light. It is all a question of dimensions that space holds, in conjunction with time, running parallel, that exist in accepting more than understanding. Trigonometry is the sole proof of such dimensions.

I have shown Π^2 to be 49 from the Earth's perspective which means Π is 7. But Π should be 3 from the law of Titius Bode. And Π^2 then have to be 6. By placing Π at 7 and Π^2 at49 it means the distance positioning came about from space in the fourth dimension $7^2 = 49$ and not time through proton development in the third dimension. $\Pi = 3$ then $\Pi^2 = 6$; $\Pi = 6$ then $\Pi^2 = 12$. The square that applies to seven is an indication of definite heat creating space and not time through proton growth alone. But one can still detect the proton growth of 4, in the accumulation of seven.

There is of course a much simpler way about to go in explaining the Titius Bode 10/7 and 7 / 10, and I do give tit in another part of the book, but knowing the purist of academics, such simplicity would go by unnoticed.

In a previous part I established the planetary relation according to space-time application of R^3 (space) and T^2 (time) holding equal value. In the following pages I wish to apply the same standards, only relating the standards to cosmic principles I already explained.

A star form time at a value of 7/10 (4 ($\Pi^2 + \Pi^2$) = 55 and the core has to be less than 56 to allow light (29,6 + 27) than 56 to allow light (29,6 + 27) ore holding a higher value than 56,6, the light will no longer escape from the inner core and as the core grows, the star will start absorbing its light (photon) production until such time it turns into a full blown "black hole". At a value of 3($\Pi^2 + \Pi^2$) the core will dissolve space to the effect that the neutron no longer has space in which to be. The neutron then will abolish the star altogether.

When a star is in a Roche limit, the value is $(\Pi/2)^2$ ($\Pi^2 + \Pi^2$) = 48, but since it is two stars it becomes 2 x 48,7 = 97. That places the space-time development of the binary in a position held by stars two era's previously. The space-time enhancement is about double to normal growth. This fact also carries a high degree of significance when I explain how the solar system came to be in the layout in which it presently presents itself.

This places the star in a space-time occupation that was valid before the "Big Bang" surpassing the cosmic value of 112. It puts the stars in the Roche-lobe in a state where the stars in the binary combines to a cosmos excluding the outside to favour the inside where the atmosphere is excluding the cosmos and does not have space separating the stars, but merely the atoms. The time relation of the atoms hold a combining value similar than that, which is 2($\Pi^2 + \Pi^2$) $(\Pi/2)^2$ = 98. The atoms will not destruct, but all space between the atoms (Π as well as 3) will diminish as the atom alone falls outside singularity, but all other space has reduced to singularity. Fusion will not occur since the presence of heat maintaining the proton / neutron will still generate through matter spin of the two objects as one, bringing about a linear locking, applying a circular contact with space-time.

With the two structures forming a cosmic dual of survival each will apply $4\Pi^2$ $(\Pi/2)^2$ where the factor of $(\Pi/2)^2$ will hold the relevancy the one has to the other. When the two structures have equal prominence the relations will be as follows. In spoken language each will regard the other as a neutron attachment of half Π and half Π^2, therefore $(\Pi/2)^2$.

$4\Pi^2 (\Pi/2)^2 (\Pi/2)^2 4\Pi$

$4\Pi^2 (\Pi^2)$ and $4\Pi^2\Pi^2$

As both hold and equal value to the neutron or second Π^2 the two will apply a joint value of $(\Pi^2+\Pi^2) (\Pi/2)^2 + (\Pi^2+\Pi^2) (\Pi/2)^2$.

This value places the space-time within the lobe to a time duration where heat was liquid, just after the forming of the neutron. In such a star, not only will the heat and space of the electron have liquid space, but also the neutrons will find itself in dense space. The combination places the space-time value in a position where it will enhance the neutrons in the space-time between the two. When the two structures, almost equal all neutron space, it will naturally become a combining or combined proton star or double Black Hole.

In the event of the two core structures being so equal and similar, it will start growth the space-time the protons hold, placing the double proton in an unnatural era. At this point I wish to indicate that the factor values indicate the space in time separating the objects or if you wish the numbers only apply to the heat value of the space keeping the structures apart. The matter part will remain apart from the unoccupied space-time gone into singularity. Saying this I have to add that with the event of the matter finding itself in space-less time, predating the "Big Bang" the matter eventually will also have to dissolve to time as the matter starts to fall out of space and going on to singularity. The matter, however, will follow another route. All these facts I introduce to prove my point that the Sun is not even a star but through the Grace of God it is there. If the Sun was a true natural star, all "planets" will incinerate with no chance to support life.

As indicated above, the Roche-lobe, from the matter's vantage point (not the space separating the two objects), the one object relates to the other object being a neutron to the proton.

$$4\Pi^2 (\Pi/2)^2 + 4\Pi^2 (\Pi/2)^2$$

Therefore $(4 (\Pi^2 + \Pi^2) \Pi^2/4) /2$
$= ((\Pi^2 + \Pi^2) (\Pi^2)) /2 = 97$

To enable any person to see how significant and out of era this is, compare this value to that which the last surviving element, Plutonium holds at 94. The true value of plutonium carrying an overload of neutrons to stabilize the element is then $5(\Pi^2+\Pi^2) (\Pi/2)^2 (3/5) = 244$.

Analysing this is as follows:

5	End of space-time
$(\Pi/2)^2$	Limit on demolishing space-time
$(3/5)$	Heat stretched to its limit. This whole ration spells one nuclear disaster waiting to occur. The space-time environment within the binary, even exceed this Plutonium.

Therefore half of the combination regards the other half as $\Pi^2\Pi$ and $\Pi^2\Pi$ is a neutron position holding the three dimensional Universe to the same value in space as that which separates matter from matter being 3. Since the 3 stands completely excluded from the atomic combination of $(\Pi^2 + \Pi^2)$ holding the second object as

The stars maintain their individual circular displacement values, as they move closer, driven by their linear space-time displacement. At a point where space-time becomes a unit, star A's linear displacement has to overcome the Roche lobe boundary of star B. This applies to star B in the same way. At the limit of the Roche lobe border, star A and star B will establish a mutual circular displacement value, limiting the other star's linear displacement. Then a situation develops, where the mutual circular displacement will replace the individual circular displacement values, which both stars had. At this point, the stars would find it impossible to move closer, because of the Roche lobe limit. As both stars share equally in the circular displacement, they have to share an equal linear displacement. This would leave them unable to break the linear displacement equality, therefore they maintain at a distance apart, spinning around a mutual axis somewhere in the middle of the distance keeping them apart. Mathematically the equation can express as follows:

In the event where the one star's space-time value would overshadow the second object's space-time value, therefore the value of $\$_c$ would locate within the boundary of the larger star's Roche lobe border. In such a case the Roche lobe, would in effect not apply and the larger system will incorporate the smaller system within the larger system's space-time.

The effect that this would leave on the two systems which are able to maintain a mutual linear displacement value, is that they would either share in a common space-time growth or the one system will destroy the other system at a certain point in space in time.

$\Pi^2\Pi$. This can only apply when the one object occupying less space-time has a proton value ($\Pi^2 + \Pi^2$) that is less than the superior object's position on Π^2 This means the one is totally dominating the other in all aspects. Some quarters of the Newtonian High Priest in High ranking made claims that the moon once formed part of the Earth. In the following elaboration I shall prove why I dismiss this claim as utter nonsense. From these facts about a binary, one can then clearly see that having two structures in a position overshooting the Binary scenario, is very much fantasy. It is just not possible because the valid space-time will exceed 112, and the structure will not have the ability to hold position in the Universe that is limited to 112.

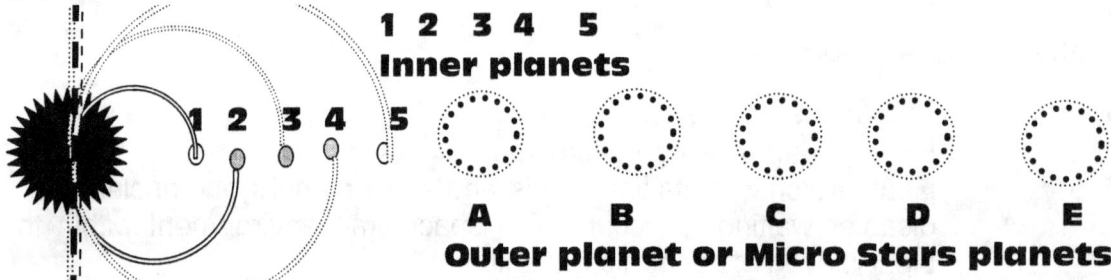

1 2 3 4 5
Inner planets

A   B   C   D   E
Outer planet or Micro Stars planets

In accordance with the Lagrangian rule there cannot be more than five objects pairing onto a centre object. However in the case of the solar system we have a double except that in the one pairing there is one devastated and in the pairing of the larger group we also find the most outer structure demolished or at least in ruin and in association with cosmic debris. With this evidence there now is a manner in which to determine how the rules were broken.

From the fragments we find that is scattered all over the solar system and the irregularities we find as I indicated up to this point we can see there is a lot of history in the story, which will unfold as we investigate. There was tussle in core battles between the Earth and the Moon, where each falls into a sequence arrangement of a proton value of the Earth is ($\Pi^2 + \Pi^2$) and it will hold the second object (the moon) at $\Pi^2 /2$. This is because the second object is in the "gravity" application of the larger object (the Earth) and the "gravity" factor of the Earth takes on a linear value, half that of the gravity factor of the Earth. The Earth will not allow any linear action to exceed 10Π and at $(\Pi^2 + \Pi^2)$ $(\Pi^2/2)$ it exceeds that value. However this can only be a result of another tussle of much bigger importance and ferocity. The debris can only formed from many disarrangements and cosmic laws being broken as a chain reaction that sprung from one specific event. From the evidence that the debris leaves it is a fact that there were many numbers more rocky structures that was left over from the Unknown star demise, but as the law would not allow more than five, it was five that was formed with one being crushed as an aligning planet. But there were more than those forming the line, but I have not the time, nor the space in this letter to go into speculating about the crushed ones since the publisher limits this letter and therefore because of the length limit the publisher placed on the book I cannot delve deeper. The moon for example too must have been one of the after thoughts of the disaster and in *Seven Days Of Creation* I speculate about this with some convincing proof as how that came about. I place a great importance on the fact that the Moon and the Earth performed as a Roche partnership and this much more than any other reason brought the development or the possibility of life developing to the Earth and not one of the other solar structures. That too I cannot share in this book because I am limited about the length of the book and it has to be accompanied by a vast quantity of proving information. Without the evidence I know I shall meat with sure rejection from our Newtonian camp and that is as sure as the Sun Shines during the day.

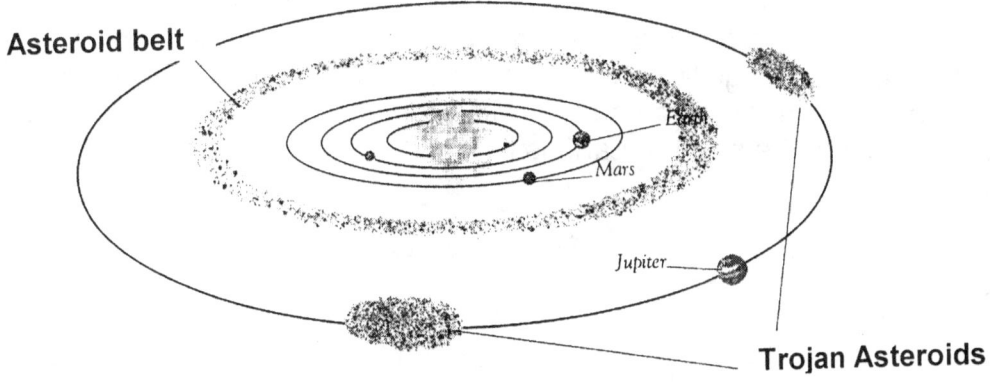

Asteroid belt
Earth
Mars
Jupiter
Trojan Asteroids

There is a great deal of pieces of hard rock floating all over the show. Some are nicely grouped in units, which one can clearly see was designated places and solid units. The only way the rocks could have brittle as they did was becoming excessively overheated. The only way it could have overheated in the manner it did was by bridging the Roche limit divide with much larger object, and there is a much larger object in the region and sharing an orbit with the immense structure. This was bullying at its best and we can see from this how cosmic plunder, thieving, brutality and theft came about as the larger murdered the little.

When the two structures go into a duel. As the two core has a dual the fragmenting of larger to smaller is a method of distributing heat by expanding. It common language we call it an explosion, space-time occupational value of $(\Pi^2 + \Pi^2)$ $(\Pi^2/2) = 97$, and the core value of the Earth is at 7/10 $(4 (\Pi^2 + \Pi^2)$ the combined value will even exceed the critical space factor of $3 (\Pi^2 + \Pi^2)$ applying to stars holding space, therefore the space separating the two objects will vanish into singularity. The reason why the Roche principle maintains core separation is that the core combinational value, seen from one or the other objective, is $(\Pi^2 + \Pi^2)(\Pi/2)^2 = 48$. The individual space-time factor of each core is 7/10 $(4(\Pi^2 + \Pi^2) = 55$, therefore the space holds less heat and therefore more space.

Lets reconsider what was apparent.

We know from the fact that there are four peculiar hard surfaced objects formed within the inner part of the sun that there was some abnormality in the solar system development. The inner objects are named planets although in the entire Universe there are no call for such object releasing from the Milky Way this early and therefore the fact of their presence relate to some irregularity that came about in the past. In order to have the four smallest planets blessed with a solid surface while the other "giant planets" all are a mushy gas structure must be because the five inner planets was part of a star much more equal to the Sun that any we now have. In order to be of significance and not be totally destroyed in the Roche battle the Sun I suppose was the size of what Jupiter at present is, the second binary was then about the size the Earth is today.

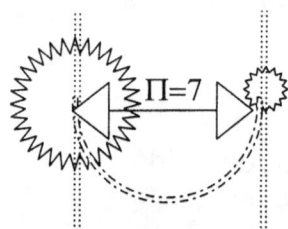

I have explained that matter is seven in relation to the Titius Bode law and that pts space in a position of 10. If 7 is Π, then Π^2 is where the Earth is because I am in the Earth and therefore I am the centre of the Universe bringing the Earth position also as the centre of the Universe. All motion spins around me making me Π to all of Π^2. If $\Pi = 7$ then $\Pi^2 = 7^2 = 49$. Therefore my position in relation to the governing singularity is 49. I am holding space in Π^6, which then is $(10)^6$, which is then 10^6 kilometres away from the Sun and with material being 49 the factor in relevancy is 49×10^6 kilometres away from the Sun.

Since I am on the third planet from the sun the relevancy my planet have with the Sun is in line with the factor relevancy, which I have with the Sun. That puts me with my Earth in relation to $49 \times 10^6 \times 3$ (since I am on the third planet from the

Sun and if the development of our solar system was normal) the allocated position of the Earth must then be 3(49 X10^6) = 147 000 000 kilometres

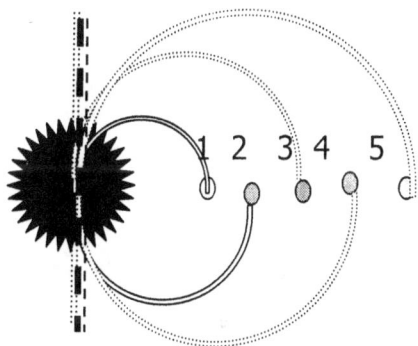

By the way that is the reason why Mercury follows such a strange pattern while circling around the Sun Mercury presents 7/10 = 10 /7 so in the case of Mercury aligning with the Sun nothing makes sense for poor old Mercury. More to the point there is evidence of unusual space –time development at the place Mercury now occupies.

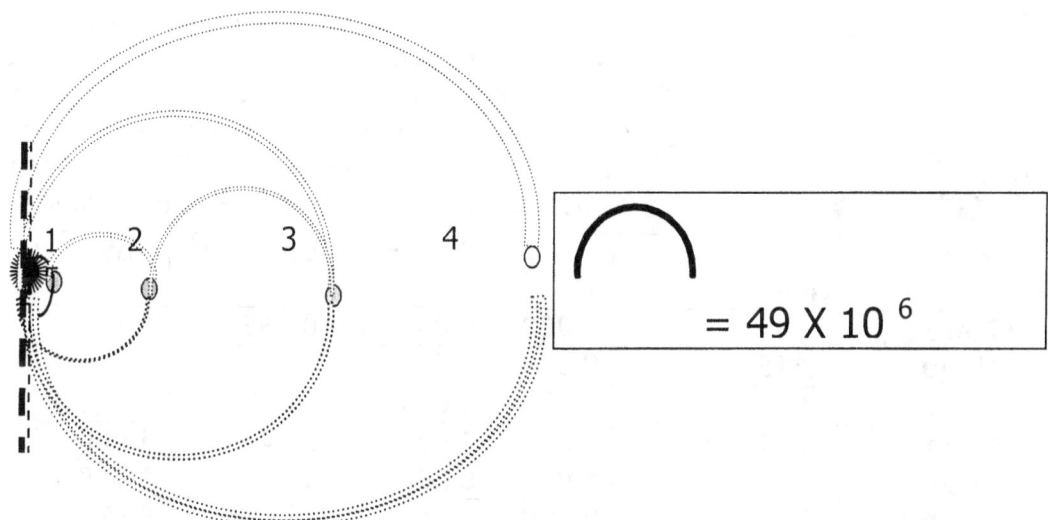

$$= 49 \times 10^6$$

There is a clear defined value, which serves as **k** in every event of every planet in relation to the spin as well as the development in accordance to the Titius Bode law. B y the square of seven representing material related to the double cube of time through which material must rotate while aligning with the sun the distance on average is one million.

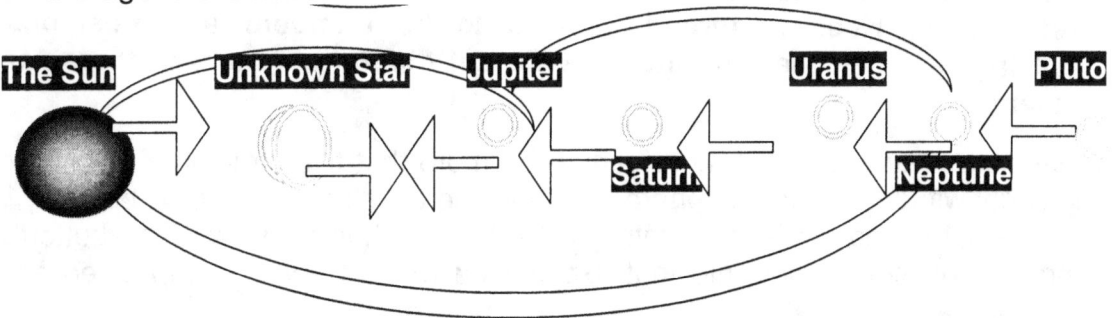

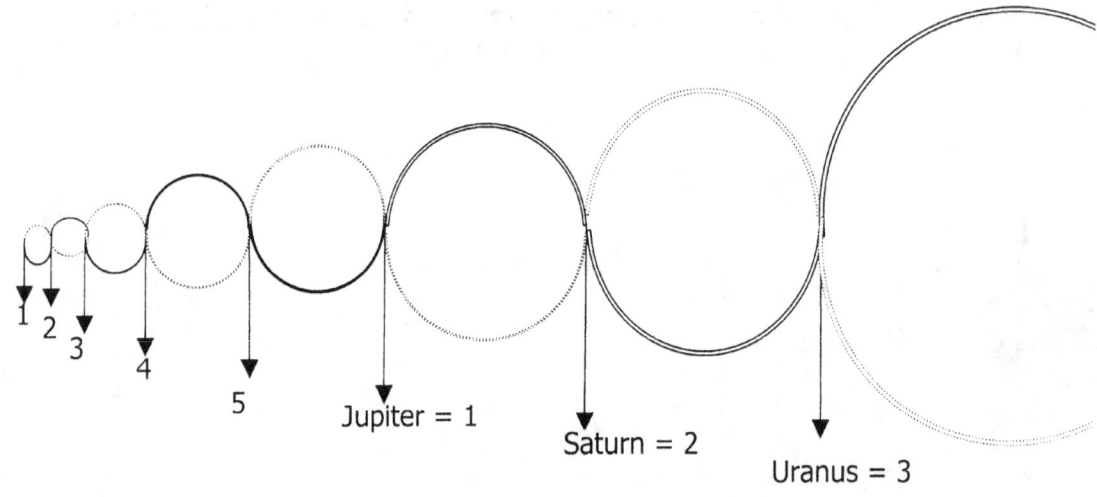

Jupiter = 1

Saturn = 2

Uranus = 3

| 1 | Mercury | 2Venus | | 3Earth | | 4Mars |

1	**Mercury**	**49 X 10^6 km**
2	**Venus**	**98 x 10^6 km**
3	**Earth**	**147 x 10^6 km**
4	**Mars**	**227 x 10^6 km**

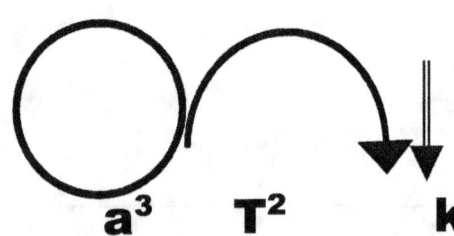

a^3 T^2 k

What Kepler saw was more of a dimensional nature than the practical mathematic symbols and values. On the one hand was a value to the third dimension, which equalled two-dimensional values one the second dimension, and one to the first dimension.

Planet	Period years	T	T^2	Distance	Space a^3	Ratio
Mercury	0.241		0.058	0.39	0.059	0.983
Venus	0.615		0.378	0.728	0.381	0.992
Earth	1.000		1.000	1.000	1.000	1.000
Mars	1.881		3.54	1.524	3.54	1.000
Jupiter	11.86		140.66	5.20	140.6	1.000
Saturn	29.46		867.9	9.54	868.25	0.999
Uranus	84.008		7069	19.19	7067	1.000
Neptune	164.8		27159	30.07	27189	0.999
Pluto	248.4		61703	39.46	61443	1.004

At the first glance, Kepler's formula seems to be numbers and positions applying between the sun and specific but different planets in the At the first glance, Kepler's formula seems to be numbers and positions applying between the sun and specific but different planets in the solar system.

Under normal development there cannot be more that five structures relating to singularity without forming a sphere. It is how the development should be. Why it is different I do not wish to go into but I will say it has to do with the butterfly diagram. However getting into that explanation would be very complicated and time consuming and for that reason I think it is best left to the Sven days Of Creation.

1	**Mercury**	**49 x 10⁶km**
2	**Venus**	**98 x 10⁶km**
3	**Earth**	**147 x 10⁶km**
4	**Mars**	**196 x 10⁶ km**

This is what it is at present so there is strong indication about some development above and beyond the normal cosmic growth that the solar system did experience as part of the Hubble growth. The normal flow would put the fragments at time relevancies as I interpret them by using the space-time rule in time in space confirming normal development in space-time. However it is not as innocent as it all seems and much evidence tells of a violent and crime filled past that shaped our solar system in becoming as unique as it eventually did with sporting a place that could host life and all.

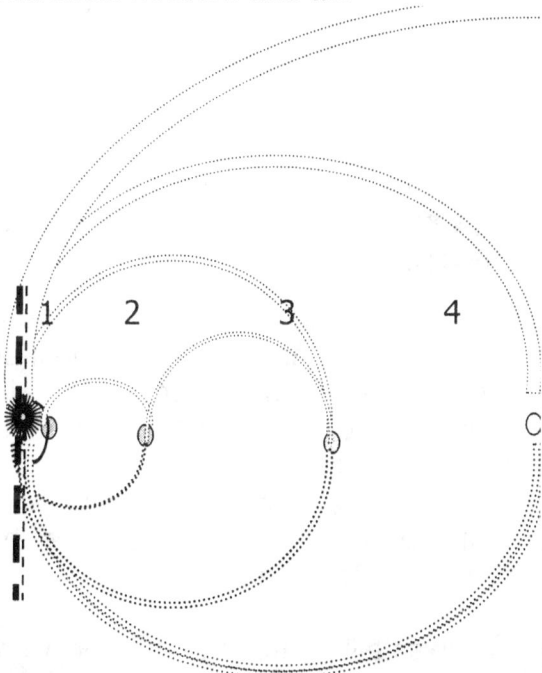

One can clearly see there was a push of the planets in the Titius Bode law moving outward towards the direction of the last structure. The plants move from one to five but five landed in a massive problem. The growth coming from the sun was defining the structure to a specific location while that location was blocked by a micro star may times the size of the planet. This Micro star was in tune with quite another singularity line because of the history the micro star had in early development with the Sun. The micro star was located as structure on in alliance with another set of developing structures and was unmoved by the oncoming dwarf. The response the micro star showed in relation with the Sun was in another frequency than the Micro star was. I guess Jupiter at the time was not as disproportionate gigantic as Jupiter seems today but being a Micro star in relation to the other micro stars such as Saturn, Uranus Neptune and Pluto (yes Pluto but Pluto again has a history which is his story and we have no time for that). With the sun forming space-time and Jupiter blocking such an advance there came trouble to paradise.

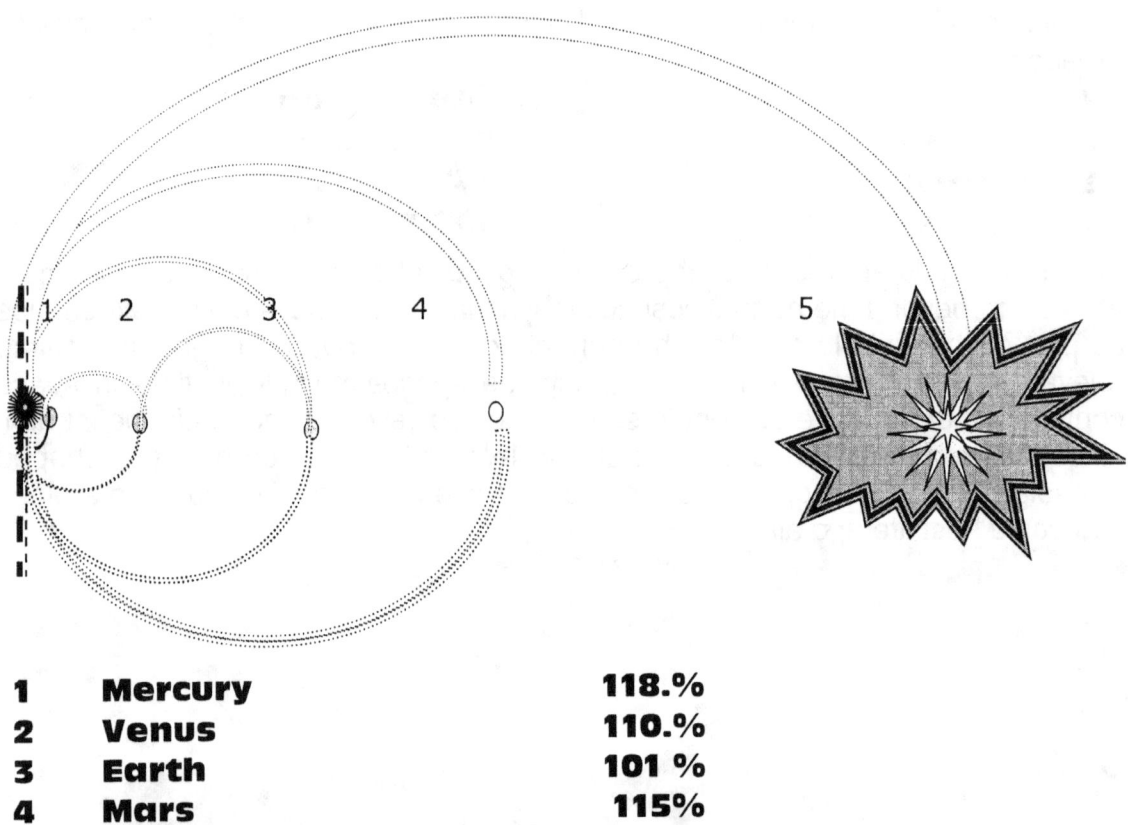

1	**Mercury**	**118.%**
2	**Venus**	**110.%**
3	**Earth**	**101 %**
4	**Mars**	**115%**

That indicates than space grew from a source to the outside of Mars and there is only one way an enormous quantity of heat could release at that point. It must be from the fragmenting of the fifth solid planet as the planet had the same fate as its mother star did and went the same rout by being forced to expand into a territory where no more expanding in such a direction is possible. Such a situation demolishes the core by exaggerating the heat load to appoint where almost the entire core liquefies and as the singularity of the little planet cannot take the bearing of the micro star, the little planet becomes wasted space-time.

The fragmenting tells how so many "moons" or satellites fragmented and were captured by the giant micro stars. Where the Sun came into the opportunity to capture some, the Sun did not use those satellites as "moons" but had them as liquid heat.

The planet went liquid but that is another bone of contention because science do not even recognise heat going liquid by forming gas as heat expands. Well, if they do they never made it part of cosmology. In the star hydrogen has a function in relation to heat. So has helium have a very special function. The oxygen in the relation to the carbon in relation to nitrogen has a specific purpose and that purpose tells to what point did the star stage of development reached. The Hydrogen has the duty to produce motion by as much duplication as the star singularity requires in maintaining. The Helium contains the captured heat, which then is charged to the carbon layer that keeps the heat in transit. With the oxygen coming into contact the oxygen transports the heat where the hydrogen transfers the heat from a solid as it was in the carbon and to an extent in the oxygen and release the heat to be served as fuel driving the core region. When looking at a fire one can clearly see what every element does with heat in liquid form. The carbon keeps the coal red while the oxygen store the heat as smoke and the hydrogen takes the heat as flames into a volatile motion. Every element stands directly related to the purpose it holds with heat and in the manner it serves the star. We can see this from the way a fire ignites and burns. The cosmos has rules and the rules will apply everywhere.

This same effect happens when a small fire starts in a large room. The heat lodges in the smoke as the smoke fills the room the smoke grows increasingly to the ceiling of the room and nestles in the smoke. When a draft enters the room the heat transfers to space and that action the scientists regard as an explosion.

Science fixes their attention on the incoming oxygen that causes the explosion. This might be true in part, but that is a small part of the whole picture. In the cloud of smoke gathers heat, confined to the smoke with the flow of direction out of the roof. The advancing heat flows from the burning fuel igniting underneath the smoke.

Material form heat at the bottom, the heat advances to the top, but the heat at the bottom remains higher than does the heat at the top.

At this point we have to look at another natural phenomenon. As heat can convert to space so can space convert to heat. We all know what happens when a compressor forces air into the air container. The container heat rises dramatically. Science call it pressure, but pressure it is not. Pump the compressor to say ten bar and leave it overnight. AT first the container will be hot. After a while (say 12 hours), the heat will reduce to room temperature. At that point one may call it pressure, but with the heat amongst the matter, it is space turned to heat. With the flow of time, the heat will return to space. TIME IS THE SPIN RATE OF HEAT IN A SPECIFIC SPACE.

To go back to the oxygen argument, one has to examine the burning oil well in Kuwait during the Iraq war. In order to distinguish the oil fire, they used blasting material to cut the heat from the oil. According to science it is to cut the oxygen from the fire, but the oxygen will flow in just as fast as it flowed out. The oxygen will return immediately, therefore the oxygen as such does not increase the fire. It is the response that the oxygen has with heat that kills the fire. By accelerating the time from burning to beyond the interaction with oxygen, that is what really kills the fire. The incoming oxygen does not transfer sufficient heat to restart the fire. What is also true is that not only does the oxygen bring heat in, but also it relieves the burning material of heat, therefore allowing more space to convert to heat.

One has to seek for this natural phenomenon as it occurs inside the star. In the case of a steel cutting however, the acetylene is the fuel, and the oxygen's role is to enhance the heat, not burn the metal. Again it is the way oxygen responds to heat and that accelerates time. The oxygen removes heat by bringing in more space to become heat, as much as remove heat to become space. We have to recognize the dual role of oxygen and not merely the fact that oxygen burns. OXYGEN DOES NOT BURN AND OXYGEN CANNOT BURN. IF OXYGEN

COULD BURN, THE EARTH WOULD NOT HAVE ANY OXYGEN LEFT BY NOW.

In fact we observe the metal degeneration as cutting but it is time enhancing on one small area of the metal. The metal is "rusting away" at one point. The rusting process speeds up by thousands of years at one specific point. It is not the oxygen but the way the oxygen responds to heat and the interaction between heat and space that cuts the metal. In other words, the oxygen only carries the heat and space to the iron.

In every galactica, a certain value of space-time was sealed in, released, as its time becomes valid. Every galactica therefore, corresponds to a different value at a different rate in as far as compensating for the time draining in the Universe, as the Universe is shrinking. We, as humans have up to now, placed a false value to the universal expansion, because we have valued the invalid aspect to the Universe, which is space.

Seen from the valid perspective, which is time, the Universe is shrinking as it loses in the valid perspective, which is time; the Universe is shrinking as it gains in the invalid component to the Universe, which is space. Therefore, the Universe is not expanding, but it is shrinking all along. I have been in so much disagreement about so many aspects of science in the way they regard the Universe, that I do not wish to split hair about such trivia. In the light of this, I only point this aspect out, but I do not wish to make it an enormous issue, because it does not really matter, from what angle you look at it. Another outcome may be where both objects maintain the claim to singularity, by pushing the space-time occupied to new levels of occupied space-time values. The result of the establishing of new individual but unequal points of singularity is the oval way objects rotate, first favouring the on in the matter part and the other in its space part and afterwards turning the points of reference around.

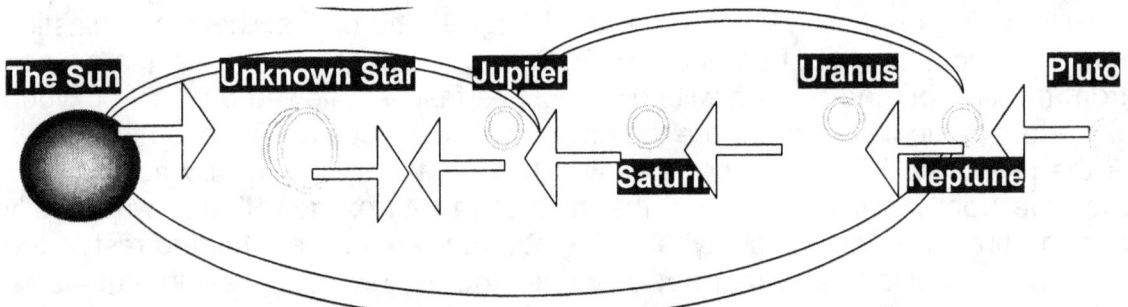

Since the First movement of time, the Roche factor was present and from the Roche factor came the Titius Bode principle. Each object seen, as well as not seen, represents a different period contained by a different specific value of space. All matter is time in a different frozen state in space and the time gives the space its particular value.

The Roche limit that came about between the Unknown star and the star we call sun contributed to the demise of the smaller star.

The Roche limit is:

The region surrounding each star in a binary system, within which any material is gravitationally bound to that particular star. The boundary of the Roche lobes is an equipotential surface, and the lobes touch at the inner Lagrangian point, L_1, through which mass transfer may occur if one of the components expands to fill its lobe. It names after the French mathematician Edouard Albert Roche (1820-83).

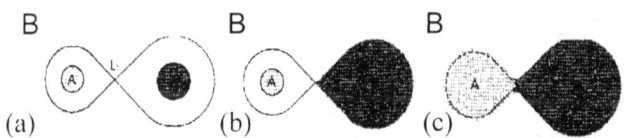

THE ROCHE LOBE: **In a binary system, the Roche lobes of components A and B meet at the L_1 Lagrangian point.**

(a) In a detached system, neither star fills its Roche lobe. (b) In a semidetached system, one massive component, B, fills its Roche lobe. (c) In a contact binary, both components overfill their Roche lobes and share a common envelope. Lets explain the importance of this Roche limit and how the Universe used the Roche factor to produce the Big Bang. That is where it all started...

Any person taking Newton seriously should at least take on the challenge and find the comets colliding with the sun, find how much the planets moved closer to the sun since the days of Newton and indicate where there is unprecedented collision between stars. Yet the closest the Universe comes to that is to show " how stars blow bubbles" in space and that is to use the precise words.

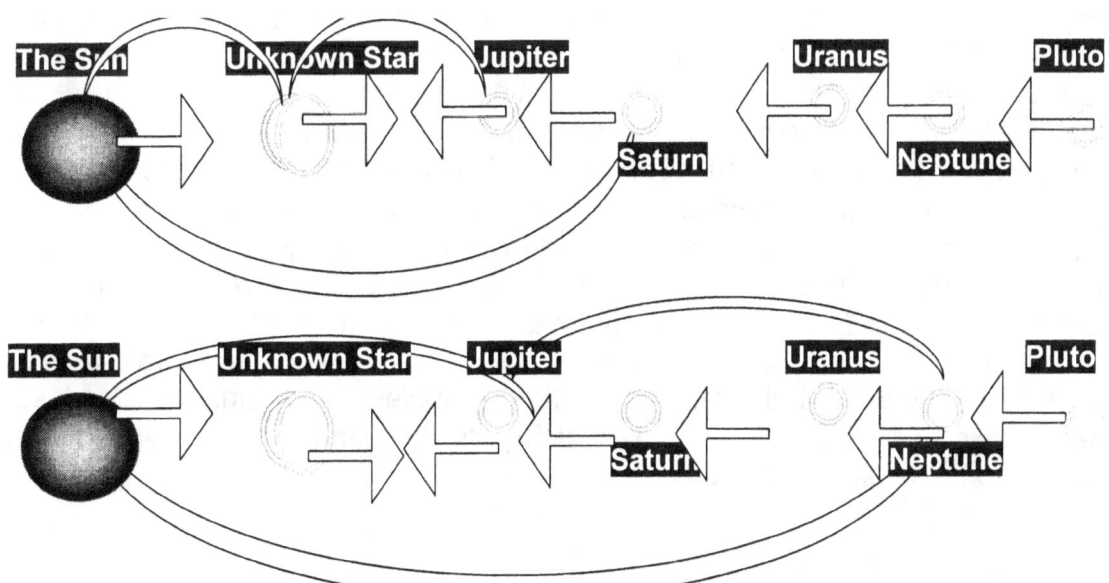

Jupiter related to Unknown star by the matter-to-matter relation of 7 /10 + 7/10 in relation to the space of 10 /7. However there the relation with the other micro Stars stopped! Jupiter was the one 7/10 marker and Saturn held its position as the other 7/10 marker. Then Saturn had the 10 / 7 relation with the Sun. From Saturn point of view in relation to singularity control the fact of Unknown star did not exist. Where the sun development pushed Unknown star out, it did not move at the rate of the Micro stars since it was a large star by individual measure. It pushed the sun Back as well and it pushed Jupiter but the other four that did not see Unknown star held firm. That brought conflict since Unknown star did not move well with the others. It had a singularity relation with the Sun but that was it.

The Micro stars had a singularity relation with the sun and Unknown star was never in the picture.

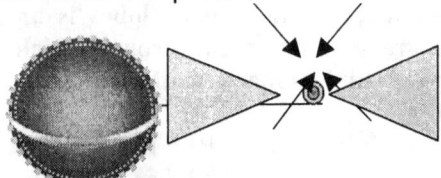

The extending of **k** in **k = a^3 / T^2** had a murderous effect on Unknown star since from the sun the star had to grow away from the Sun but five other Micro stars had now addition about the intentions of Unknown Star and therefore Unknown star became something like Johnny –in –the- middle.

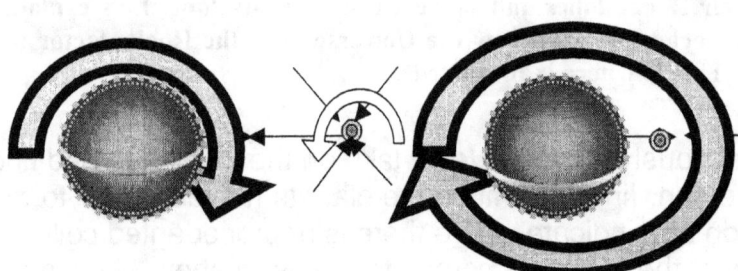

From having separate to having a joint Π^2 as the lesser star core is dissolved to heat.

The war came to rage as a battle for supremacy began to emerge because Unknown star and the Sun came into a Roche limit. How the actual figure places the position that Unknown star had I will not get into because that is fairly complicated and it is time consuming. In the Roche limit there is an intense flow increase in liquid time and it is the minor structure that cannot take the contraction. The advances star puts the flow Π^2 it receive from the spin that forms the space Π^3 to the value of Π. Then as the Roche limit is bridged the time value $4\Pi^2$ is in conflict with the time difference limit $\Pi^2 / 4$ and since the time ratio of $4\Pi^2 / \Pi^2 / 4 = \Pi^0$, which in effect brings about that the singularity controlling the better developed singularity also takes control of the lesser developed space-time since both then relate to Π^0 and is in relation to **k** = Π because **T^2** =Π^2. However there is no way that the lesser core can stand the heat since the better developed core increase the flow of liquid extending the flow to the lesser star and then the lesser star is taxed with the burden of Π which puts the same value of space Π^3 in relation to the lesser developed core.

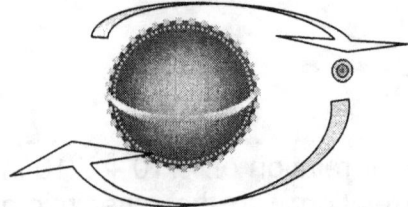

It is well advised to remember that it is not the Universe that grows but it is the Mterail in the Universe that expands by the margin Kepler mentioned as **k = a^3 / T^2**. Therfore in all the expanding of space that took place and did not take place the diameters of the Sun as well as Unknown star did expand. As they expanded so di the distance parting them not respond by the same measure because the Micto stars was blocking the route into expanding as far as the well being of Unknown Star goes. Soon time arived that introduced serious conflict in Paradise.

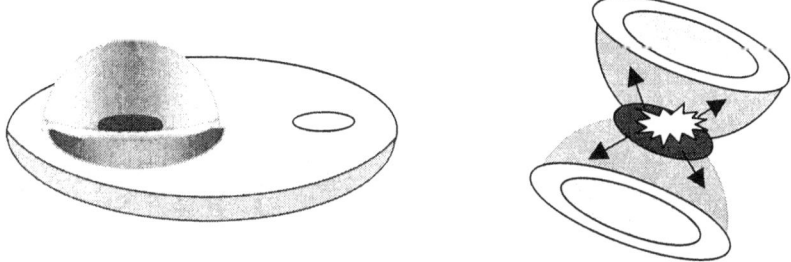

The Roche limit is evidently widely applying throughout the entire cosmos but Newtonian rules do not explain the phenomenon. The heat increased condemned the fate of Unknown star and the core of Unknown star expanded as it brattled and blew. Five chunks remained as well as many fragments, which became comets. The Hydrogen gas now at present forms what is known as Kuiper belt.

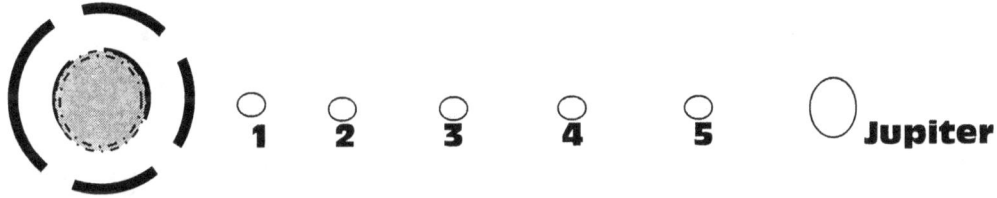

Then there were five totally unrelated fragments on the inside of five Sun related micro stars. This spelt disaster in many languages. The five now formed did not correlate to the five micro stars and therefore the five micro stars still did not at all align with the newly fragmented hard rocks.

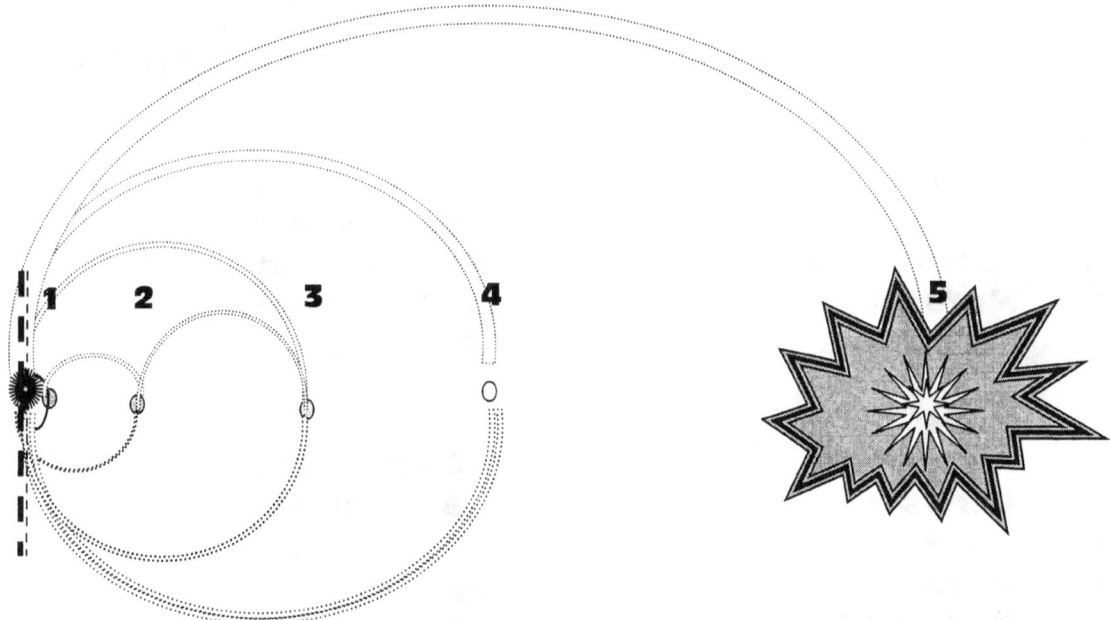

We have a ring of debris and cosmic junk clustered around where a fifth planet should be on the very edge of the inner planets. Logic tells us that too must be part of a solar disaster but how can one prove that? I once again do not propose the following as proof but my main attempt is only to show there are methods one can use when applying my cosmic code to find answers to answerless questions.

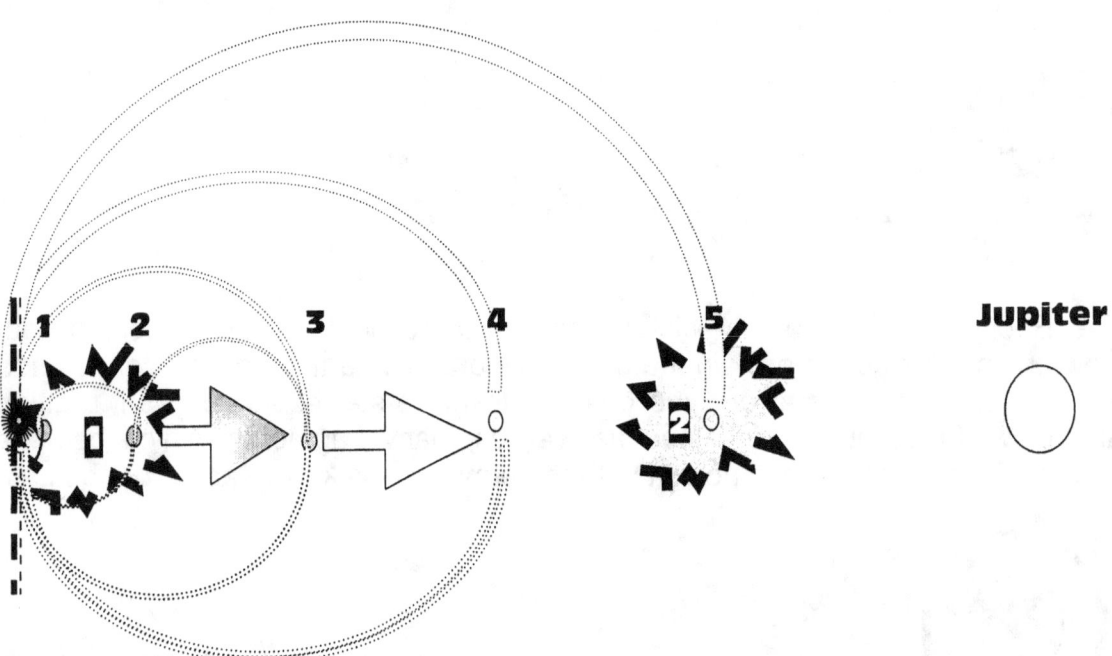

The evidence we see is in the numbers we find. There was this event that produced an enormous quantity of heat near or close to the sun. This we find in evidence where Mercury expanded 18 % more than that which should be gauged as normal. The sympathy decline in space development as the planets developed at a further distance from the event. In the case of Venus we find an exaggeration of 10 % and in the case of the earth there is an exaggeration of 1 %. This is to be expected since the demise of one little star In a Roche limit is hardly expected to change the face of the entire Universe at large. By the tome the development approached Jupiter there should be no evidence of growth since Jupiter now did not connect with a page object and the growth could hardly constitute to influence a micro star.

Is:

1	Mercury	49 X 10^6 km
2	Venus	98 x 10^6 km
3	Earth	147 x 10^6 km
4	Mars	227 x 10^6 km
5	Fragments	413.1952

Should be:

1	Mercury	49 X 10^6 km
2	Venus	98 x 10^6 km
3	Earth	147 x 10^6 km
4	Mars	196 x 10^6 km
5	Fragments	245 x 10^6 km

1	Mercury	118.%
2	Venus	110.%
3	Earth	101 %
4	Mars	115%
5	Fragments	168%

However when we include a Quantity of debris at a location where the Titius Bode law does indicate the presence or position of a structure, the significance change considerably.

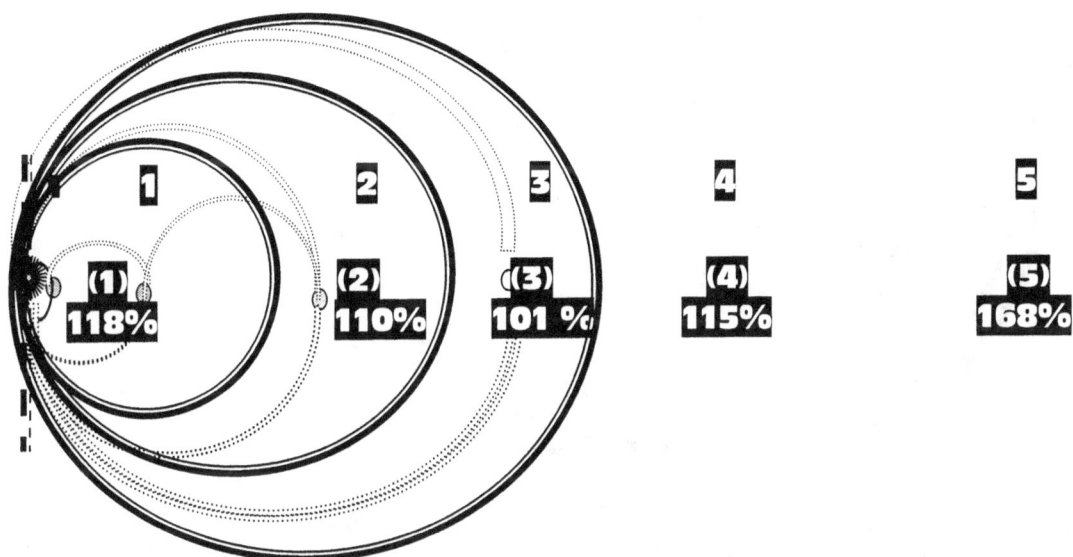

Where the reducing should start nullifying the space growth, which is past Mars there we find an increase that is most astonishing. We find that there is an increase of as high as 68 % where there should be no traces of any growth left. This can only be the result of a heat release of gigantic proportions in the manner of a (very little) Super Nova spectacular.

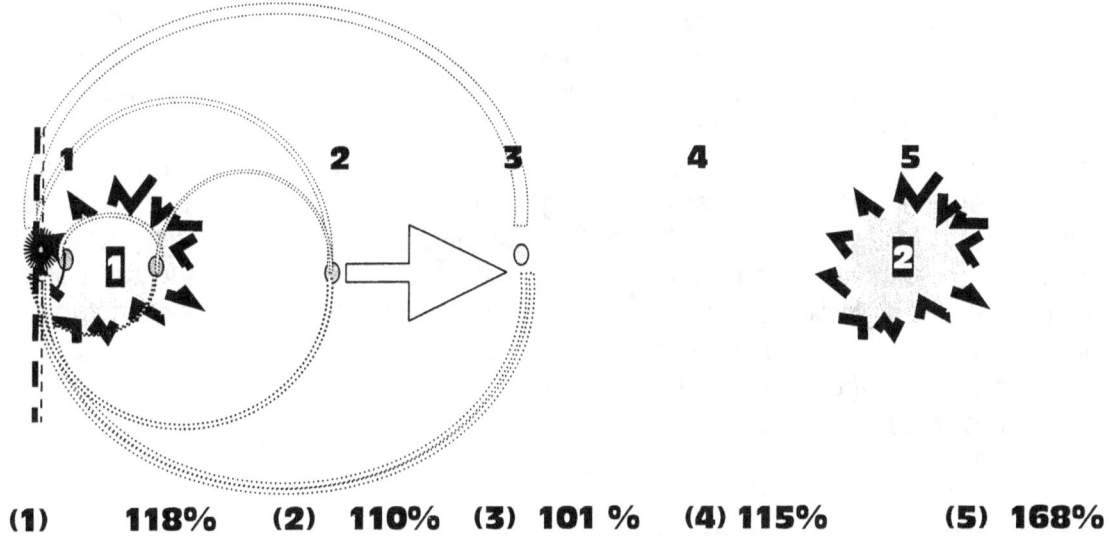

(1)   118%   (2)   110%   (3) 101 %   (4) 115%   (5) 168%

Jupiter released the same fate on the fifth planet as that which befell the fifth plants mother star old Unknown star. I the fifth planets did no fragment (and believe me there was no alternative solution) the alignment between the planets and the micro stars would not have realised because then there was no linking the micro stars and the planets in relation to the using of the Titius Bode law.

Any application to use this method in gauging the development of the micro stars would bring no clear results since we and the micro stars are not connected in the manner we connect with the four planets

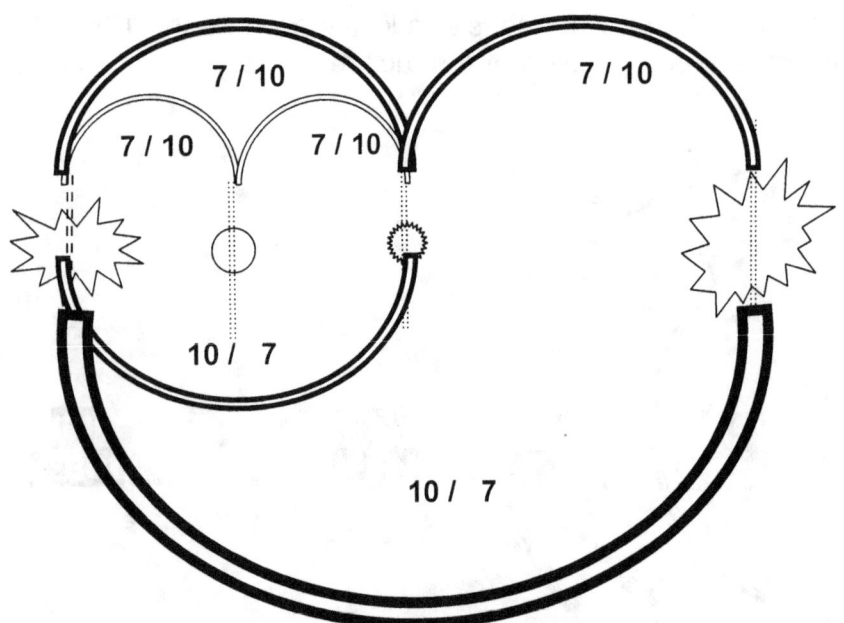

The significance and implication that the application of the Titius Bode principle holds on the Hubble constant reflecting on era to come as well as era gone past turns the cosmos from a small piece of vacuumed holding a few atoms to a vastness no computer irrelevant of size can ever determine.

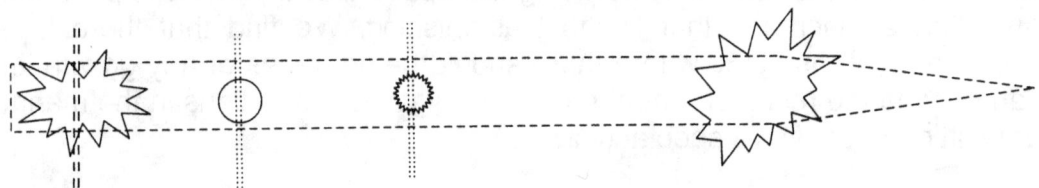

In view of the Titius Bode time depletion, time in flow creating space has a far more complicated arrangement than Xepted science can even produce on a chart. To be honest every person knows that Xepted science cannot even place the planets on a chart, depicting true distance to size, but they WILLFULLY never mention that information when the chart they show is as false as a three-dollar bill. In a sense it does no harm leading people down the ally in such a way, because others in my class of mental insignificance in society is far to un-intelligent to realize the correct way and will therefore not understand the correct way in any event. Now you ask "So what about Neptune and why does Neptune not fit the pattern…"well that storey is more complicated which I therefore reserve for *"Seven Days Of Creation"* because the explaining is as simple after I produced a much wider vision with a lot more explaining to make it as simple as this…

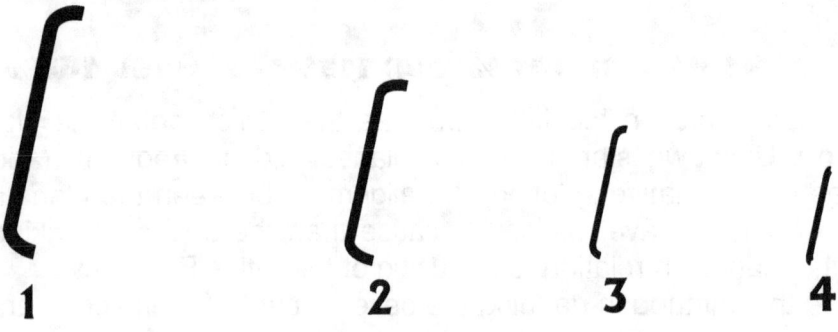

1 SINGULARITY HOLDING MATTER TO OUR FUTURE IN SPACE-TIME
2 SINGULARITY HOLDING MATTER IN RELAVANCY TO SPACE-TIME
3 SINGULARITY HOLDING MATTER AT THE SPEED OF LIGHT
4 SINGULARITY HOLDING MATTER BEYOND THE SPEED OF LIGHT

That is a mistake with a stinging tale. It is as dangerous as a scorpion to Xepted science. In an introducing article, I named Anglo-American Mythology, I pointed out how misconception feeds society, favouring the lies and untruths and blatantly ignoring the truth. In the past, since time began the powerful used this on the brainless masses, and for a period where that civilization lasted, got away with that strategy. The next civilization that came to power, followed the same methods applying the same dogma, and in the end paid the same penalty because their greed, lust for power, and sublimations gave them control over the masses for a while. The misconception those in favourable positions forced onto the masses, made the very people in power so shortsighted, their course on vanity lasted but a few generations. This is achieved because our Earth environment is tolerant and can buffer a lot, to save life in the end for life's contamination on Earth. They wish to extend life's connection to Earth, as being a connection to the cosmos at large and will be in effect as long as life remains in the cosmos. When "going abroad" to our "next door planet" that holds all supporting evidence of life carrying organisms, the connection to the cosmos remains and connecting to the Earth is of little consequence. That bluffing must stop. I realize no one on Earth will ever take note of what one sod (like me) on Earth is shouting, but misery awaits our Martian Colonists.

Enough of this for I shall never halt the madness, so why bother with trying to in any case. Suffering will be the reward for the fools attempting to catch the bounty of "fame, riches and glory" on behalf of the All Powerful Dollar and the dollars absolute true benefactors. Those with eyes, let them see, those with ears let them hear and let the rest self demolish.

I admit, man derives all the above-mentioned lies mainly from corrupting the Bible. The Roman Catholic Church started the trend 1 500 years ago and still maintains it as best it can. You can prove whatever you wish to prove from facts you take from the Bible. The Bible is in support of whichever standing you may support. This is not because the Bible is incorrect, it comes from our insignificance to appreciate the full content of the whole Bible. The Bible promote only truth, man takes from that what man wants, divert the truth to suit his need by corrupting the lot.

I know presenting evidence as well as I do, will not change the course of science. I know too, that I am too small to pinpoint conclusively whether Authentic Author referred to the creation of the first cosmic period, or the first solar period. This is not because of inaccuracy on the part of the Bible. This inaccuracy comes from the human interpretation of Authentic Author on his vision, and then my insecurity to interpret his interpretations. It is human error bringing on misinformation. The Bible's recollection CAN and DOES apply to either period. Therefore, to respond in containing human misjudgement on my part, I shall re-apply the vision of Authentic Author in the context of the first solar day. I showed how it fits to the first cosmic day already.

Binary stars, spinning to self-destruction will produce significant heat. Heat create space, space forms winds. That is facts that the Bible present and is indisputable. Where the Earth was, was still a void, containing a sphere of circular displacement and this will reduce linear displacement to zero.   Linear displacement is space and circular displacement is containing heat for matter survival.

Binary Star Minor overheated. That is why the core brittle and fragmented. This action will release tremendous contained heat, the heat will produce magma flowing in space like water in space and this eruption of heat space that created winds. Once again the recollection fits the scenario. Releasing the heat and producing space will establish space-time and fill the void where the Earth should fit. This is fact and if anybody even tries to dismiss this will be because of abstinence on his or her part. I did not prove the Bible correct. The Bible told the truth and in such correct detail, it is beyond human comprehension, but sublimation on the part of Newtonians and science before them, disallowed their ability seeing it.

At this point I invite you, the reader to go back and read about the Newtonian version of cosmic structure forming. Compare that FORCE applying MAGIC to the Biblical portrait of events and see where the fools hide. When comparing notes about Newton's view and the Bible's view, judge for yourself who in the end understood Newton and who did not. All I wanted was to find some one that could look past the mechanic and appreciate the work he is representing. I never seek prominence, I only wished to introduce my view and let another more educated, more significant and more wise man take the reigns from there. I always knew I am not the man to do the job. My field of knowledge is too small, too limited and above all, too insignificant.

I found no one that could look past me and see my formula $R^3/T^2 = 1$ and $\$T = (\Pi^2 \, x\Pi^2) \, (\Pi^2\Pi) \, 3 = 1836$ which is the relevance of the cosmos. By not finding a person that could see past me, I knew that person will not be able to look beyond "a burning Sun and see the frozen state in which the Sun is. Without noticing such crucial evidence, the rest goes lost. That person that sees me and not my formula will never see the cosmos for what it is.

While the one proton connects to space in singularity (Π^2 going to singularity) and connects space Π (in singularity) the other proton brings time Π^2 directly to space Π, re-uniting space-time as a unit to singularity (Π^3). We may call this re-unification unifying the " gravity - motion" in order to identify the one proton unifying time with space, while still in contact with the other proton holding (Π^2) time to (Π) space in as much as being occupied by matter (Π^2) and unoccupied heat Π forming 3.

This explains then the absolute value of time as a square, with the square having both a circular value of R^2/T multiplied by the linear value of R/T producing the link to singularity in time and singularity in space.

That is what an insignificant formula $R^3/T^2 = 1$ where $R^2/T \times R/T = 1$ represents space-time in singularity as well as space-time in densified, occupied and unoccupied format. That means the everything of the whole lot, or as we say in Afrikaans, the "Heelal" meaning Universe.

When comparing notes about Newton's view and the Bible's view, judge for yourself who in the end understood Newton and who did not. All I wanted was to find some one that could look past the mechanic and appreciate the work he is representing. I never seek prominence, I only wished to introduce my view and let another more educated, more significant and more wise man take the reigns from there. I always knew I am not the man to do the job. My field of knowledge is too small, too limited and above all, too insignificant.

I found no one that could look past me and see my formula $R^3/T^2 = 1$ and $\$T = (\Pi^2 \times \Pi^2)(\Pi^2\Pi)$ 3 = 1836 which is the relevance of the cosmos. By not finding a person that could see past me, I knew that person will not be able to look beyond "a burning Sun and see the frozen state in which the Sun is. Without noticing such crucial evidence, the rest goes lost. That person that sees me and not my formula will never see the cosmos for what it is.

While the one proton connects to space in singularity (Π^2 going to singularity) and connects space Π (in singularity) the other proton brings time Π^2 directly to space Π, re-uniting space-time as a unit to singularity (Π^3). We may call this re-unification unifying the " gravity - motion" in order to identify the one proton unifying time with space, while still in contact with the other proton holding (Π^2) time to (Π) space in as much as being occupied by matter (Π^2) and unoccupied heat Π forming 3.

This places Kepler's formula at a relevance s $\mathbf{a^3/T^2 = k}$ where $\mathbf{k^1}$ will hold a relation to heat with time in eternity, while AT THE SAME MOMENT infinity also apply, giving $\mathbf{k^0}$ the value of singularity. This is all to do with the Universe remaining in contact with singularity while keeping the Universe in matter and heat, in the dimensions of space and time. This explains then the absolute value of time as a square, with the square having both a circular value of R^2/T multiplied by the linear value of R/T producing the link to singularity in time and singularity in space. That is what an insignificant formula $R^3/T^2 = 1$ where $R^2/T \times R/T = 1$ represents space-time in singularity as well as space-time in densified, occupied and unoccupied format. That means the everything of the whole lot, or as we say in Afrikaans, the "Heelal" meaning Universe.

THE COSMOS IS NOT OUTSIDE; IT IS INSIDE, INSIDE EVERY ATOM. THE ATOM CANNOT DISAPPEAR, AS IT CANNOT VANISH. After all it is energy.

In an effort to convey what I see as time I wish to convert the Cosmic Calendar to some Cosmic Time Scale, This scale does not name events, but rather use relevancies, running time and space as the event unfold from singularity. The singularity comes from the point where all matter occupied and unoccupied confirmed one line with out space as space was infinite and time was eternal. As purely an indication of physical time found in the cosmos applying at this moment I wish to introduce an illustration in bringing a better comprehension of time flow. There is a possibility of many other starts to the flow of singularity, but this point holds most valid significance to the theme we explore

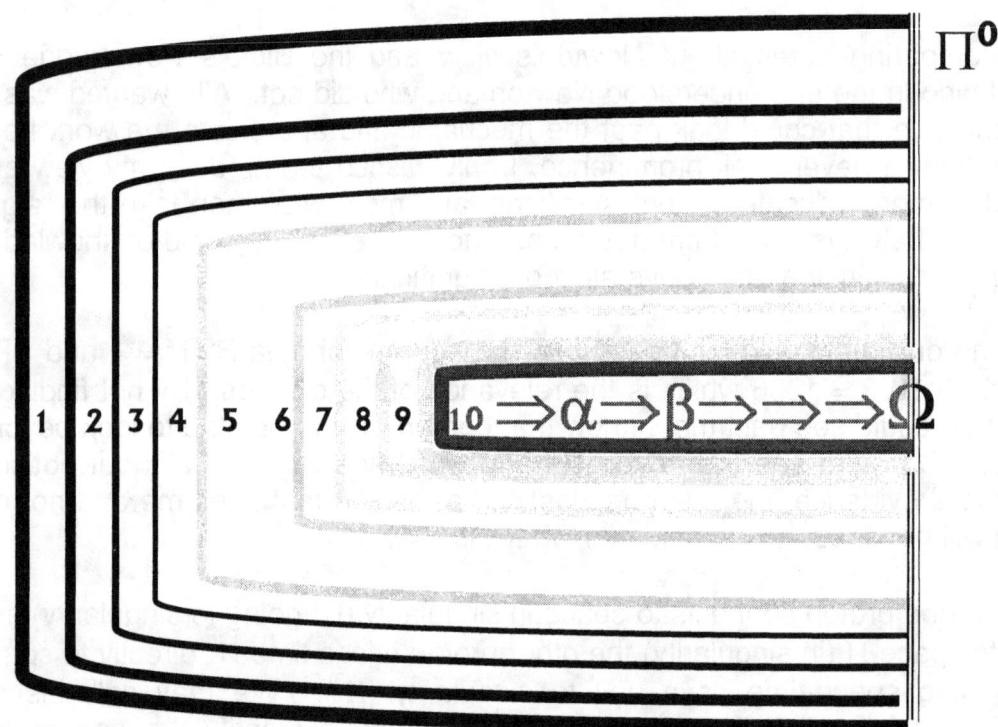

Newtonians tell about a Big Bang explosion that included everything there is.

$t = 10^{-43}$ seconds
the very first instant, the entire Universe were much smaller than a neutron and the temperature was $\approx 10^{32}$ K

$t = 10^{-34}$ seconds
The Universe underwent an increase in pace expansion growing in size with a factor of 10^{30}. The Universe becomes a soup of quarks and leptons at a temperature of $\approx 10^{27}$ K.

$t = 10^{-4}$ seconds

Quarks combine to form protons and neutrons and their anti particles. The Universe cooled down to such a slow pace electrons no longer can breakdown the particles remaining. Particles of matter and anti matter collide and annihilate each other. There is a slight excess of matter not finding annihilating partners, surviving to form the world that we know today.

t = 60 seconds

The Universe has by now cooled down enough to form protons and neutrons and with colliding can stick together to form the nuclei of low mass elements ^{2}H, ^{3}He ^{4}He and ^{7}Li. The predicted relative abundances of these nuclides are just what we observe in the Universe today. There is plenty of radiation around, but light cannot travel before it interacts with nucleus. The Universe is opaque to its own radiation.

t = 300 000 years

The Universe has now fallen to $\approx 10^4$ K, and electrons can stick to bare nuclei when they collide forming atoms. Because light does not interact appreciably with (uncharged) particles such as neutral atoms, the light is free to travel great distances. From this light comes background radiation Atoms of hydrogen and helium can hold together under the influence of gravity, and begin to clump up forming galactica and stars. In every small human mind we try to find time, which we know and trust. The cosmos holds time much to the properties the Creator describes in the Bible. I shall not be blasphemous and say it is the time the Creator refers too, because the cosmos is time the Creator created with all other aspects, therefore the Creator refers to time at a pace that puts our vision of eternity in the same class as we find an explosion to be.

To us the future is dark because it holds more space to less light. On the other hand the past is bright because it holds lighter to less space. The space we see lacks luminosity, because there are much more space to hold light. Stars that came before us cannot vanish for they are matter, holding matter to occupy space of matter. Matter (and space in the form of unoccupied matter or heat if you wish to call it that) is energy and energy cannot destruct, vanish disappear or leave the point of singularity. Again we are facing the situation the Bible warned us about. We are thinking of the heavens in terms of Earth instead of thinking of the heavens in terms of heavens.

On Earth we humans, connect time to human relation. Today become yesterday with the event of tomorrow. We think of today disappearing, as tomorrow is dawn. In the Universe that may not be the case, because where true cosmic time holds space, we humans shall never enter. In contrast to Popular Newtonian belief we will not run down the corridors of some Black Hoe to another Black hole and in the process mesmerize time. Such is not for us fitted with carbon-holding life in a position of singularity.

In stars one year is as eventful as a million years because of space holding time away from singularity. On the moon the next million years will be as eventful as the previous million years with nothing to report in newspapers. Slightly of the point but still relative: ever seen the Newspapers come out and say nothing much happened? There is always news only the relevancy may change from day to day. Well, this does not even happen on the moon because on the moon life will not bring news as it happens; it does not happen!

Looking at the sky we do not see space we see time. The photon is space travelling through time and that is the only space we see, that which the photon hold and that the photon occupy as much as bring to us. Space is light in utter darkness. Space is a ray of photons in magnitude not directed to our singularity but in countless other directions where singularity manifest space in time. Space is heat and heat is photons of lesser implication or not directed our way. To my view (what it may be worth) the Authentic Author did not move back in time, he merely moved foreword in space. By moving in space he reached Moment-Alfa. Moment-Alfa is in eternity and eternity never ends therefore he moved to eternity in space eluding time by having a vision. He could therefore see what he reported because he was in that space but not in that time. If he were in that time he would still be there, he would be there eternal and never come back to report on what he saw. There is much rumour of a Big Bang, however I am inclined to think the biggest bang that ever was also became the smallest bang there ever can be. It was the instant when heat parted from cold. The line started in infinity because the line was continues but being continues it never was. The very instant followed the previous instant identically and the instant was so identical it remained the same by never moving while always moving uninterrupted eternity upon eternity. Then came the entire Universe when infinity broke free from eternity. It was when darkness broke into light. It was when whatever possibly can be became a possibility to be. It was when the first number mathematically arrived and from the one became two by gong 1^0 to 1^1. That infinity is so small it houses everything there is in the entire Universe. The entire Universe still is in a spot that formed a dot. The spot has no outside but it only has an inside while it is inside all that spins as it generates all that can spin. Yet, by spinning it brings motion into being. The spin creates a drive that keeps the Universe mobile. Still the first forming of the dot from the spot came about to the inside and not the outside, which makes the Universe shrink and not expand. It is the smaller things that come into relevance as the larger things were placed in relevance when time began. The Universe is shrinking into the oblivious since the Universe never had anywhere to expand to. Never once did one Newtonian sit back and consider their laughable proposal of an expanding Universe with nowhere to go when it is expanding.

When the cosmos came to motion, motion was not yet defined. When the cosmos brought about motion, the first motion was relevancies. Cold parted from hot. Eternity parted from infinity. Motion parted from motion absence. Infinity broke the laboriousness of eternity for the duration of infinity. The spot became

The Spot

becoming

the **Dot**

This was the era of distinction, when separation brought an all-possible new Universe

From what the spot was to what the dot now is might be just a mathematical implication of going from 1^0 to 1^1 but in reality that first motion was the creating of and establishing of an entire Universe with all possibilities now in it. Never again can that much growth become a reality, although to us the growth is beyond what we ever can notice. But it is because the growth is so massive and we are so small that we are unable to notice such almighty growth.

COSMIC DAY   SINGULARITY
1     $(\Pi^3)^2 = 961.$
SINGULARITY Forming
2   time   space
$\Pi \times \Pi^2 \times \Pi^3 / 5 = 192$ $\Pi^2 \times \Pi^2 \times \Pi^2 / 5 = 192$

3
$10/7\pi^2/2(\pi^2 + \pi^2) = 139$ $7(\pi^2 + \pi^2) = 138$ $7/10 \ \pi^2/2(\pi^2 + \pi^2) = 136$

4
$2\pi(\pi^2 + \pi^2) = 124.$ $2(3)(\pi^2 + \pi^2) = 118.$ $10 \div 7(4(\pi^2 + \pi^2)) = 112.$

5
$\$T = \pi^2(\pi^2 + \pi^2) = 107.$ $3\pi((\pi^2 + \pi^2) = 102$ $3^2(\pi^2 + \pi^2) = 98.$

6
$10 / 7\pi(\pi^2 + \pi^2) = 88.$ $10 / 7(3(\Pi^2 + \Pi^2) = 84.$ $4(\Pi^2 + \Pi^2) = 78.$

7
$10/7(\pi/2)^2(\pi^2 + \pi^2) = 69$ $7/10(4((\Pi^2 + \Pi^2) = 55.$ $2((\Pi^2 + \Pi^2) = 39.$

8
$10/7(2((\Pi^2) = 28.$ $10/7(\Pi^2 + \Pi) = 18$ $7/10(2(\Pi^2 + \Pi) = 14$

9
$2(\Pi^2) \ 19$ $10 / 7(\Pi^2) = 14.$ $7 / 10((\Pi + \Pi) = 5.$

10
$7/10(\Pi) = 2.$ $(\Pi/2) = 1.57$ $7/10(\Pi/2) = 1$

What is the Universe?

The Universe is what became visible when the invisible found a means to generate motion into the immovable and by splitting the inseparable that parting brought a division within the undividable. That which has no limits found a centre when that which has no inside started moving apart from that which has no outside. The separation of the undividable formed a line running in the centre of all spinning matter connected to that which allows spin to all spinning matter which holds space align with time while it keeps time synchronised with space in motion. The Universe has no sides, is undetectable, is only found outside our spectrum of reality yet is, controlling what it is generating the unreal from where we are part of the unreality outside singularity. We call it singularity yet it creates by generating space- time in time in space without being in form inside the Universe. It creates the Universe in time so fragmented we will never understand.

The information in this book I purposely made easy to red, and easy to follow. The content however, is only a drop compared to a bucket of information contained in all seven parts of "MATTER'S TIME IN SPACE The Theses". All information presented to you, in the first part of the book is an introduction to the second part. In the latter part of the book the conclusion comes. With out a detailed introduction in the first part, the information in the second part will be of little value, as the information is very conclusive about the first part. I have been researching cosmology since 1978, on a part time basis. The conclusions may seem simple; that however is in retrospect. Some of the arguments took me up to six months to arrive at since I do not have all the information on hand.

Facts I advise the reader to become acquainted is the following:

aanplasing, verplasing, versnelling and inperking
As this book is a translation from an Afrikaans originally, some terminology and expressions I had to revise to accommodate my Ideas. Where I could I used modified English words to express a thought or an idea. One such a term is gravity. I had so much criticism about the word, which I feel I do not deserve. There is a certain notion clinging to the idea represented by gravity. Gravity links to a force that is all compelling, but I do not agree with SUCH AN COMPELLING FORCE, such as the word gravity implies. Gravity I introduce, works on two principles, but gravity to Newtonian standards is a single force. When I refer to gravity the normal reaction is that I am referring to the force I deny. By declaring that gravity THE FORCE CONTROLING ALL OF THE UNIVERSE AS A STANDARD CONSTANT IS NON EXSITING, I bring the wroth of the scientific world upon me. When I make the statement that there is no gravity, every person considers me mentally unstable.

Of course there is a movement of energy keeping all objects attached to the Earth, but gravity implies work, and with that work principle I disagree, because that is NOT WORK, THAT IS A COSMIC BALLANCE THAT STARTED AT A POINT OF ETERNITY AND WILL END AT A POINT OF ETERNITY. ONLY LIFE, STANDING ALONE AND DETACHED FROM THE COSMOS AS THE ENERGY THAT MANIPULATE SPACE-TIME CAN COMMIT WORK, THE REST IS A BALALANCE RUNNING CONCURRENT FROM TIME'S BEGINNING TO TIME'S END. I HAVE UNBELIEVEBLE DIFFICULTY IN RELAYING THE DIFFERENCE BETWEEN LIFE WITHIN THE COSMOS, AND THE COSMOS WITHOUT LIFE.

In the entire Universe there is no work, it is a balance running concurrent through time and space. The balance within stars shift in some cases to favour space and in other cases more to favour time. But in all of that shifting, a continuous balance strikes every aspect of space-time. This applies new ideas never brought to light before and the new concepts clashes with the conventional names that science applies to current ideas. I had to divorce the science ideas from those I introduce and the only way was with new etymology. I have to start implementing the newly created terminology, which will apply to the rest of this book. This stems from my lack in ability to find suitable words in the English language that would define the concepts as they are, in order to establish the difference in meaning from the current words, which convey the existing misinterpretations (or if you wish, to my view incorrect applications).

Firstly, we start with the word **densified**, which is not a normal English word, but a word I had to produce in order to make a comprehensible statement. The correct word that applies is concentrated, that much I do know about the English language. But concentrated has not the correct meaning or the expression that I would like to bring over. Concentrated can apply to any substance, be it gas, liquid or solids where one of the ingredients become more than the rest of the ingredients. In that way, matter as a solid substance produced from the eternal substance which is heat, cannot be concentrated. Nothing in the entire Universe can compare with the density of pure heat that spins at a rate in which that very heat can produce a value and which has a density far beyond anything else. Therefore, I chose to use the concept of concentration in a position where it makes a lot more sense. A star is concentrated space-time, but there is a huge difference between a star's concentrated space-time and the value of pure matter. When a star does therefore become densified space-time, it can only be at the end of the Big Crunch eternity , witch I prefer to call moment Omega; that is when space becomes eternal and time becomes Zero. In this light I chose to call matter densified space-time. Densified space-time should therefore be in a definition where matter or substance has reached a point in density that will last one eternity, but has no limit. Concentrated space-time, on the other hand does have a limit, which is at the point where it becomes densified space-time.

The second word I created is **Aanplasing,** which is the ongoing redirection of heat as in matter to heat as in time and that connects to a circular deepening of the separation that matter undergo transforming to time as it discards heat for the cold of fusion. Later (I hope) it will be clear enough for every reader to comprehend and to distinguish between the various factors that bring about **aanplasing** as should the reasons be clear why I prefer to have created this new word.

In this case however, there was no English word that could merely be altered and then be re-applied. A more suitable word that relates to a better meaning in the case where I brought in "densified" would have been the Afrikaans word "verdigting", where "verdigting" stands in relation to "konsentrasie" (concentrated). The fact of the matter is that I am not wilfully forcing Afrikaans down the throat of the Anglo-American and in the case of density I was able to modify a known English word that could adopt a new concept. By using an English word, would mean that there is no liberation from the "misleading" focus that depends on gravity, nor can it liberate the feature of this "misconception".

As for the Afrikaans words: **aanplasing, verplasing, versnelling** and **inperking**: there are no such words or concepts in existence that the precise meaning can derive from the English written or spoken language. Should any such words exist, the misconceptions that remains connected to the original English words, would not bring justice to the concept which I wish to apply to convey the meaning that lies behind the correct value of the thought. If I stuck to the word "gravity", the concept I wanted to introduce would forever remain confused with Newton's application. To that end the new realization would then never come across in the way I intend it to be.

Inperking: This stands apart from the idea of curtailment because in curtailing something or someone, means that object's or person's movement or moveable motion is deprived. This then brings over the misconception in the accepted notion of an expanding Universe. In due course I shall explain the concept, but inperking involves the same value that was there at first and will be there in the end, only the location in the balance shifts to favour one or the other part of the same coin. Because of the fact that none such a thing applies when space-time "accelerates" (versnel), and where this brings about inperking, it does not apply. Instead, all functions and factors still apply when inperking becomes valid, therefore the meaning of inperking becomes more applicable and this word describes the process much better. Inperking relates to time, where the duration of time extends, but not the value of time as such, as time applies in the cosmic sense.

One should realize that the entire atom, as well as its surroundings including all other surrounding atoms are reduced in space-time volume, so the atom is not actually curtailed, nor is its surrounding which means the word curtailment does not really apply. All aspects of occupied and unoccupied space-time are in reality, re-focused down in the true sense and above all, remains to the precise relative relation value it had for one entire eternity where the relation between such times, only refocus. However, scaled down would neither apply, because that would not refer to the time involvement, which lies at the hart of this revaluation. Where less time applies, inperking would be more severe and where more time applies, less inperking will apply.

Versnelling: It carries exactly the same meaning as acceleration, but the meaning or concept connected to acceleration implies to matter as the matter increases its own positional change in space and time. That is not the impression I wish to relay when referring to versnelling, because it is exactly the opposite of that meaning. In this, the actual meaning is more applicable to the true connection. These I must explain carefully, not to convey confusion. When a person stands outside an explosion of some sort, the time laps seems instantaneous, quicker than the senses can relate to. However, inside the explosion, time is almost standing still. Any person, who is inside such an explosion, would relate to time on the outside as being instantaneous. Whether this statement is accepted or not, the truth is that a person in an explosion cannot die, although his body is shattered in a million pieces. Such a person is sealed in a period separated from the period he and we lives in. This I explain at a later stage. The time duration slows down immensely, but from the outside, it accelerates immensely. Therefore, time versnel to the outside of where ever one relates to.

I wish to bring over the fact, as just been said, that the concept we have, is quite the opposite. Versnelling implies that the motional increase lies with the transfer of space-time, regardless whether matter occupies it or not. As aanplasing (not gravity) and versnelling bring about inperking (not curtailment as the body remains free to do as it wishes) the space-time that the body occupies and the surrounding sphere are in constant state of versnelling. The increase in motion has an effect on the matter, but the matter stands weightless as its specific density applies the time in that particular space.

Verplasing: This word is preferred to that of displacement, because although the matter in motion is displaced, time and space are implicated in the process.

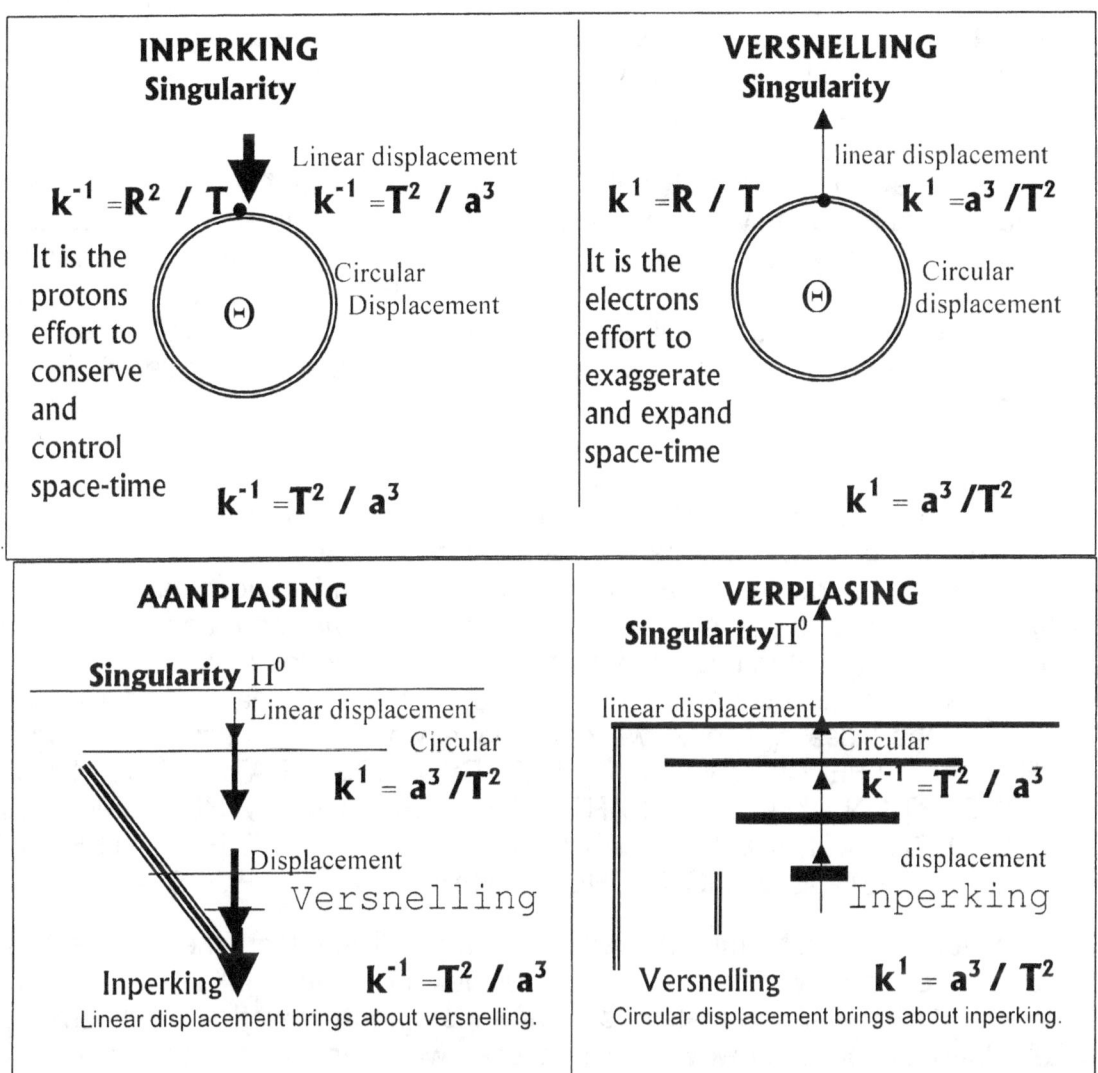

Verplasing is in fact the transferring of newly created magnetic space-time by matter, as a body composed of atoms has to replenish the space-time it occupies in order to maintain its position, place and structure in space-time in time in space, according to its geodesic positional allocation within the star's space and in time. Verplasing comes in effect as matter progresses in position, but the time-affect of verplasing that it has on matter, comes into real effect when an object reaches Mach $_3$ depending on its shape and altitude. In short: **Aanplasing** is relatively where matter is in a geodesic motionless position as space-time carries the motion component of the two values. This means that aanplasing is relative to positive space-time displacement.

Verplasing on the other hand has to do with the motion being with the newly created space-time in relation with the matter and the geodesic space-time remains relatively motionless. In both cases inperking and versnelling is a consequential result of the process. The difference is in the application of the time component itself.

A practical example of the difference between aanplasing and versnelling is as such: a body in **aanplasing** is in example a skydiver is falling towards the Earth and **verplasing** is where a body, such as a rocket is on a trajectory path as it fires into space. Both bodies will comply with the linear and circular displacement, but the circular displacement will relate oppositely in each event.

In the past, I have been severely criticized by academics for the view I hold and to my judgment rather unduly. The largest complaint by academics comes because of my critic about Newton's findings. There is a vast difference to the cosmos and the Earth surrounding us filled with life, the part we humans take fore granted, but the only place we will find life will be on this planet. Compare Earth to the moon, and the moon represent the real cosmos. It is dead, except for matter that grows with cosmic time.

To understand the cosmos the cosmos and life has to stand apart, the two on they're own and only then does the cosmos appear as it is. A rocket launched is not part of the cosmos, but is part of life's extending energy. Life is the ability to occupy and manipulate space-time occupied and unoccupied. In the manipulation, there is a very strict limitation to what man can and cannot accomplish. It is as if man has become drunk with his supposedly unlimited ability. This will lead to tragedies man still has to meet. There is no reason sending brave men off to their death, if only we can find a way to recognize the dangers. NOTHING IN THE COSMOS CAN DIE, BECAUSE DEATH IS A HUMAN MISCONCEPTION. ALL ARE PART OF DIMENSIONS WHERE WE ARE INCLUDED OR EXCLUDED. THIS CONCEPT ABOVE ALL SEEMS TO EVOKE THE MOST REJECTION AMONG THOSE I HAVE COME ACROSS.

The last four five wrote full time and with which I went on the Internet. The persons I wish to draw to my book do not seek information on the Internet and those who do look for information, cannot understand my work. The commercial route I tried to follow; however no publisher would touch it because it is not recognized science. That reason forced me to seek an academic route, in order to find recognition for my work.

To the correctness in my work, I have no doubt; however there may be parts I still have not clarified sufficiently. I am more than willing to meet such challenges because I know the proof is there, I was just not yet able to recognize the question.

Being human we tend to form concepts through culture and such concepts we then translate to the cosmos in an attempt to fid meaning moreover about our worth than the facts relating to the cosmos. We find ourselves physical apart and as much linking and with that we then wish to link and find meaning. What links and what divide is the stumbling block because we place man in the cosmos through which we then wish to produce a link. Such a link does not exist but for the tiny speck of cosmic dust which we call either home or Earth. The main issue to divide what should be divided and then accumulate what should connect and doing that is the purpose of this book.

Stars will never collide because stars can never collide.

The only absence In the cosmos is zero and without zero there cannot be an end to eternity but only an everlasting cycle that breaks to start one more eternity now and then. With the cosmos created minute by minute from no space within the cosmic centre, the cosmos is ruled from a position with every thing but nothing is but where we know God must be. By accepting singularity and the rule there of brings into the cosmos things physics are unable to explain, mathematics are unable to calculate and man is unable to dismiss. If you accept physics you have to accept God because you cannot except one proving singularity without the other coming through singularity.

If it is that simple then why is it complicated.

BEST WISHES,

PETRUS. (PEET) S. J. SCHUTTE

There is more about this in other books with the titles:

AN OPEN LETTER TO
SELECTED ACADEMICS" Divorcing Kepler from Newton and ending the three hundred year unlawful marriage
MATTER'S TIME IN SPACE THE THESIS ISBN 0-9584410-8-1 starts with
A Cosmic Birth...Dismissing Nothing I. S. B. N.   0 - 620 – 31609 – 8, which is the part one that ends with part seven, which is an open letter about The Seven Days Of Creation.

Part one to seven of MATTER'S TIME IN SPACE: THE THESIS ISBN 0-9584410-8-1 is
A Cosmic Birth... Dismissing Nothing.
An introducing of my theory by applying the principles to nature.
AN OPEN LETTER ON
XEPTED MISTAKES
Mostly an abbreviated version of book taking physics into what physics brings to life and the developing of the mind
AN OPEN LETTER ON
INTER GALACTICA SPACE TRAVEL
(Explaining the sound barrier in relation to centre of the Earth and how that affects intergalactic space ventures)
AN OPEN LETTER ON
CORRECTING COSMOLOGY
Forming the basic allowing a better understanding of the basics of cosmology.
MATTER'S TIME IN SPACE:
THE HYOPOTHESIS
(Concerning the way the universe started and how galactica develops)
AN OPEN LETTER ON
" STARSSTUFFN'
(Concerning the way the universe started in as much as stars developing)

" SEVEN DAYS OF CREATION"
Is about the forming of the planets and gas structure that are stars to be in the solar system and matching the events with seven solar periods or days that follows the Biblical lines of development)
MATTER'S TIME IN SPACE: The Thesis
(As a combination of the above)

FOR other related information, PLEASE VISIT THE WEB SITE, FOR YOUR CONVENIENCE

To find more about the book please visit
gravity @tlantic.net

This was the prologue letter announcing
MATTER'S TIME IN SPACE: THE THESIS
ISBN 0-9584410-8-1
FROM THE ORIGINAL AFRIKAANS: "MATERIE SE TYD IN RUIMTE" I. S. B. N.   0 – 620 – 27041 - 1
WRITTEN BY PEET SCHUTTE
© KOSMOLOGIESE EN ASTRONOMIESE TEGNIKA

You may also contact me by land mail at:

Po Box 1093,
Ellisras
0555
REP. of South Africa.

THIS WAS

An open letter

TO SELECTED ACADEMICS

ISBN 0-9584410-9-X